Mechanisms and Machine Science

Volume 14

Series Editor

Marco Ceccarelli

For further volumes:
http://www.springer.com/series/8779

Vijay Kumar · James Schmiedeler
S.V. Sreenivasan · Hai-Jun Su
Editors

Advances in Mechanisms, Robotics and Design Education and Research

Springer

Editors

Vijay Kumar
University of Pennsylvania
School of Engineering
  and Applied Science
Philadelphia, Pennsylvania
USA

S.V. Sreenivasan
Mechanical Engineering
University of Texas
Austin, Texas
USA

James Schmiedeler
Aerospace and Mechanical
  Engineering
University of Notre Dame
Notre Dame, Indiana
USA

IIai-Jun Su
Mechanical and Aerospace
  Engineering
The Ohio State University
Columbus, Ohio
USA

ISSN 2211-0984   ISSN 2211-0992   (electronic)
ISBN 978-3-319-00397-9   ISBN 978-3-319-00398-6   (eBook)
DOI 10.1007/978-3-319-00398-6
Springer Cham Heidelberg New York Dordrecht London

Library of Congress Control Number: 2013935696

Foreword

We are delighted to present a select compilation of papers that represent recent advances in kinematics, mechanisms, design and robotics research and education. The papers celebrate the achievements of Professor Kenneth Waldron who has made immeasurable contributions to these fields in the last fifty years. His leadership and his pioneering work have influenced thousands of people in our community. During this period, he has mentored over 35 doctoral and 30 masters students, all of who owe their success to Professor Waldron's tutelage and mentorship. In addition, Professor Waldron's doctoral students themselves have advised more than 50 of their own doctoral students, who have in turn gone on to mentor nearly 20 of their own doctoral students to date. Therefore, his influence continues to grow and spread through the ongoing work of these "academic grandchildren" and great-grandchildren of his, with more generations to come.

The book has four categories of papers that serve to illustrate the impact of Professor Waldron's work. In *Historical Perspectives*, Waldron and Waldron, Chase *et al.*, Vohnout, and Velinsky offer retrospectives on Professor Waldron's life, research, and service to the community. Davidson, Su *et al.*, Ananthasuresh, Midha *et al.*, Sugar and Holgate, Meissl *et al.*, Ge *et al.*, Brassitos and Mavroidis, and Raghavan all describe novel contributions to mechanisms and machine theory in *Kinematics and Mechanisms*. The section on *Robotic Systems* consists of papers by Flores and Kecskeméthy, Mulgaonkar *et al.*, Long and Cappelleri, Zhou *et al.*, Rovetta, Vertechy *et al.*, and Notash that provide insight into fundamental research problems in robotics. In *Legged Locomotion*, Wensing and Orin, Abdallah and Waldron, Schmiedeler and Funke, Tsai *et al.*, and Schache *et al.* present recent advances in research on legged robot and human locomotion. Finally, in the *Design Engineering Education* section, Srinivasan, Zielinska and Kedzior, Lilly, Hirose *et al.*, and Yan describe new exciting efforts to invigorate our educational programs.

We thank the authors for their contributions and the editorial staff at Springer for their assistance in compiling and editing this monograph. Finally, without the support of the Department of Mechanical and Aerospace Engineering at The Ohio State University, this book would not have been possible.

Professor Kenneth Waldron's Graduate Students

University of New South Wales

1973 B.R. Seeger Ph.D.
    K.W.A. Hui M.Eng.Sci.
1974 B.P. Yeo   Ph.D.
1976 J.E. Baker   Ph.D.

University of Houston

1978 R.T. Strong   Ph.D.
    K. Kheyrandish M.S.
    S. Nadkarni   M.S.
1979 A. Ahmad   Ph.D.
    J.H.W. Sun   Ph.D.
    J. Lott   M.S.
1980 J.C. Huang   Ph.D.
    A. Kumar   Ph.D.

The Ohio State University

1981 V. Cooley   M.S.
    M. Moldovan M.S.
    V. Vohnout   M.S.
1982 T.M. Jou   M.S.
1983 J.Y. Chuang   M.S.
1984 S.M. Song   Ph.D.
1985 G.J. Wharton M.S.
    S.J. Bolin   M.S.
    S.K. Agrawal   M.S.
1986 R.D. Cope   Ph.D.
    S.L. Wang   Ph.D.
    M.J. Tsai   Ph.D.
    J. Reidy   Ph.D.
    R. Geoghegan M.S.
    B.L. Lilly   M.S.
1987 V. Kumar   Ph.D.
    J. Gardner   Ph.D.
    A. Burkat   M.S.
    R. Mandavilli M.S.
    N.N. Murthy   M.S.
    K. Ramani   M.S.
    T.J. Ward   M.S.
1988 M. Huang   Ph.D.
    S. Nair   Ph.D.

1989 S. Coe   M.S.
    A. Thenamkodath M.S.
    V. Varadhan   M.S.
1990 V.S. Murthy   Ph.D.
    S. Bhargava   M.S.
    L. Brandolino   M.S.
    R.D. Dixon   M.S.
    T. Gozali   M.S.
    S.R. Hargro   M.S.
    S. Sreedhar   M.S.
1991 S. Mukherjee   Ph.D.
    P. Srikrishna   M.S.
1992 P. Nanua   Ph.D.
    J.E. Chottiner   M.S.
1993 M. Husain   Ph.D.
    P.C. Chin   Ph.D.
    Y.K. Chou   M.S.
1994 S.V. Sreenivasan Ph.D.
    W.Y. Chung   Ph.D.
    J.C. Yu   Ph.D.
    Y.R. Wang   M.S.
1995 R. Rachkov   M.S.
1997 S.C. Venkateraman Ph.D.
1998 C.J. Hubert   M.S.
1999 P.H. Yang   Ph.D.
2001 J.P. Schmiedeler Ph.D.

Stanford University

2005   J.G. Nichol   Ph.D.
2006   S.P.N. Singh   Ph.D.
2007   M.E. Abdallah Ph.D.
2009   W. Fowler   Ph.D.
       S.K. Kim   Ph.D.
2010   A.D. Perkins   Ph.D.
2011   P.J. Csonka   Ph.D.
       C.M. Wong   Ph.D.
Current C. Enedah   Ph.D.
       D. Jacobs   Ph.D.
       K. Najmabadi Ph.D.
       L. Park   Ph.D.

My Professional Career: A Summary

Kenneth J. Waldron

In 1964 I was studying for the degree of Master of Engineering Science in the Department of Mechanical Engineering of the University of Sydney. This was a course work plus thesis master's degree. I took a class from Associate Professor Jack Phillips on kinematics of mechanisms. Like many others before, and since, I was fascinated by the proposition that one could predict the movability of an assemblage of rigid links connected by joints by simply counting the numbers of members and joints, and noting the number of degrees of freedom of each joint. I was further fascinated to learn that there were anomalous mechanisms that had more mobility than predicted by the constraint criteria. I wrote up some ideas that really amounted to treating each closure separately. Abe Soni had recently published a paper about constraint analysis in ASME Transactions, I think it would have been J. Engineering for Industry. In those days journals published discussions on the papers they published. Ken Hunt came to visit Jack in this time frame, so the three of us wrote a discussion on Abe's paper. That was my introduction to research in this technical area. At Ken's suggestion I contacted Bernie Roth, and a couple of others in the U.S. about doing a Ph.D. Bernie came up with a Research Assistantship, so I took myself off to Stanford in the middle of 1965.

At Stanford I took classes on kinematic synthesis and spatial mechanism theory. However, I had the constraint analysis problem in my back pocket, and that wound up being the path I followed in my doctoral thesis. Bernie supported me throughout on his grant. I don't know how that worked. I was still very naïve about such things.

I had the idea of applying screw system theory, as re-formulated by Phillips and Hunt, to the constraint analysis problem [1]. That worked, and I was able to generate a number of new instances of overconstrained mechanisms, more or less by inspection based on the geometries of the relevant loop screw systems [2, 3, 4].

The limitations of applying screw system theory to elucidating overconstrained linkage geometry soon became apparent. Observing that the axes of all joints lie within a given screw system only proves instantaneous mobility, unless one can also prove that the screw system is invariant to motion about the linkage joints. Also, for some linkages, like the Bennett mechanism, the relationship between the screw system geometry and the linkage geometry is so deep that it cannot be understood without extensive analysis.

At Bernie's suggestion, I developed a method of formulating closure equations for a spatial loop. He had introduced me to the concept of representing a position of a body in space by the coordinate transformation relating positions of points in the body to a reference frame embedded in it to their positions relative to a fixed frame. I used a version of the Hartenberg and Denavit specification of kinematic members and joints. The original H&D formulation focused on the joints, logically enough. However, that meant that one of the parameters characterizing the geometry of a link was numbered differently to the others. I found that annoying and modified the notation so that all the parameters that characterized a link were numbered synonymously with that link: effectively shifting the focus from the joints to the link. Anyway, I was able to demonstrate that the order of the screw system of a loop corresponded to the number of independent closure equations that could be written for that loop, and that overconstraint implied dependency among the closure equations [5, 6].

I have continued to use that version of the H&D parameters throughout my career without further thought. Relatively recently I was accused of originating the version of H&D that most people now use for robotic position equations. This is an example of something I regarded as a minor issue that has turned out to have enduring significance.

After marrying Manjula in 1968, I taught Bernie's courses during the 1968-69 year while he was on sabbatical. My thesis sat on the corner of the desk mostly untouched during the year. I then had to scramble to finish it up since I had an offer of a lecturer position at the University of New South Wales, starting at the beginning of the 1970 academic year.

During that year, I was the nominal supervisor of Vic Scheinman, who was designing the Stanford Arm. That was my first formal involvement with robotics.

At U.N.S.W. I made a foray into biomechanics with Barry Seeger, and Boon Ping Yeo. This was my first attempt at understanding the mechanics of legged locomotion.

With Eddie Baker, I took up the overconstrained mechanism issue again. We developed an approach to including screw joints in the closure equations that rested on the transcendental characteristics that this introduced into the equations. In hindsight we should have presented this work in the language of set theory. Our colleagues still would not have understood it, but maybe they wouldn't have said that it was "not real mathematics". Eddie continued with this work throughout his career, demonstrating the existence of many new types of overconstrained linkage [7].

In 1973 Manjula and I took a mini sabbatical, travelling in Europe and the U.S. I visited my former roommate, Fazle Hussain, at the University of Houston, and discovered I was being interviewed for a job. For multiple reasons Manjula and I decided to take up this opportunity to move back to the U.S., so we moved to Houston in 1974.

At Houston I picked up a problem that had been in my mind since being introduced to rational synthesis in one of Bernie's courses at Stanford. Burmester theory had been implemented in computer software through the work of George Sandor with Ferdinand Freudenstein. Bernie had batch processing code that plotted circle point and center point curves given a set of four discrete design positions of a lamina. I had noticed that, although picking points on the curves gave cranks of invariant length in the design positions, combining those cranks into a four bar linkage seldom gave a usable solution linkage. There were several reasons for that, in fact there were three major ways things could go wrong.

One of these was that, if the linkage was to be driven by a crank rotating in a uniform direction, the coupler might move through four, or five design positions in the wrong order. This problem related only to the choice of driving crank. I was able to demonstrate that the circle point curve could be considered to be divided into segments. Choices of circle points on the same segment would yield driving cranks that would move the coupler through the design positions in the same order. The segments were bounded by the six image poles associated with pairs of the four positions. Thus, it was possible to define which segments of the curve would yield driving cranks that would move the coupler through the design positions in the desired order [8].

The next issue is that for a given position of the driving crank of a four bar linkage there are two possible positions of the coupler and driven crank. For a Grashof Type I linkage, there is no way to get from one of these configurations to the other without disconnecting and reconnecting one of the joints. Thus, if the solution linkage passes through some design positions in one of these configurations, and the others in the other one, it is not a valid solution. I noticed that it is possible to discriminate between the solution configurations by looking at the sign of the angle between the driven crank and coupler. Based on this, I worked out a graphical construction that could be used when selecting a circle point for the driven crank to ensure a solution linkage passed through all design positions in the same configuration. Unfortunately, I soon discovered that Elizabeth Filemon had preceded me on this.

There is another part to this problem: the so-called change of branch problem. Depending on the choice of the driving crank circle point, there may be no available choices of the driven crank circle point for which the angle of rotation of the coupler relative to the crank is less than 180°. In that case, it is not possible to find a solution that doesn't change branch. Thus, it is necessary to restrict the choices of driving crank circle point even further, beyond that needed to avoid the order problem. We worked out a graphically based method to do that [9].

There is yet another problem that constrains the solution space for finitely separated position problems. We usually want to drive the mechanism by means of continuous rotation about one of the joints, usually, but not always, a center point of one of the two cranks. That means the linkage has to have at least one joint that can be rotated continuously. That means it must be a Grashof Type I linkage. Although the Grashof inequality is simple, it does not lend itself to reduction to a graphical method like the solutions for the order and branch problems. Robert Strong worked with me on an algebraic solution to this and on ways to implement the branch solution numerically [10]. By this stage we had a pretty complete theory for problems with two, three or four design positions. Jo Chuang took a look at the implications for five positions, in which case only a small number of discrete solutions are possible.

Yet another variation was to look at the implications for spatial linkages. Bernie Roth had developed a spatial version of the Burmester theory while I was a student at Stanford. Working with John Sun we took a look at applying our theories for the order and branch problems to spatial linkages, with partial success [11].

Finally, we took a look at the implications for synthesis of adjustable planar linkages. That was the work of Anees Ahmad [12].

While at Houston I also picked up my interest in manipulator design again. I worked with Alok Kumar on characterizing serial chain workspaces. One idea we formulated was the dexterous workspace. This is the set of positions that the hand reference point can assume in which the hand can be placed in any orientation, if one ignores joint motion limits. This did not make much in the way of a ripple at the time: it was only an ASME conference paper, but I notice the concept is alive and well in recent literature [13].

By 1979 things were not going so well in Houston so I did a little looking around. There was a vacancy at Ohio State University. We moved to Columbus in Fall 1979.

I had met Bob McGhee at the first RoManSy conference in Udine in 1973, where he presented a paper on the work he had done with Andy Frank on the Phony Pony, the first computer coordinated walking machine. I hoped that I might have an opportunity to work with him at OSU, and quickly became involved with the OSU Hexapod project group. In 1980 DARPA came to Bob with an interest in pursuing a practical scale walking machine, and in 1981 we took up the initial contract in what was to become the Adaptive Suspension Vehicle project. Bob was the principal investigator, and I was co-principal investigator responsible for mechanical engineering work. David Orin and Charles Klein from the electrical engineering department were also co-investigators. At various times during the ten year course of the project quite a number of other faculty members participated. These included Said Koozekenani, William Olsen, and Fusun and Umit Ozguner from electrical engineering, Gary Kinzel, Krishnaswamy Srinivasan, and Necip Berme from mechanical engineering, and Bruce Weide and Karsten Schwann from computer science.

I also continued my work on linkage synthesis after moving to OSU, in collaboration with Gary Kinzel, and his students. We put together the RECSYN program that integrated the theories I had developed on partitioning the solution space to eliminate the order and branch problems, with software to generate circle and center point curves, etc. The initial implementation was on a Vax mini-computer with Tektronix storage tube displays – very primitive by modern standards. The program actually worked well, and it evolved over several generations, adding new features. However, as computers became rapidly more powerful it became practical to simply generate solutions throughout the solution space, test them, and eliminate those that were of the wrong Grashof type, or had order or branch problems, thereby eliminating the need for our approach. It also became apparent that the precision position approach was too inflexible to be a practical design tool.

Given the size of the walking machine project, that became my primary research focus over the next nine years. I started with Shin-Min Song trying to apply our linkage synthesis techniques to the design of walking machine legs. We decided that we needed to generate a straight-line ankle path relative to the body of the machine, so we could walk on the level without raising and lowering the center of mass of the machine. It was necessary to be able to do this at a variety of leg lengths to accommodate walking over obstacles [14]. While we did generate some viable looking designs, it became apparent that the seven bar and above solutions we were looking at were too complex to be practical, so we changed our focus to the two dimensional pantograph rotating about a longitudinal swing axis, which was the design we ultimately used. By this time Vince Vohnout was on board as a research associate, after completing his MS. He had a big influence by finding practicable ways to implement slides and integrate hydraulic actuation cylinders, as well as leading the structural design of the machine [15].

Working with Simon Song, we developed the overall dimensions of the machine, and critically, the working volumes of its legs [16]. Shih-Liang Wang also contributed to this work as part of his master's thesis. We also developed a theoretical proof of the proposition that wave gaits maximize velocity for a given minimum longitudinal stability margin [17]. This had been postulated by Bessonov and Umnov, and validated by a numerical search. Our work on this was immensely complex, and conceptually difficult, and ultimately nobody cared. Simon and I reworked his doctoral dissertation with additional material that I wrote into the monograph Machines that Walk [18].

In my original conceptualization of the machine, presented at the first project review meeting in 1981, I had realized that we would need to use hydraulic actuation, but that a conventional, valve controlled hydraulic system would be very wasteful of energy. For this reason I thought about using a hydrostatic actuation system with each actuator directly coupled to a variable displacement pump: what is now referred to as displacement control. After thinking some more about this, and after we had settled on a six legged configuration with eighteen degrees of freedom I went cold on this concept because of the weight implications of eighteen variable displacement pumps. We looked at dual pressure hydraulic systems, rather like the concept Boston Dynamics is now using on Alpha Dog, before becoming discouraged by that complex horror. Fortunately, Vince and his brains trust did a comparative analysis, and concluded that the displacement control concept was workable, and we decided to go with it. Despite its success, the fluid power industry has been very slow to take advantage. I can report that displacement control is now on the cusp of commercialization. Direct comparison tests conducted by Professor Monika Ivantysova at Purdue University have demonstrated fuel savings of the order of 40% for typical construction equipment tasks.

Around this time, in late 1983, we were under contract to build a full-scale prototype. It became apparent that it would be extremely difficult to deliver on this working through the established university shop facilities. Bob and I discussed the situation and were in agreement that it would be better to work through a company outside the university for this purpose. We discussed it with Clint Kelley, the

DARPA project monitor, and it was apparent there would be no obstacles from that end. At the time, as noted above, Vince Vohnout was a research associate in mechanical engineering committed full time to the project. Dennis Pugh and Eric Ribble were in similar situations in electrical engineering. We sat down with them and suggested they incorporate themselves, and we would give them a subcontract to cover construction and testing of the machine. In due course that came to pass creating Adaptive Machine Technologies Inc. The company is still doing business under the name AMT Systems.

Around the same time, it became very apparent that we did not have suitable space to embark on such an ambitious construction and testing program. I had secured laboratory space in back of the north wing of Robinson Laboratory, which was the largest space available within the then department facilities. We had built a full-scale prototype leg, and operated it on a test stand in that laboratory. It had the two-dimensional pantograph geometry, and slides implemented as roller guides that we later used on the ASV, after redesign to improve integration. We learnt a lot from that prototype, such as the importance of having a means of controlling the foot attitude, and how to shape the shank.

Anyway, assisted by the College of Engineering we embarked on a search throughout campus for more space. We were rather apologetically shown a large shed on West Campus. The interior of this was rather sad. It was unoccupied, and very decrepit. It was just what we needed! The interior had been partitioned into cubicle type offices. We needed to get rid of the partitions and clear the main part of the building for use as high bay space. To do this playing by the rules and going through the campus facility office would have taken forever, and probably cost a lot. Fortunately, in our decrepit building on West Campus we were not very visible to the university bureaucracy, and I adopted the philosophy that what they didn't know would not hurt them. We hired a gang of undergraduates for a few days and equipped them with sledge-hammers and pry bars, and had the unwanted partitions cleared out in no time. The AMT people were more than capable of setting up the electrical system so that no codes were grossly violated. They also had to do some roof repairs, and other maintenance to get the building in shape, but it subsequently served us well.

In 1985 Bob McGhee went on sabbatical leave at the Naval Postgraduate School in Monterey, and I became, effectively, the director of the project. In 1986 he formally retired from Ohio State University to take up a permanent position in the Department of Computer Science at the Naval Postgraduate School, and I became, officially, the principal investigator of the project. Three years later, in 1989, we had a new contract form DARPA and were planning the work to be done. I was on vacation in North Queensland, when I was contacted because things were going amiss with the contract. It took some time to work out what was going on, but a new director had taken over at DARPA and deemed our contract to be a waste of money. He wanted to void it in its entirety. We called in the cavalry in the form of Senator John Glenn's office and were able to keep the first year's funding, but the project had to be moved into shutdown mode. I remember standing at a public telephone in the middle of the night in Cairns, Queensland to talk to someone from the OSU research office about the situation with the contract.

Discussions in project meetings were vigorous and productive. One issue was allocating contact force among the feet in contact with the ground. We knew, after earlier work with the OSU Hexapod, that we had to control the feet in stance in force, but the question was what force should be commanded at each stance foot when there could be anywhere from three to six on the ground at any given time. Chuck Klein and David Orin figured that this could be formulated as a linear programming problem with a relatively minor approximation. The approximation is that the limiting friction force is directly proportional to normal force when projected into the xz and yz planes. In other words the friction cone became a friction pyramid. Chuck successfully pursued this approach with one of his students.

I thought about another approach: I thought that it would not be productive for the feet to work against each other. This meant that the components of the contact forces at any pair of feet along the line joining the contact points should be the same. I recognized that this condition was similar to that enforced on the velocities of any two points in a rigid body. Since that condition resulted in a helically symmetric field of velocity vectors, it seemed that the contact forces should also be distributed in a helically symmetric field, which would make it easy, and efficient to calculate them if one knew the field axis and its intensity. Vijay Kumar and I took this up [19]. We were able to work out a way to find the field axis and intensity given the resultant wrench that the contact forces were to equilibrate. This meant we had a very efficient, closed form algorithm for computing the force to be commanded at each foot in stance, given the wrench acting on the machine that was to be equilibrated. That was, itself, not so straightforward. The machine was massive enough, and fast enough, that we had to model its inertia and estimate the inertia wrench in each computation cycle. Rather than using the basic idea of maintaining static stability by keeping the projection of the center of mass within the support pattern we used the simple scheme of computing the wrench to be equilibrated: inertia wrench plus weight, computing the force allocation and checking that none of the commanded contact forces were less than zero, or actually a minimum threshold value.

There were a number of other theses that came from the ASV project. John Gardner took a more control oriented look at the force allocation problem. His work was co-supervised by Cheena Srinivasan [20]. Working with Mingzen Huang, we took a look at the relationship between speed and load carrying capacity in walking vehicles [21]. In contrast to wheeled vehicles, larger loads can be carried at slower speeds because the average number of legs in support increases. Satish Nair took a look at energy flows in the actuation system of a walking vehicle. His work was co-supervised by Raj Singh.

During the decade in which we were working on the ASV project we were also still pursuing research on serial manipulators. Indeed, it is notable that people who did their master's theses on manipulator work moved onto the walking machine project, and vice-versa. Thus, Vijay Kumar did his master's work on manipulator workspaces, before moving on to the multi-limbed force allocation problem cited above. Shi Liang Wang worked on obstacle crossing gaits for six-legged walking machines for his MS before shifting to workspaces and singularities of serial manipulators for his doctoral work. Others who produced doctoral dissertations on various aspects of serial manipulators during this period included Ming June Tsai and John Reidy. Another paper that seemed to make little impact at the time, but has lived on was Waldron, Wang and Bolin: "A Study of the Jacobian Matrix of Serial Manipulators" that introduced a compact formulation of the algebra that has been widely adapted [22].

During this period I also had another whack at the muscle recruitment problem in human walking with Ralph Cope. Marcus Pandy looked at biological quadrupedal locomotion with support from the ASV project under Necip Berme's supervision [23]. We were interested in how animals cross obstacles, particularly how they do it without looking at their feet.

While on sabbatical at Stanford I worked on Oussama Khatib's Artisan manipulator system, including the macro and mini manipulator wrist system. Together with Bernie's student Madhu Raghavan we explored the kinematics and coordination of mixed serial-parallel manipulation systems. That resulted in a paper in the ASME Transactions on Dynamic Systems Measurement and Control that is the only journal article I have co-authored with Bernie [24]!

On January first 1988 I began a five year term as technical editor of the *ASME Transactions Journal of Mechanisms, Transmissions and Automation in Design*. This was one of the two daughter journals that had resulted from the fission of the *Journal of Mechanical Design*. The other was the journal of *Vibrations, Acoustics, Stress and Reliability in Design*. The laundry list titles reflected the technical committee structure of the ASME Design Engineering Division at the time. *JMTAD* was struggling, with a weak subscription base. Tom Conry was the technical editor of *JVASRD* at the time. We talked about ways to improve both journals. He felt the stress and reliability in design material was a poor fit with vibrations and acoustics, and that it should be moved to my journal. I was apprehensive about this because the page number limits imposed by the ASME publications committee were tight. Tom told me that he was, in fact getting very few papers in these areas, which turned out to be true. We realigned the journals and I proposed that the name of my journal revert to *Journal of Mechanical Design*. The changes were approved effective January 1st 1990. The change of name solved the subscriber base problem!

Despite the name of the journal we were neither receiving nor accepting papers on design theory or methodology. There was growing interest in that area, but authors felt they had nowhere to publish, because papers in these topics fared badly in the review process in traditional engineering journals, including *JMD*. I tried to create a climate in which quality papers would be welcomed and published by appointing Erik Antonsson as associate editor with a brief to handle papers in design theory and methodology.

Around this time I also was active in research in design methodology myself in collaboration with Manjula. That resulted in the book Mechanical Design: Theory and Methodology [25] that we co-edited, and in which we co-authored several chapters.

Shortly after the end of the ASV project, Ken Hunt came to spend a part of a sabbatical leave at OSU. I had been tinkering with some theoretical ideas emanating from screw system theory, and my experience with the problem of coordinating multi-limbed systems, and found him to be a helpful sounding board. This collaboration resulted in the paper entitled "Series-Parallel Dualities in Actively Coordinated Mechanisms" that explored the duality between screw systems and wrench systems, and the mechanical consequences of that relationship [26]. It implied that, for any serial chain there was a dual parallel chain such that the screw systems of the one were isomorphic with the wrench systems of the other, and *vice-versa*. Further exploration of such relationships was to form the basis of much of my research over the next few years. The works of Vasudeva Murthy [27], Muqtada Husain [28], and Pie-Chieh Chin, in particular, followed this line of research.

I was also interested in applying what we had learned about coordinating multi-limbed systems to other types of system, notably multi-fingered hands. Vijay Kumar had already done some work on that problem. Sudipto Mukherjee [29]and Wen-Yeuan Chung [30] took the idea further. At about the same time I made my first foray into dynamically stable locomotion with Prabjot Nanua [31].

In 1993 I took up the responsibility of chairing the department. That did divert much of my energy and attention, but I did continue to work on research problems. I thought that the ideas we had developed for coordinating multi-legged, statically stable robotic systems could also be useful for wheeled systems with independently driven wheels. Those ideas were taken up by Shankar Venkataraman and S.V. Sreenivasan [32].

Another idea I was very fond of at that time (and still am) was trying to emulate nature and do force control of actuator arrays by recruitment. Successfully implemented, it would finesse the issue of wasting energy through control devices like hydraulic valves. That was the inspiration for Pohua Yang's work [33], and was continued at Stanford by Rocco Vertechy, where we attempted to use arrays of polymer actuators. Unfortunately, it is extremely hard to construct large arrays of simple actuators using currently available technologies.

I also got involved in design for manufacturability issues, co-advising Jason Yu with Kos Ishii. His work was on an alternative approach to design for robustness (rather than Taguchi's methodology). The central idea is that true optima often do not give good designs because they may be very sensitive to variations in the design parameters. It is better to seek an area of the design space that gives good performance but is not very sensitive to parameter variation.

Finally, David Orin and I received funding to pursue a dynamically stable quadruped design. This ultimately became KOLT. We started by trying to develop an appropriate compliant leg design. That was the core of Jim Schmiedeler's work [34].

I had been active in IFToMM: The International Federation for Promotion of Mechanism and Machine Science for many years. At the World Congress in 1999 I was elected to be President effective January 1st, 2000. I served two terms as president until December 31st 1997. That activity created professional relationships, and friendships with people throughout the world.

In 2000 I was completing my second term as department chair at Ohio State University and decided that it was time for me to step down. I decided that I wanted to focus on doing research, and working with postgraduate students for what was left of my career. I felt that a change of institution might be beneficial. I talked with my many friends in the design group at Stanford. It took a while, but they were able to set up an appointment as professor (research). Consequently, I left Ohio State University at the end of the 1999-2000 academic year and headed west.

The grant for the dynamic quadruped project was still in place, so I moved part of the money to Stanford. David and I continued to collaborate on the project using video-conferencing, and exchanging students. We also got a successor grant. Jim finished his work, and Jamie Nichol and Surya Singh came on board. Jamie did much of the mechanical design of KOLT, while Surya focused on sensing and data fusion issues [35]. David's students Darren Krasny and Luther Palmer worked on control. During this period Joaquin Estremera came for a year as a visiting scholar from Spain. He actually led most of the testing and data collection [36].

Muhammad Abdallah had been working with me on an actively suspended wheeled vehicle problem, but chose to do his doctoral work on dynamic legged locomotion, and developed a powerful design approach [37]. His work actually represented a transition to dynamic bipedal locomotion, that became the TRIP project. Alex Perkins pursued the dynamic behavior and control strategies for this device, primarily using simulation [38]. Paul Csonka pursued the hardware design, including a novel hybrid actuator [39].

I had also been working with the SUMMIT unit in the School of Medicine on a large NIH supported project with a focus on applications of virtual reality techniques in surgical training and diagnosis [40]. A project aimed at using haptic feedback for dermatological diagnosis grew out of that collaboration. Chris Enedah pursued this project. We were able to demonstrate an ability to detect and transmit skin texture information, but the overall project was multifaceted and much more challenging than initially meets the eye.

In March 2009, my close colleague Professor Kosuke Ishii unexpectedly passed away. I was best situated to supervise his advanced doctoral students through the remainder of their projects. Thus, I acquired four additional doctoral students working in various areas of design for manufacturability. Sun Kim and Whit Fowler were close to finishing, but Karthik Manohar and Jenny Wong would take a year or two to finish up. Karthik developed a design decision support system, including market feedback, for very complex products. Jenny worked on the evaluation and reduction of manufacturability risks arising from use of new manufacturing technologies. I also have my own design for manufacturability project. Kioumars Najmabadi, who works for Boeing Commercial Aircraft, has been working with me on evaluating and managing design risks in very complex products, with equally complex manufacturing environments.

Around 2006 Jim Schmiedeler and I were approached by Oussama Khatib and Bruno Siciliano to contribute a chapter to the Handbook of Robotics. David Orin was to be the sectional editor for what turned out to be the opening chapter of this award winning publication [41]. We recently revised the chapter for inclusion in the second edition.

One of the things we learned from the KOLT and TRIP projects was the importance of the foot, and of understanding the mechanics of its impact with the ground. We can run over very diverse substrates with no significant change in our gross running action, and without bouncing and slipping between the foot and the ground. Dan Jacobs recently completed his dissertation on modeling foot impact. Following on from his work, Linus Park is completing a study of the mechanics of the metatarsal joint, and of the effects of tendons connecting the foot segments to the thigh and shank.

Starting in 2007, I have been spending half of each year in Sydney, Australia. I have a half time appointment at the University of Technology, Sydney. I am actively participating in several projects there. One of these is the development of a robot to perform inspection of steel bridges and other ferrous infrastructure. Older steel bridges can be very complex in structure, and were protected by lead based paints. It is very difficult to comprehensively assess the condition of the paint, and the presence or absence of corrosion. The device we are exploring for this purpose is a seven-degree of freedom inchworm robot with magnetic feet on both ends.

We are also working on exoskeletons both for rehabilitation and for industrial applications. Marc Carmichael is finishing his dissertation on the use of a biomechanical model of the torso and arm in the control of an upper limb exoskeleton. I am co-advisor of his project with Professor Dikai Liu.

One important project that doesn't fit into the above, roughly chronological account is the text Kinematics, Dynamics and Design of Machinery that Gary Kinzel and I co-authored [42]. This was a multi-year effort for the first edition that appeared in 1998, not to mention the second edition that appeared in 2003. We are now working on a third edition with Sunil Agrawal joining us as third author.

This is a necessarily brief summary of a satisfying and eventful career; one that is still in progress, although at a reduced level of activity. There are many things that I have consciously, or inadvertently left out. In compiling an account like this one looks for landmarks around which to organize it. I have used the many doctoral projects that I have supervised, or co-supervised as those landmarks, together with other significant efforts such as books, and major projects. Likewise, by and large, I have chosen to cite only journal articles and books. There were many masters' theses that I do not have space to mention, and many conference papers, some at least as important as the items cited.

As I look back on all this, there is one essential truth: an academic career is about the students. The greatest reward is to see them progressing in productive careers of them own. I have been blessed to work with a great many very able people, and I appreciate the honor of having contributed to their lives in even a small way.

References

1. Waldron, K.J.: The Constraint Analysis of Mechanisms. J. Mechanisms 1, 101–114 (1966)
2. Waldron, K.J.: A Family of Overconstrained Linkages. J. Mechanisms 2, 201–211 (1967)
3. Waldron, K.J.: Hybrid Overconstrained Linkages. J. Mechanisms 3, 73–78 (1967)
4. Waldron, K.J.: Symmetric Overconstrained Linkages. J. Engineering for Industry 91(1), 158–164 (1969)
5. Waldron, K.J.: A Study of Overconstrained Linkage Geometry by Solution of Closure Equations, Part I: A Method of Study. Mechanism and Machine Theory 8(1), 95–104 (1973)
6. Waldron, K.J.: A Study of Overconstrained Linkage Geometry by Solution of Closure Equations, Part II: Four-Bar Linkages with Lower Pair Joints Other Than Screw Joints. Mechanism and Machine Theory 8(2), 233–248 (1973)
7. Baker, J.E., Waldron, K.J.: The C-H-C-H Linkage. Mechanism and Machine Theory 9, 285–297 (1974)
8. Waldron, K.J.: Note on the Order Problem of Burmester Linkage Synthesis. J. Engineering for Industry, Series B 97(4), 1405–1406 (1975)
9. Waldron, K.J.: Elimination of the Branch Problem in Graphical Burmester Mechanism Synthesis for Four Finitely Separated Positions. J. Engineering for Industry, Series B 98(1), 176–182 (1976)
10. Strong, R.T., Waldron, K.J.: Improved Solutions of the Branch and Order Problems of Burmester Linkage Synthesis. Mechanism and Machine Theory 13, 199–207 (1978)
11. Sun, J.W.H., Waldron, K.J.: The Order Problem of Spatial Motion Generation Synthesis. Mechanism and Machine Theory 17(4), 289–294 (1982)
12. Ahmad, A., Waldron, K.J.: Synthesis of Adjustable Planar 4-Bar Mechanism. Mechanism and Machine Theory 14, 405–411 (1979)
13. Kumar, A., Waldron, K.J.: The Workspaces of a Mechanical Manipulator. J. Mechanical Design 103(3), 665–672 (1981)
14. Song, S.M., Waldron, K.J., Vohnout, V.J., Kinzel, G.L.: Computer-Aided Design of a Leg for an Energy Efficient Walking Machine. Mechanism and Machine Theory 19(1), 17–24 (1984)
15. Waldron, K.J., Vohnout, V.J., Pery, A., McGhee, R.B.: Configuration Design of the Adaptive Suspension Vehicle. IJRR 3(2) (1984)
16. Song, S.M., Waldron, K.J.: Geometric Design of a Walking Machine for Optimal Mobility. J. Mechanisms, Transmissions and Automation in Design 109(1), 21–28 (1987)
17. Song, S.M., Waldron, K.J.: An Analytical Approach for Gait Study and its Applications on Wave Gaits. IJRR 6(2), 60–71 (1987)
18. Song, S.M., Waldron, K.J.: Machines That Walk: The Adaptive Suspension Vehicle. MIT Press, Cambridge (1988)
19. Kumar, V., Waldron, K.J.: Force Distribution in Closed Kinematic Chains. IEEE Trans. on Robotics and Automation 4(6), 657–664 (1988)
20. Gardner, J.F., Srinivasan, K., Waldron, K.J.: Closed Loop Trajectory Control of Walking Machines. Robotica 8, 13–22 (1990)
21. Huang, M.Z., Waldron, K.J.: Relationship Between Payload and Speed in Legged Locomotion Systems. IEEE Trans. Robotics and Automation 6(5), 570–577 (1990)
22. Waldron, K.J., Wang, S.L., Bolin, S.J.: A Study of the Jacobian Matrix of Serial Manipulators. J. Mechanisms, Transmissions and Automation in Design 107(2), 230–238 (1985)

23. Pandy, M.G., Kumar, V., Waldron, K.J., Berme, N.: The Dynamics of Quadrupedal Locomotion. J. Biomechanical Engineering 110(3), 230–237 (1988)
24. Waldron, K.J., Raghavan, M., Roth, B.: Kinematics of a Hybrid Series-Parallel Manipulation System. J. Dynamic Systems, Measurements and Controls 111(2), 211–221 (1989)
25. Waldron, M.B., Waldron, K.J. (eds.): Mechanical Design: Theory and Methodology. Springer, New York (1996)
26. Waldron, K.J., Hunt, K.H.: Series-Parallel Dualities in Actively Coordinated Mechanisms. IJRR 10(5), 473–480 (1991)
27. Murthy, V., Waldron, K.J.: Position Kinematics of the Generalized Lobster Arm and Its Series-Parallel Dual. J. Mechanical Design 114(3), 406–413 (1992)
28. Husain, M., Waldron, K.J.: Position Kinematics of a Three-Limbed Mixed Mechanism. J. Mechanical Design 116(3), 924–929 (1994)
29. Mukherjee, S., Waldron, K.J.: Multi-Fingered Manipulation of Loaded Objects. J. Applied Mechanisms and Robotics 2(1), 21–29 (1995)
30. Chung, W.Y., Waldron, K.J.: An Integrated Control Strategy for Multifingered Systems. J. Dynamic Systems, Measurement, and Control 117, 37–42 (1995)
31. Nanua, P., Waldron, K.J.: Energy Comparison Between Trot, Bound, and Gallop Using a Simple Model. J. Biomechanical Engineering 117(4), 466–473 (1995)
32. Sreenivasan, S.V., Waldron, K.J.: Articulated Wheeled Vehicle Configuration with Extensions to Motion Planning on Uneven Terrain. J. Mechanical Design 118(2), 219–311 (1996)
33. Yang, P.H., Waldron, K.J.: Massively Parallel Actuation. In: Proc. AIM 2001, Como, Italy, July 8-11 (2001)
34. Schmiedeler, J.P., Waldron, K.J.: The Effect of Drag on Gait Selection in Dynamic Quadrupedal Locomotion. IJRR 18(12), 1224–1234 (1999)
35. Nichol, J.G., Singh, S.P.N., Waldron, K.J., Palmer, L.R., Orin, D.E.: System Design of a Quadrupedal Galloping Machine. IJRR 23(10-11), 1013–1028 (2004)
36. Estremera, J., Waldron, K.J.: Thrust Control, Stabilization and Energetics of a Quadruped Running Robot. IJRR 27, 1135–1151 (2008)
37. Abdallah, M.E., Waldron, K.J.: The Mechanics of Biped Running and a Stable Control Strategy. Robotica 27, 789–799 (2009)
38. Perkins, A.D., Waldron, K.J., Csonka, P.J.: Heuristic control of bipedal running: steady-state and accelerated. Robotica 29(6), 939–947 (2011)
39. Csonka, P.J., Waldron, K.J.: Characterization of an Electric-Pneumatic Hybrid Prismatic Actuator. J. Mechanisms and Robotics 2(2), 021008-1–021008-16 (2010)
40. Dev, P., Montgomery, K., Senger, S., Heinrichs, W.L., Srivastava, S., Waldron, K.J.: Simulated Medical Learning Environments on the Internet. J. Am Med. Inform. Assoc. 9, 437–447 (2002)
41. Waldron, K.J., Schmiedeler, J.P.: Kinematics. In: Siciliano, B., Khatib, O. (eds.) Handbook of Robotics, ch. 1, pp. 1–26. Springer (2008)
42. Waldron, K.J., Kinzel, G.L.: Kinematics, Dynamics and Design of Machinery. John Wiley and Sons, NY (1998) (Second Edition, October 2003)

Contents

Part I: Historical Perspective

Part II: Kinematics and Mechanisms

Part III: Robotic Systems

1 Design for Inclusivity:
Meaningful Collaboration with Differences

Manjula and Kenneth J. Waldron

Abstract. The ubiquitous availability of the internet for global networking has made collaborating by differences necessary. This is challenging because of the human history of tribal organizations. In this paper we draw from our 45 years of personal and professional journey together to elicit the skills that have helped us to successfully engage across racial, cultural, and academic differences. It required designing new rules of social and academic engagement that changed how we related with the "other". We rely on these techniques daily to inform us on how to facilitate collaboration in any situation capitalizing on the diversity of thought, body, experience, belief, and/or training while focusing on the similarity of our journey.

1   Introduction

"Differences in religious beliefs, politics, social status, and position are all secondary. When we look at someone with compassion, we are able to see beyond these secondary differences and connect to the primary essence that binds all humans together as one."--- **Dalai Lama**

The late and important techno-social change agent, Steve Jobs, said in his 2005 commencement address at Stanford [1]. "…You can't connect the dots looking forward. You can only connect them looking backwards, so you have to trust that the dots will somehow connect in your future. You have to trust in something— your gut, destiny, life, karma, whatever—because believing that the dots will connect down the road will give you the confidence to follow your heart, even when it leads you off the well-worn path, and that will make all the difference." He was talking to graduates, and using his unique life journey as an example to

Manjula · Kenneth J. Waldron
ME Design Group, Stanford University, Stanford, California, U.S.A.
e-mail: mwaldron@stanford.edu

V. Kumar et al. (Eds.): *Adv. in Mech., Rob. & Des. Educ. & Res.*, MMS 14, pp. 1–16.
DOI: 10.1007/978-3-319-00398-6_1   © Springer International Publishing Switzerland 2013

make the point that one cannot predict the future. We must seek and allow our faith in our destiny to guide us knowing that we have the capacity to rise should we fall, finding sustenance should we be needy, and willing to be vulnerable and foolish should we become full of ourselves [2]. This is a profound human ability: to connect and engage something beyond ourselves to liberate us by unlearning beliefs that are not serving us and engaging differently with those around us, thus changing the very fabric of the social structure of which we are part of.

That we, an Australian white male and an Indian dark skinned female, chose to marry in 1968 was sheer foolishness in the eyes of many of those who knew us. It was Ken's personal integrity and Manjula's courage to be different that held us together thus far through our social vulnerability [3]. As we look back over the forty-five years of our collaborations both personally and professionally, there were lessons learned from the wisdom of our folly that have guided us to collaborate in a meaningful way with people who are different in status, nationality, disciplines, styles, gender, age, thoughts, values, race, and ethnicity. It has been a wild journey, but the essential process is tractable, trainable, and lends itself to redesigning collaborations from an unusual and whole life perspective.

It is this knowledge that helped Manjula to create and offer an undergraduate design course called Design for Diversity with the Associate Dean for diversity and "first gen", Tommy Woon, at Stanford.

Perhaps we have been closet social designers throughout our lives. We contributed to the social change of the sixties, and in our sixties we continue to affect it. What do we mean by that?

Both of us have lived off the beaten path, following our hearts even though the paths we took before we met were as different as night and day. However, there was something enduringly human in our life engagement that brought us, and has kept us together even when social curve balls were thrown at us. These issues had the effect of amplifying our differences and testing our resolve and commitment for a meaningful social coexistence.

As designers [4] we have used our design and networked thinking to devise durable products and collaborative relationships that are inherently inclusive. Did we know it consciously going forward? Of course not! But as we connect the dots backwards, the dots that seemed random and disconnected form a perfectly designed and orchestrated wonderful, unique, and strong tapestry of life that we were crafting.

In our paper, we will retrospectively use the design process wisdom from our personal and professional journey to draw out basic human qualities that make collaboration across differences meaningful and productive. As engineering designers, we examine these qualities and abilities deeper and not fall prey to resorting to the social clichés of "it was just blind love". We introduce and use the term "*rules of engagement*" to mean how we relate to those who we encounter in our social environment. Some of those rules are given to us unquestioned in our upbringing. They form part of our unconscious personality. Others we acquire through conscious reasoning.

This article therefore complements Ken's professional career summary also presented in this book. We hope that it will be engaging and useful reading, and provide a guide in designing deep, lasting, and inclusive collaborative alliances in a globally diverse and digitally networked world.

We organize this paper by first setting the stage, then following a brief journey of personal, professional collaborations across differences, and eliciting the lessons we have learned as keys for design for diversity, creating rules of engagement that facilitate collaboration across differences in academic and social settings.

We conclude by crowd-sourcing our readers: extending an invitation to contribute their thoughts on what would it take to design a Collaboratorium where robust and creative collaborations can emerge capitalizing on the diversity of thought, body, mind, social experiences and belief, and/or training in a globally networked digital world.

2   Foundations: The First Quarters of Our Lives

2.1   Manjula — From Sukker to Stanford

One could say I have been a closet diversity engineer most of my life, engaged in designing a world that empowers those who were socially marginalized and different to find their voice on the table [5].

My grandparents and parents were freedom fighters and committed Gandhi followers. They consciously rejected the Indian caste system, religious differences, and social violence to win India's freedom. As the family legend goes my grandfather in the naming ceremony "havan" for his daughter, put Allah, Raam, God, Prabhu on its four corners. This offended the Hindu priest. He wanted Allah removed. When my grandfather refused, the landlord threatened to evict them. My grandfather, not to be deterred, found a house for sale nearby, found someone who will loan him the money, and then bicycled 50 miles that night to purchase the house. Next day they moved their meager belongings into it and then, true to his Gandhian beliefs that all religions represented the same truth, performed the ceremony as he had wanted to the night before.

I was born in 1943. My parents lived in Sukker then, in what is now Pakistan. India's freedom was being negotiated. Mahatma Gandhi had rejected Bose's suggestion that they use violence against the British. In 1947, The India-Pakistan partition was the price that was paid for it. Its violence was a tremendous blow to our family tenet of nonviolence. Our family fled what is now Pakistan as refugees, to escape the carnage [6]. In independent democratic India policies were created and enforced for a secular modern India, despite recurrent episodes of communal violence that would, ultimately, cost Gandhi his life. Legal safeguards were provided for the socially deprived. For the first time they could find educational opportunities and thus become part of the larger dialog.

I had friends from different castes and all parts of India. Their first language was often different from mine. I had Hindu, Muslim, Christian, and atheist friends. Because of my father's work with the Indian Railways we moved a lot, and lived in different parts of India exposing me to many of the diverse languages and cultures represented within India. In fact the first language I learned to read and write was Gurmukhi, and not Hindi. This experience helped me to learn to befriend those different than I at many levels.

Nevertheless, independent behaviors of young women were socially restricted. This attitude curbed the growing feminist expressions of the younger, post freedom, set. For example, I was coerced into wearing a sari; into pleasing and not openly questioning those with higher social status (no matter how they violated my personal value system). Contact with the opposite gender outside the family was "verboten". Women's sexuality could not be expressed openly and freely. Those in power strictly enforced their authority. It gave me plenty of fodder to protest and be angry about the gender unfairness, inconsistencies, and to question its legality. It got me into trouble at home and in school. Nevertheless, fear of punishment was not sufficient to quench my internal fire of seeking to be an independent educated professional woman.

Diversity was in the very air I breathed. My parents, who were themselves, both university graduates, were committed to educate me alongside my brothers. At the time, women were few and far between in science. In 1962, I was the only woman in a class of about one hundred to graduate in Physics Honors at the University of Delhi. However, since I enjoyed design and technology, the appeal of theoretical physics had diminished by the time I graduated. I applied and was selected on national merit as one of three women to attend and graduate in electrical (communication) engineering from the Indian Institute of Science, Bangalore. After graduation, IBM rejected me for employment because of my gender. I worked as an engineer in Bombay (now Mumbai) for Tata Hydroelectric Power Company. I was accepted to attend Stanford in the Fall of 1966. Until then I had never been out of India.

My family was probably relieved to get me off their hands. In fact they were very progressive to allow me to leave India as an unmarried woman, despite the protests of many in their community that this transgression on their part would reduce my prospects of an arranged marriage.

In 1966 the Stanford campus was very white, and the engineering student body was all men. In January 1967, around 200 students took the Ph.D. qualifying examination in electrical engineering. I was the only woman who passed it, along with about 100 men. Engineering at Stanford was a bleak and lonely place for me as a woman [7]. The rules of engagement that people around me used appeared to be different to those I had encountered in India. Of course, they were unwritten social rules to which I had no access. It made for a difficult journey. But for the support from my parents and advisor, I would probably have left.

Over the course of my college career I clandestinely dated many men from different walks of life. It was not easy and there was constant tension between what was sanctioned, and what I desired. It seriously challenged my personal integrity. To set it in perspective, within my extended family most people had arranged marriages within our caste, with very few exceptions. At some level I knew that was what was expected of me. Therefore the dating scene at Stanford, although exhilarating, was very disorienting for me.

2.2   Ken: From Sydney to Stanford

I grew up in paradise! When I was six years old, my parents moved to a new house on the headland between Freshwater and Curl Curl beaches. Freshwater beach is the next one north of Manly on the north side of Sydney Harbor. It was, and is, a gorgeous little beach with good, relatively safe, surfing conditions.

Of course, I didn't appreciate the marvelous setting in which I was living until much later. Initially, I did not know how to swim, and the surf could be pretty intimidating when the wind was propelling big waves on shore. Other times the conditions were such that the waves were breaking very close to shore and, if you were a little kid, they would pick you up and drive you into the hard wet sand.

Harbord, which is what the area was called, was in transition from being an impoverished fishing village to a suburb of Sydney. There were quite a few rough boys who came to school without shoes. It was relatively remote from the city of Sydney. When I finally got to university I had a prolonged trek each morning by bus to Manly wharf, ferry to Circular Quay, and another bus up town to the University of Sydney, with the reverse in the evening. I got a second hand, beat up, VW beetle in my second year and would commute by road, but that was not any better since the only reasonable way from Manly to the city was over the Spit Bridge. That road became thoroughly choked with cars in peak hour, so I reverted to commuting by ferry most days. However the car taught me a lot about practical mechanical engineering.

Growing up so close to the ocean, I did develop a fascination with the prolific marine life I could explore just by hopping around the rocks on the headland near my house. That fascination is with me to this day. I sometimes wonder why I didn't become a marine biologist. I don't think it ever occurred to me that one could do that for a living.

I don't really know where I got the idea of becoming an engineer. Most likely it was the vocational guidance counselor in high school. They gave us all a battery of tests to indicate what we were best suited to pursue. I did well in all of them and was told I could do whatever I wanted, but that maybe I should consider engineering. I was a little primed for that anyway since my grandfather was working as a draftsman at the Water Board in those days, and he would give me the company journal that was full of technical articles on construction of dams and pipelines.

Someone at school gave me an application form for a Steel Industries Traineeship, which I filled out. Then, as now, there was a public examination administered by the state of New South Wales at the end of high school. The exam was administered at the school, but was centrally graded, and the results were published in the newspaper. It was a tense time since university admissions and scholarships rode on that result. When the results were published I had come eighty-fourth in the state. I would attend the University of Sydney with a Commonwealth Scholarship, meaning my tuition would be covered. I had also won a Steel Industries Traineeship, which meant that I would work at the Port Kembla steelworks in my vacations, and earn a little money. I must have put either mechanical or electrical engineering on the form, because when I showed up for interview by an officer of the personnel department he asked me to decide then and there which it was to be. I felt more comfortable with the visible cause and effect of mechanical systems, so I said mechanical engineering.

I would be the first member of my family on either side to attend university, so I had to find my own way. In fact, to this day I'm the only one (other than our own children) who holds a postgraduate degree. My sister is four years younger than I am, and I guess she wasn't prepared to cope with undergraduate life. A talented musician, she dropped out after one year and focused on playing the cello.

Australia at that time was a very homogeneous society. We were under the sway of the, now infamous, white Australia policy. There was an influx of white immigrants from war-devastated Europe. There were Dutch and Italian boys at my high school. They had a rough go of it. I remember the Dutch boys did not understand any English, at least initially. The Italians were not so linguistically challenged, but because of their darker skin, they were called derogatory names. I had very little experience with people of other ethnicities. In first year chemistry at Sydney University we had laboratory in a huge room with rows and rows of benches. There were one or two students from Nigeria, and a few Indian students from Fiji, including girls, who stood out in their colorful flowing saris. I didn't interact with any of them, choosing to stick with my friends, who were mostly boys from my high school.

After completing the bachelor of engineering degree with first class honors, and a stint working at the steel works, I was awarded a Commonwealth Scholarship to pursue the degree of Master of Engineering Science. This was unusual for Australia at the time by virtue of being a course work plus thesis degree. While pursuing that degree I took a course on mechanism kinematics with Jack Phillips and became fascinated by the problem of mobility in linkages. I wrote up some of my ideas. Jack introduced me to Ken Hunt, so that when I decided to come to the U.S. to pursue a Ph.D. Ken advised me to contact several universities, one of which was Stanford. Bernie Roth offered me a research assistantship, and I came to Stanford in Summer of 1965.

At the time, most Australian students who wanted to go overseas to complete their education headed to the U.K. That included most of my peers at the University of Sydney. A few stayed on to do doctorates in Australian universities. Looking to the U.S. was a bit radical. My parents always had a subscription to National Geographic magazine. I read them voraciously. Probably I had read so much about places in the U.S. that I wanted to go see for myself. It was just as well because there was very little going on in mechanism theory in Britain at that time.

I think I was well informed about the diversity of people in the world through my reading of many issues of the National Geographic. The civil rights movement was in the news, so I was aware of the racial inequality issues in the U.S.A. However, to me all this was "theoretical". I did not think that it did, or could ever apply to me.

At Stanford I was sharing a room in Crothers Memorial Hall. There were quite a few other international students in the dormitory. I also attended some functions at the International Center, and was interacting with other international students. Sultan Bhimjee and Fazle Hussain were sharing a room nearby. They were both Muslims from Pakistan but were, in fact, from very different cultural backgrounds. Sultan was from West Pakistan. Fazle was from what was then called East Pakistan, now Bangladesh. After the academic year ended I rented an apartment off campus shared with Fazle, Sultan and Scott Williamson, who was from Edinburgh. That was summer 1966. Manjula arrived on campus for the start of classes in Fall. We had a party in our apartment, and she came. We danced together and I drove her home, but we did not start to date until January 1968.

3 Collaborating Globally

3.1 Summer of Love

In January 1968 we took the same Modern Algebra class. We had seen each other around, and attended functions as international students. At some level we were both aware of our social differences and neither of us was equipped to learn how to bridge them. Yet, close proximity had a magical effect as we realized that at some deep level we were not all that different. Before we knew it we had committed to live the rest of our lives together. Thus, the Australia-India-US triangle of negotiations commenced. In those days snail mail was the only means of affordable communication. Sending a letter and receiving a reply took two or three weeks. Finally, we convinced our families that giving us their blessings was the only sane thing to do. At a human level we could get them all to see that our family values were common human values and therefore in perfect alignment. We were married in a California style Christian ceremony in our friend's house, vows that were reiterated in a Hindu ceremony at my grandfather's house in India and a reception in Ken's parents' place in Sydney in the Summer of 1968.

3.2 Paradise Lost

Ken was on an exchange visa and had to leave after completing his degree, and a one-year stint as acting assistant professor at Stanford. He had been offered a position at the University of New South Wales. Manjula put completing her Ph.D. on fast-forward and became ABD by the time we left to move to Sydney.

At this time we encountered the difficulties of securing a visa for Manjula to be a resident of Australia. Being of Asian birth, she was not then eligible to become an Australian citizen, despite our relationship. What Ken didn't understand was that it was an open question that she would be able to live in Australia at all. We filled out the application form, but one requirement was a copy of a police record from the community she lived among in India. We knew that no such record would be obtainable, she had no police record. After several telephone calls to the consulate in San Francisco we were summoned for an interview. We must have convinced the consular official that we were harmless, because the necessary visa was provided, in the nick of time.

When in Sydney Ken introduced Manjula as his wife to some of his old university friends and, at least in one case, was roundly snubbed. In others no further interest was shown to continue the friendship.

Manjula was pregnant with Andrew [7]. He was born in December 15th, 1969. Both of us were full-time junior faculty members. With no childcare or support available for working mothers it made for a rocky start.

A year or so later we found out that Andrew was profoundly deaf. Lalitha was born on December 3rd 1971. It was a difficult time for us with the demands of learning about educating a deaf son, engineering our parenthood, and with both of us being ambitious, struggling young academics. To compound it all the white Australia policy was still in place and that meant that there was strong institutional prejudice against Manjula's Asian heritage, and therefore against our mixed race children. International travel for our young family became a nightmare. In the early 70's, for Manjula being a woman engineer, who was colored and a conscientious working mother of a handicapped child was not socially acceptable in Australia.

In 1972, the conservative federal government that had ruled Australia since the Second World War, and had formulated and maintained the white Australia policy was defeated. The new Labor government set about a program of reform, including dismantling many of the provisions of the white Australia policy. By 1974 Manjula was eligible for Australian citizenship. Ironically, we had already decided to leave.

Our social and academic life had taken on an international complexity that we were ill-equipped to handle. The social diversity of our family was now on steroids with no relief in sight within the local society. Like true academics we took a sabbatical to see if there was somewhere else we could make our family life more workable. Reluctantly, although of necessity, we decided to move to Houston, Texas in 1974. When we arrived there it seemed that we had leapt from the frying pan into the fire. It was hard to believe that Texas was part of the same country as Northern California, which was our only other experience with the US.

Ken, as a white male, fitted the mainstreamed engineering scene in the US, while Manjula's career, as a colored female, was more precarious. We became aware of social biases that existed in the engineering academe. This was further compounded by our mixed-race marriage and our wanting to be and live together. We barely survived as a family and decided to move once again when an opportunity came for Ken to move to Ohio State University in 1979. We decided to try the Midwest. Paul Ravi was born on Aug 7th only a month before we were due to move to Columbus.

Once again Ken fitted the mold of an engineering professor and Manjula's career hiccupped along. The same factors that lay behind our professional/personal difficulties in Houston seem to have followed us to Columbus. It did not compute. Columbus was north of Mason Dixon line and not the south. It was 1980, fifteen years after the civil rights law was passed. Our commitment to stay together was gravely tested. Manjula's parents came to help us so she could work towards restoring her career to become tenured. This was accomplished. However, with no diversity tools to work with, and with international, intergenerational, interracial, and disability issues under the same roof, our life together created its own social toxicity. Out of necessity, Manjula became active in multicultural and diversity issues at OSU.

We learned that changing the social structure is not easy, but is doable if one is adaptable and willing to learn to change. The consequences of defying social hierarchies that set the rules of engagement were never taught in our science and engineering training. We began to realize that the game of life is played very differently when you step out of the well-trodden paths laid out over centuries. Thinking out-of-the-box through understanding social realities that surround us becomes essential. It requires redesigning the rules of engagement that make the society operational. In mid 90s Manjula attended a multicultural conference at Stanford organized by Tommy Woon, then a multicultural director at Stanford. Thus began a long collaboration to understand and address the issues of collaboration by differences that continues to this day.

3.3   *Paradise Gained*

In the early eighties, with a grant from the newly established Design Theory and Methodology program at NSF we began interdisciplinary research in design. At a DTM conference that we organized in Berkeley in 1987, it was clear that there was more commonality to engineering design methods than our silo existence led us to believe. There was something inherently human in the process of designing that transcended disciplinary boundaries. As Bernie Roth said, one learns design by designing. It is inherently a human integrated experience. We realized that as engineers, and bioengineers our knowledge of humans and their social interactions was incomplete, and the assumptions on which it was founded needed re-examination. The science for designing integrated dynamic systems, like that of life-cycle design of biomedical products was complex, and reductionist science was quite inadequate to solve the complex problems that confronted us. We introduced concepts from the science of complexity in education and research. Our design thinking research was showing us that designers were human first and their design processes had human dimensions [8]. This work led us to actively collaborate with our business, psychology, and humanity colleagues to create the

Center for Integrated Design (CID) at OSU in 1995 in which we engaged faculty from the entire University, including Law and Medicine.

In 1996, our personal family medical crisis exposed us to the limitations of the biomedical reductionist scientific paradigm and put it up close and personal. It was our non-engineering colleagues from the humanities, business, and law who exposed our blindness and opened our eyes to the dimensions we were missing in our bioengineering understanding. We thus embarked on searching for the missing cards in our playing deck. We both learned the tenets of holistic coaching and have learned to practice and change our learned beliefs to now produce different results in mind, body and spirit. We have acquired many tools that help us, and those we interact with. As a result, Manjula has trained as a chaplain resident at Stanford, she practices yoga and Chi-Gong, is certified in mind-body medicine, holistic wisdom coaching, and energy medicine, and teaches resilient aging and diversity classes based on holistic health.

In Yoga there is a concept of 5 koshas [9] or sheaths that hold our truth. These are the Mental (M), Physical (P), Social-spiritual (S), Intuitive (I, integrative) and our Emotional (E) and Life (L) forces are held together through this truth. The acronym SIMPEL™ works well for us. We have access to which of these principles are guiding us at any moment, if we take time to be mindful. We have the capacity to change if we stay present in the information accessible to us through acceptance of the moment. We have found the tools that allow us to regain balance in our SIMPEL™ house are very beneficial to collaboration across differences. Some of these tools like meditation (relaxation response), moderation in food, movement, positive attitude, and in psycho-social interactions were known a long time ago but were lost and distorted over time, and lost relevance.

In the last 20 years these principles have been researched and re-contextualized through modern science and medicine [10]. Other tools have been developed through necessity brought on by technology. To paraphrase Herbert Simon from The Sciences of the Artificial [11]— many things may use natural components in today's world but what we have around us is not natural, it has all been designed. None of us humans exist in our natural state. We inhabit a world that has been created to suit our needs. We agree and believe that we can design a social system of which we are part that facilitates positive interactions across diverse ways of being. It requires giving up our habitual ways of being through mindfully changing the beliefs instilled in us during our upbringing. Aggregating by similarities is one of our natural neural traits [12] around which tribal societies were created. Hence we have a long history of tribalism. Some may even argue for it to be natural. However, social science is increasingly showing that this is not so. Race, ethnicity, languages, cultures are all social constructs that foster tribalism [21].

Network technology is increasingly challenging societies based on tribalism, and our short, collective history of designing a caring diverse culture, is pointing in the right direction to create a peaceful world. Even though the media may have us believe otherwise—the world is much more peaceful now through social design than ever before [13, 14]. Research is showing that peace is measurable, designable and executable if we are willing to change our rules of engagement to be based on globally agreed upon values [15].

4   Key Learnings That Promote Collaboration across Differences

As we reflect on the story narrated above of our extraordinary journey, here are some key points that we have learned to successfully collaborate.

The human qualities we have found useful are: Active listening and understanding, caring, sharing, trust, value, commitment, empathy, common humanity, compassion, understanding, letting go of fear and negativity, forgiveness, receiving and giving love, acceptance, safety, reframing, gratitude, faith, charity, hope, open mind and heart, adaptive—willing to change the rules of engagement.

How do we design a life with such qualities so as to relate to the other especially if they are different from us?

Here is a list that captures our learning based on our journey:

1. Connect to the breath

This is most important to bring calm in the moment. Take a deep and relaxing breath, focus on it and be grateful for it (put a hand on the chest if needed). Breath is the essence of life and it is given to us and connects us to the other unconditionally. This awareness alone bridges the difference in the present. Despite its simplicity it is not always easy to implement, especially if we have grown up in a narrow tribal belief system because being with the other may evoke a state of fear within us. Our body is the best indicator of fear and connecting with it allows us to ground and center ourselves and harness its power by connecting to our breath [16].

2. Develop personal integrity

Confront the real source of anger, so as to be present for the other who is different. It requires that we can accept and actively understand our own feelings so that we can be present for the other on an as-needed basis [17]. This means getting in touch with the internal and external realities that we are encountering; understanding the frustration and anger we are feeling within us by becoming aware of it, its source, and committing to letting it go from within us. In other words "fess up" to our own reality, diffuse it from within us. It is not about the other and once the anger is understood it becomes innocuous. Our resolution to never go to bed angry with the other helped. It requires time for reflection and willingness to create a mutually understood environment of safety. Taking this time helps us.

3. Acknowledge and respect the other's feelings with humility and honesty

Engage in active listening and understanding: to our inner self as well as the other so that the situation can be explored more deeply to gain a better understanding of the issues at hand. Mindfully examine the source of our own feelings. It requires gaining calm so as to not retaliate mindlessly in the heat of the moment, but embracing what is broken within us that we are projecting onto the other. It requires respect and caring for the wellbeing of the other. Since emotions are contagious it requires practice to learn to pause when aroused and examine the situation. We are getting better at it.

4. Be willing to change our point of view and say sorry from the heart

This requires opening up to new possibilities. It is important to know what is due to something we have no control over and something that is learned from our environment growing up. It requires changing the story that takes over our mind out of habit and telling it differently and seeing how it feels. However, the very willingness to say sorry with genuine feeling gives us a window to pause, breathe, accept, and self reflect on our own feelings for the other. This we find takes reflection and requires patience, commitment, and caring.

5. Use power of commitment

When we feel the stress of different ways in which we engage with the other, it is important to negotiate and design and implement a new functional rule of engagement so we accommodate the other in a positive light. Be aware and practice until this new engagement becomes a norm. Instead of perfection we have developed a Pareto rule of successful engagement. If we succeed 80% of the time we have done well and it deserves celebration. If the rule works 80% of the time in the relationship then the rule of engagement is worth keeping and adapting around. This has helped us most through our low points.

6. Build trust with the other

Build trust so that they can be who they are in our presence and vice versa. This is the hardest when deep social prejudices are acculturated within us growing up that create the mistrust of the other. The insidious part of this is that we are consciously unaware of this tribalism within us. Building this bridge is essential for meaningful collaborations. It requires cultivating a new belief system. The fast emotional brain that jumps to older unconscious behavior patterns needs to change so that the stories we tell ourselves change. This is a challenging but rewarding achievement [18]. Understanding, caring, and empathy are essential for this. For example the deep racial divide that we grew up with still stumps us at times and by acknowledging its source and challenging its validity through breathing and active understanding of its source helps.

7. Create a common set of values

Common values are needed for daily living so meaningful relationships can develop [19]. Fortunately for us our core values were mostly aligned. However, there were many ways in which we mindlessly acted that did not square with them. We had to negotiate to come to commonly held values. It is something we still work on.

8. Have open dialog and communication

Establish communication through mutual acceptance and caring. It is willingness to be open and to discuss what matters to us and discover and acknowledge the positive qualities of the other [20]. This is challenging especially in a hierarchical tribal system. Bringing it into the open in a thoughtful caring manner can be difficult but is doable.

9. Be willing to bring the functional, relevant, and shared expectations from the past

To create a meaningful future, be willing to recognize, let go of, and reframe past behaviors or rules of engagement that are irrelevant. One way is to use regenerative questioning, also known as killer questions, to reflect on what is coming in the way of collaboration: what and who is important in this going forward, what is the need here, why is it important to bring about the change, and how will I achieve it going forward. We have found it very helpful especially in sticky situations.

10. Connect to common humanity when differences are amplified

When the mind focuses on the differences that separate us, taking a moment to connect to our common humanity of caring and sharing helps us to connect and collaborate. "He/she is just like me" helps to connect and bridge the difference. This is very useful in letting go of old grudges and historical wrongs and hurts.

11. Expose and let go of conscious or unconscious judgment

Let go of judgment and competition to level the playing field so that both can save face and win. Acknowledge and embrace what is coming in the way and then consciously let go of the past to create a new present. This is especially true in hierarchies where there are status differences. It requires defocusing and looking at the larger picture where both are active participants.

12. Make room for gratitude in the moment and for the moment

Suspend unnecessary criticism of self and others by letting go of grudges, regrets, and resentments through compassion, care, and connecting to the underlying love of humanity. Instead, appreciate what the other has done and accomplished. Focus on what works instead of being consumed with what doesn't. It requires us to connect to the larger picture and put the current interaction in perspective. Going to bed and waking up with a gratitude list is very helpful. Above all enjoy the moment for it is truly a gift.

5   Conclusions

Unbeknownst to us our decision to marry and create a family was a crucible for learning about living with diversity and designing inclusivity.

To live our life in harmony, we consciously practice using these learned tools that are applicable when our differences make it hard for us to include the other. For example, when we feel stressed in approaching someone who is different in status and it is hard for us to communicate openly with them. We become aware of this stress in our body. We connect with the breath, know it is based on fear of authority from the past. We reach out to the other. It helps us to be aware of it, feel it and know that it is not real in the moment. It has nothing to do with the person we are approaching in the moment, but is a learned rule of engagement from the past that is not serving us in the now. By spending some time in this felt sense and reframing it as the sense of safety that, in reality it is, we are able to let go of misperception and relax. So when we actually encounter this person we can be more present in our communication and engagement and enjoy the experience.

However, do we always succeed in relating when our differences make an alliance difficult and it is difficult to include the other collegially in our endeavor? No, especially if we approach the situation based on long standing behavior patterns and rules of engagement or unconscious beliefs that have not yet been reexamined. However, we get better at their use every day and succeed 80% of the time. That brings peace in our relationship and suffices until further stressful encounters occur such that we are not able to handle our differences collegially. Hopefully these tools will be all we will need in the last quarter of our life.

Our children are grown and are professionals in their own right, doing well. We can see that they and the course of their lives have been instrumental in teaching us the twelve techniques listed above. Likewise the families and cultures we grew up in, have informed us with their openness to adapt and change in response to challenges. Our students, and international colleagues, have each one taught us something about how to adapt our rules of engagement to include them. We take this moment to express our gratitude to all that have contributed to our learning of how to be inclusive.

5.1   Open Invitation

We have shared our experiences living with diversity that have helped us to live and love through learning how to relate with each other by being inclusive. We would like, you the readers, to reflect on your own journey and examine interactions in diverse settings. Have you used any of the techniques we have identified yourselves and have they been helpful? Are there some techniques that work for you effectively that we have not included that facilitate your collaboration across differences? Please let us know at manjulawaldron@gmail.com. We will acknowledge your response and add to our list for future use.

References

1. Jobs, S.: (2005), `http://www.npr.org/blogs/thetwo-way/2011/10/06/141120359/read-and-watch-steve-jobs-stanford-commencement-address`
2. Hellman, M.: The Wisdom of Foolishness. Stanford School of Engineering Her lecture (January 29, 2013)
3. Bhushan, A.: The World under One Roof. Rachna Publications New Delhi India (2008) ISBN 81-88820-23-7
4. Waldron, K.J.: Secret confessions of a designer. Mechanical Engineering-CIME (November 1992)
5. Waldron, M.B.: My life: Weaving a Creative Tapestry to Empower Socially Disempowered People. Sp. ME 311b designing the Professional class handout (2010)
6. Waldron, M.B.: Gandhi-eternal (2007), `http://www.indiacurrents.com/articles/2007/11/20/gandhi-eternal`
7. Waldron, M.B.: Different Engineer, `http://www.boloji.com/index.cfm?md=Content&sd=Articles&ArticleID=3556`
8. Waldron, M.B., Waldron, K.J.: Mechanical Engineering: Design theory and Methodology. Springer (1996)
9. 5 Koshas (2009), `http://beingwithyoga.blogspot.com/2009/03/5-koshas.html`
10. Gordon, J.S.: Unstuck: Your Guide to the Seven-Stage Journey Out of Depression. The Penguin Press, New York. CMBM.org (2008)
11. Simon, H.: The Sciences of the Artificial. MIT Press, Boston (1996)
12. Holland, J.: Hidden Order: How Adaptation Builds Complexity. Helix Books (1995) ISBN 0-201-40793-0
13. Kelley, K.: TED Talk (2005), `http://www.ted.com/talks/kevin_kelly_on_how_technology_evolves.html`
14. Pinker, S.: The Better Angels of our Nature: Why violence has declined. Penguin Books (2012) ISBN 978-0-670-02295-3
15. Institute for Economics and Peace, `http://en.wikipedia.org/wiki/Institute_for_Economics_and_Peace`
16. Gunn, A.: Fear is Power: Turn your fears into Success. Hardie Grants Books, Victoria (2006) ISBN: 9-7781740-664080
17. DeLong, C.C.: Achieving Personal Integrity: A psychiatrist's insight. iUniverse Inc., Bloomigton (2012)
18. Kahneman, D.: Thinking Fast and slow. Farrar, Strauss, and Giroux, NY (2011) ISBN 978-0-374-27563-1

19. Gentile, M.: Giving Voices to Value: How to speak your mind when you know what is right. Yale University Press, New Haven (2010) ISBN: 978-0-300-16118-2
20. Rosenberg, M.B.: Living Nonviolent Communication: Practical Tools to Connect and Communicate Skillfully in Every Situation. Sounds True (2012) ISBN 978-1604077872
21. Smedley, A., Smedley, B.D.: Race as biology is fiction, racism as a social problem is real: Anthropological and historical perspectives on the social construction of race. American Psychologist 60(1), 16–26 (2005)

2   Computer Aided Mechanism Synthesis: A Historical Perspective

Thomas R. Chase, Gary L. Kinzel, and Arthur G. Erdman

Abstract. The age of computer aided mechanism synthesis began in the late 1950's, as Freudenstein & Sandor published the first paper on the topic [14]. Many exciting developments occurred over the next 60 years, resulting in the development of multiple intriguing mechanism synthesis packages at several leading research institutions.

This paper provides an historical overview of the developments in computer aided planar linkage synthesis in the time window of 1955 to the present. The origins and legacies of those packages are reviewed. Key contributions to the field by Waldron and his associates are recognized.

1   Introduction

The design of many machine elements is accomplished by developing the input-output equations and solving for the design parameters by inverse methods. When linkages are involved, however, the solution space is usually so nonlinear that it is difficult to develop viable solutions simply with inverse techniques. Therefore, special approaches to linkage synthesis problems which incorporate the constraints directly into the synthesis equations have been developed.

The majority of linkage synthesis problems can be classified in one of four categories: function generation, motion generation, path generation, and crank-rocker synthesis [43]. Of these four types of problems, function generation and crank-rocker synthesis can usually be approached using relatively simple special purpose

Thomas R. Chase · Arthur G. Erdman
Mechanical Engineering Department, University of Minnesota, Minneapolis, MN 55455
e-mail: {trchase,agerdman}@umn.edu

Gary L. Kinzel
Department of Mechanical and Aerospace Engineering, The Ohio State University,
Columbus, OH 43210
e-mail: kinzel.1@osu.edu

V. Kumar et al. (Eds.): *Adv. in Mech., Rob. & Des. Educ. & Res.*, MMS 14, pp. 17–33.
DOI: 10.1007/978-3-319-00398-6_2   © Springer International Publishing Switzerland 2013

programs or they can be recast as special case motion generation problems. Therefore, the most difficult problems tend to be motion generation and path generation,
and this is where much of the effort in developing robust computer-aided design
(CAD) programs has been concentrated.

Since computers became available in universities in the late 1950's, much time
and effort has been expended by engineers and computer scientists to develop design
software that will simplify the linkage design process. Most of the early work was
done by relatively young faculty members who had an intense interest in kinematics
and were intrigued by the new tool that computers offered. A few of these efforts
led to the development of software packages that were relatively widely used and
even commercialized.

This paper will provide a historical overview of the development of computer
aided mechanism synthesis programs. The scope is limited to planar linkages. Although path generation will be mentioned because it is covered by some of the
software packages, the main emphasis in the paper will be on CAD approaches to
motion synthesis.

The methodology of synthesizing linkages using precision precisions is contrasted with optimization methods in Section 2. Problems that can arise during precision position synthesis are also introduced. Early linkage synthesis programs were
all developed at research institutions; they are reviewed in approximately chronological order in Section 3. More recent efforts are typically developed as extensions
to existing commercial CAD software; they are described in Section 4. Some spec-
ulations on the future of computer aided linkage synthesis programs are offered in
Section 5. Conclusions are in Section 6.

2 Technical Approaches Used in CAD Software

Motion generation has been approached using two fundamentally different approaches. In the first, a large number of positions of the moving plane (coupler)
are specified, and the best linkage which moves the coupler through the positions
in an approximate sense is determined through a mathematical optimization process
(for example, see [25]). In this approach, the coupler is unlikely to pass through any
of the positions exactly. This approach is based more on optimization concepts than
on kinematic concepts.

The second approach, which is emphasized in this paper, is based on precision
position synthesis. In this approach, the linkage is designed such that the coupler
passes through a modest number of prescribed positions exactly. This approach usually results in multiple solutions. Optimization may be used ultimately to select the
best linkage from the domain of possible linkages; however, optimization constitutes a secondary process.

In the precision position synthesis approach to motion generation, 2-5 positions
of the coupler relative to the reference link can be specified. The two position problem yields three infinities of solutions. Even with five positions, multiple solutions
can result.

The definitions of circuits and branches of linkages proposed in [5] will be adopted in this paper: A circuit is defined as all possible orientations of the links which can be realized without disconnecting any of the joints. If a circuit contains stationary configurations, a branch is defined as a continuous series of positions on the circuit between two stationary configurations. Using these definitions, four-bar linkages satisfying the Grashof criteria have two circuits while those that do not have one. The single circuit of a non-Grashof four-bar linkage has two branches. A crank-rocker or double-crank has two circuits, but since they do not contain stationary configurations, branching is irrelevant. The two circuits of a rocker-crank or Grashof double-rocker both contain two branches.

Solutions generated using precision position based synthesis methods are not useful if the precision positions fall on different circuits. They are often not useful if the precision positions fall on different branches. In addition, solutions may be defective because they pass through the precision positions in the incorrect order [39]. In some situations, the designer is interested only in solutions that have fully rotatable driving cranks (for example, [16]). Basic precision position synthesis methods leave it to the user to determine whether a solution suffers from any of the possible kinematic defects. Some of the sophisticated synthesis packages described in Section 3 are programmed to automatically sort out desirable solutions from defective ones.

In the majority of cases, even after eliminating solutions with circuit, branch, order, or crank rotatability defects, the designer must still choose among multiple solutions. She may do this by explicitly identifying additional constraints or by using objective or subjective techniques for selecting among the various choices.

One of the main features of CAD software is to help the designer choose the most desirable solutions. The main objective of all software that is to be used by technician level designers is to provide an environment that will allow the designer to obtain a good or near optimum solution quickly without needing an in-depth knowledge of theoretical kinematics. As will be discussed when individual software packages are described, programs provide this assistance by graphical interfaces that guide the designer through the process, by incorporating sophisticated mathematical optimization routines, or by incorporating pattern matching and/or knowledge based systems that narrow down a large number of solutions to a small number (perhaps one) that the user can easily evaluate.

3 The Early Years of Software Development

Freudenstein and Sandor [15] were the first to publish a paper which utilized a "digital computer[1]" to synthesize a linkage. Their program was set up to design four-bar linkages for path generation with prescribed timing for five precision positions. Their program was written for a specific computer, the IBM 650, but it likely would have been adaptable to any similar machine which used the same programming language.

---

[1] At the time of publication, analog computers were commonplace.

Freudenstein and Sandor's pioneering synthesis program was also the first to attempt to identify the best of multiple synthesis solutions. Up to four dyad solutions exist for the five-point synthesis problem. These can be combined to create up to six four-bar linkage solutions, which can be extended to twelve by constructing cognates of these linkages[2]. Their program automatically selected the best solution based on a quality index comprised of the ratio of the shortest link length to the longest link length times the range of the driving crank rotation. The program also performed a displacement analysis of the solution for evenly spaced increments of the driving crank, establishing a precedent that would be followed by several later programs.

Freudenstein and Sandor [15] provide several examples of how the synthesis program presented in [14] can be utilized to solve several related synthesis problems. Specifically, they demonstrated the synthesis of a four-bar function generator, a geared five-bar linkage, and a two-degree-of-freedom seven-bar linkage. They also provided a more detailed theoretical derivation of the five precision position solution. They suggested generalizing Burmester theory to the case of observing the motion of one moving plane relative to another, thereby anticipating the formulation of triad synthesis methods [4].

Kaufman pioneered mechanism synthesis using interactive computer systems [26]. His "KINSYN" program was the first to utilize an interactive input device, a data tablet, and an output display, a dynamic cathode ray tube (CRT), to enable a user to interact with the program while it was running (see Fig. 1). The early versions of the program described in [26] utilized a custom hardware system, so it was only operable at its development site (MIT). It featured an impressive list of synthesis capabilities, including motion generation for two, three, four and five precision positions. The program was capable of designing linkages with slider joints in addition to revolute joints. It was capable of analyzing tentative solutions to determine their Grashof type, circuit, branch, order of traveling through the prescribed positions, transmission angle, and acceleration. It could animate solutions on the display device, including multi-loop extensions to the basic four-bar solution.

A later version of KINSYN, "Micro-Kinsyn", was re-designed to run on an Apple IIe personal computer augmented with a custom input module [18]. Unfortunately, it did not prove feasible to keep the program current with the rapid pace of computer hardware development at that time.

Erdman and associates [11] developed another early interactive mechanism synthesis package, the Linkage Interactive Computer Analysis and Graphically Enhanced Synthesis Package (LINCAGES)[3]. LINCAGES overcame the need for specialized hardware by utilizing either a commercially available "storage tube"

---

[2] Up to 18 solutions were created if the user chose to release the prescribed timing constraint.

[3] The third author recalls that the LINCAGES project was initiated only because KINSYN was not available outside of MIT in its early days, so the creation of LINCAGES was necessary to expose University of Minnesota students to Kaufman's groundbreaking interactive synthesis strategy.

Fig. 1 The KINSYN III hardware. The human user was utilized as an integral part of the synthesis procedure. The user observed the current state of the design on the CRT screen and input directions for continuing the synthesis by way of a data tablet. As such, KINSYN may have constituted the first interactive computer aided engineering program. (Published by ASME, FIGURE 1 from [26], Journal of Engineering for Industry, Vol. 99 No. 2, by Rubel, A. J., and Kaufman, R. E., 1977.)

graphics display[4] or a teletype for both input and output to a mainframe computer operating in time sharing mode. While the teletype option was slow and had poor resolution, it made the program accessible to venues where linkage synthesis tools had been previously unavailable. The early LINCAGES program had the capability to synthesize four-bar motion, path and function generators for three, four or five precision positions, although the four point capability was developed more extensively than the other options.

Both the centerpoint curves and circlepoint curves were displayed for four point solutions, and the user could interactively select from either one. Solution dyads were parameterized according to the rotation of one of the dyad vectors between the first and second precision positions, β_2 (for example, β_{2A} in Fig. 2). While not ideal by way of intuitive understanding[5], this parameterization enabled the user to explore the entire domain of solutions associated with four point synthesis. This was done by creating a table of tentative solutions where a range of β_2 values for driving dyads was

---

[4] Storage tube displays were popular from about 1977-1987. They were relatively affordable interactive displays. Once a line or character was written to the screen, it would remain on the screen until the entire screen was erased, usually a matter of a minute or more.

[5] The matter is further complicated by the fact that each β_2 value has two different dyads associated with it; for example, see [3].

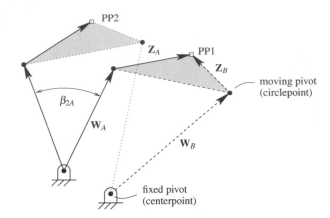

Fig. 2 Four bar motion generators are synthesized by combining two dyads, $(\mathbf{W}_A + \mathbf{Z}_A)$ and $(\mathbf{W}_B + \mathbf{Z}_B)$. β_{2A} represents the rotation of vector $\mathbf{W}_A$ from precision position 1 to precision position 2. LINCAGES utilized the β_2 value of a dyad to parameterize all solutions to the four precision position synthesis problem.

represented in the rows and a range of β_2 values for follower dyads was represented in the columns (see Fig. 3). The minimum transmission angle at the precision positions, if the solution was free of the branch or circuit defect, and the maximum link length ratio was calculated for each combination of two dyads. The user could then identify an attractive solution by selecting a solution from this table.

Filemon [13] authored a seminal paper on identifying portions of Burmester curves for four precision position synthesis which would produce linkages which have kinematic defects. Specifically, she identified sections of the curves where the precision positions could be reached in the correct order by continuously rotating a selected crank link, and where the follower link would not change position from above the ground link to below the ground link. The latter would lead to either a circuit or branch defect in the solution.

Filemon did not develop a computer based synthesis program. However, her pioneering work inspired Waldron and his associates to greatly extend and refine her work to the point that it could be utilized for computer assisted linkage synthesis. Waldron coined the term "solution rectification" to describe methods to eliminate spurious solutions in an a priori manner. Ultimately, he developed techniques for solution rectification for four-bar and slider-crank linkages for 2-5 precision positions. Rectification of the circuit and branch problem is addressed in [37, 40, 44, 45]. Identifying linkages that traverse the prescribed positions in the correct order is addressed in [39, 40, 44, 45]. Identifying linkages with a specified Grashof type is addressed in [38, 42, 34]. Controlling the transmission angle at the design positions is addressed in [35, 46].

The Rectified Synthesis, or RECSYN, program of Waldron and associates [6] implemented their solution rectification methods in a powerful four-bar synthesis program. While set up for synthesis for motion generation, path and function

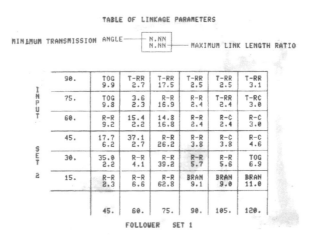

TABLE OF LINKAGE PARAMETERS

MINIMUM TRANSMISSION ANGLE ———┌─ N.NN ─┐
                              └─ N.NN ─┘ ——— MAXIMUM LINK LENGTH RATIO

		90.	TOG 9.9	T-RR 2.7	T-RR 17.5	T-RR 2.5	T-RR 2.5	T-RR 3.1
I N P U T		75.	TOG 9.8	3.6 2.3	R-R 16.9	R-R 2.4	T-RR 2.4	T-RC 3.0
		60.	R-R 9.2	15.4 2.2	14.8 16.8	R-R 2.4	R-C 2.4	R-C 3.0
		45.	17.7 6.2	37.1 2.7	R-R 26.2	R-R 3.8	R-C 3.8	R-C 4.6
S E T		30.	35.0 2.2	R-R 4.1	R-R 39.2	R-R 5.7	R-R 5.6	TOG 6.9
2		15.	R-R 2.3	R-R 6.6	R-R 62.8	BRAN 9.1	BRAN 9.0	BRAN 11.0
			45.	60.	75.	90.	105.	120.

FOLLOWER   SET 1

Fig. 3 The "TABLE" function in LINCAGES enabled the user to quickly explore the entire solution space for promising solutions. The solutions in the matrix correspond to selecting a driving dyad with a β_2 value shown in the left column and a follower dyad with a β_2 value in the bottom row. The top number in each matrix entry represents the minimum transmission angle at the precision positions, if one exists. The lower number indicates the maximum link length ratio of the solution. A top entry of "R−R" indicates that the solution has changed branch between precision positions, while "BRAN" indicates that the solution has changed circuit.

generators could also be synthesized by applying kinematic inversion. The original RECSYN was designed for storage tube type interactive displays, and it exploited an evanescent imaging option[6] to dynamically "rubber band" links corresponding to each synthesis step to the interactive selection cursor.

RECSYN had a very well developed three point synthesis option, as Waldron recognized the common need for three point solutions to practical problems. The user was guided to select a circlepoint defining a follower dyad prior to a driving dyad. Regions where placing a circlepoint would lead to a relative rotation between the coupler and follower greater than 180° were automatically deleted, as this would lead to solutions having a circuit or branch defect. Waldron created a method which he called the modified Filemon construction[7] to identify portions of the plane where circlepoints for the driving dyad could be selected without causing the transmission angle to change sign. Once a follower dyad was selected, the graphics display was updated to present the results of this construction, and the user was guided to select a driving dyad circlepoint in the remaining allowable regions. The Grashof type of a grid of sample solutions was also displayed in the allowable region. A final useful feature of the three point option was the display of the slider point circle. In addition

---

[6] The evanescent image was a dim dynamic image superimposed on the static background written to the storage tube display.

[7] The method was inspired by a similar construction applied to centerpoints by Filemon [13].

to its utility for designing slider-crank linkages, this circle also enabled identifying portions of the circlepoint plane where link length ratios tended to be poor.

The four point solution was equally well developed (see Fig. 4). The display for selecting the follower circlepoint would remove portions of the circlepoint curve where the relative rotation between the coupler and follower would exceed 180°. Once a follower link was selected, Waldron's modified Filemon construction was applied to identify portions of the circlepoint curve where moving pivots for driving dyads could be selected without causing the transmission angle to change sign. Finally, the circlepoint curve was subdivided so as to indicate the order of rotation of the driving crank as it passed through the precision positions[8]. RECSYN also included a five point option, but performing a comprehensive check of the small number of five point solutions proved more efficient than attempting to adapt solution rectification to this problem.

RECSYN was later extended to include several additional useful features [31]. Two position synthesis was added, where Waldron's modified Filemon construction was performed to help select a driving circlepoint. The program was modified to handle solutions for parallel precision positions where possible. A unique enhancement consisted of augmenting the modified Filemon construction to include a "starburst" of lines through the selected circlepoint that indicated the highest deviation angle reached at all the precision positions. Other enhancements included the addition of optimization techniques to choose the best linkage for 2, 3, and 4 position synthesis based on the link length ratio and transmission angle [1, 28, 33]. The optimization approach was later extended to allow the two middle positions in a four-position problem to vary within a given tolerance range to extend the range of the design parameters in the optimization process [36].

The MECSYN program [23, 30] was developed in the same time frame as KIN-SYN and LINCAGES. MECSYN is notable for two reasons. It constituted the first program to be capable of designing multi-loop mechanisms. The program had the ability to kinematically invert[9] basic dyads, which enabled it to synthesize Stephenson six-bar linkages and other mechanisms more complex than simple four-bar linkages [22][10]. Second, it was the first program capable of synthesizing linkages for multiply separated positions[11]. Four or five multiply separated position problems could be solved.

The SOFBAL program [32] originated with the same group that developed MEC-SYN. SOFBAL constituted a blend between Burmester theory based synthesis methods and synthesis by optimization. Burmester theory was used to generate circlepoint and centerpoint curves for four positions. However, the user did not directly select solution linkages from the Burmester curves. Rather, the curves were

---

[8] Features varied between versions. The version illustrated in Fig. 4 does not appear to implement this feature.

[9] Kinematic inversion refers to the ability to change the link which is assumed to be attached to ground.

[10] MECSYN is cited as being a work in progress in the conclusion of this paper.

[11] Multiply separated positions enable specifying velocity and other higher derivatives at specified precision positions.

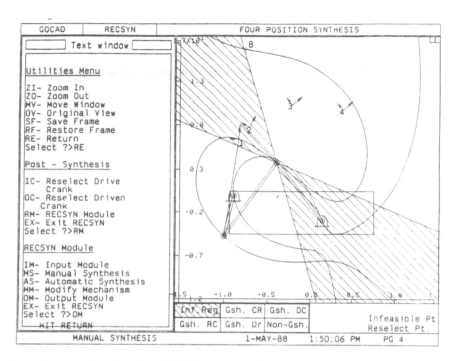

Fig. 4 A screen shot from the RECSYN program for synthesizing a motion generator for four precision positions. Both the centerpoint and circlepoint curves are shown. The user can select points from either curve. The small coordinate systems represent the four precision position definitions. The crosshatched areas represent regions that are forbidden for selecting circlepoints for the driving dyad, determined by Waldron's modified Filemon construction. They were added to the display following selection of a driven link. Point "B", labeled on the circlepoint curve, represents Ball's point, corresponding to a driving slider. Users were advised to avoid selecting circlepoints near, but not on, this point because they would tend to produce solutions with poor link length ratio, as slider solutions correspond to drivers of infinite length. As the user selected enough points to complete a linkage, its Grashof type was indicated by lighting up the appropriate box at the bottom of the screen. After the solution linkage was specified, RECSYN animated the linkage throughout its feasible range. The minimum and maximum values for the transmission angle and link lengths were also summarized. (From [27].)

parameterized and a grid of solutions mapping the entire possible solution space was constructed in a manner similar to the "TABLE" command in LINCAGES (see Fig. 3). A quality score was then assigned to each solution in the grid by applying a user-controlled objective function. The user then interactively refined the search by manually zooming in on promising portions of the grid. SOFBAL shared the multiply separated position capability of MECSYN. Unfortunately, neither MECSYN nor SOFBAL ever became widely available.

The SIXGUN program [2], authored by the group which created LINCAGES, was designed specifically for synthesizing multi-loop mechanisms[12]. The computational engine of the program could generate Burmester curves for four precision positions for either dyads or triads. Relative precision positions, described in [4], were used to implement kinematic inversion for synthesizing triads. The program would establish the topology of the mechanism being designed by reading files that were external to the program. As a result, any mechanism that could be modeled with free vectors, dyads or triads could be designed, including all the Watt and Stephenson six-bar linkages [10]. SIXGUN was never released in its most general form, since a non-expert user could potentially define a nonsensical combination of synthesis components. The LINCAGES-6 package [24] addressed that problem by modifying the original program so that it was limited to synthesizing a catalog of pre-defined six-bar linkage topologies.

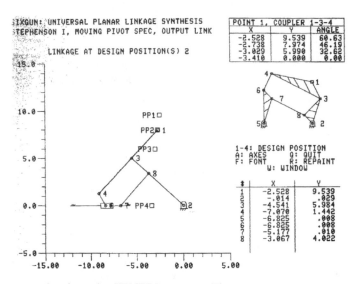

Fig. 5 A screen shot from the SIXGUN program. The program was set up to synthesize a Stephenson I six-bar linkage in this example. The generic topology of the linkage being designed was illustrated in the figure in the right column. This topology was defined by a file separate from the program itself. The numbering of the pivots indicates the order in which they were selected. In this example, the position of point 1, the angle of link 1-3-4, and the angle of link 5-6-7 were defined by precision positions input from the user. The user was then guided to select either pivot 2 or 3 from a set of Burmester curves. A free vector was then used to set the position of pivot 4 relative to pivot 1 at the first precision position. The user could then select either pivot 5 or 6 from a new set of Burmester curves generated using the placement of pivot 4. A final set of Burmester curves was generated for selecting pivots 7 and 8 by internally computing a set of relative precision positions from the earlier input and selections. Note that the solution shown utilizes a slider point; i.e., pivot 5 is at infinity.

---

[12] SIXGUN began as an attempt to codify the methods for synthesizing all six-bar mechanisms defined in [12], but it quickly transformed into a more generic synthesis tool.

One problem associated with designing multi-loop linkages using precision positions is that solution mechanisms are more likely to suffer circuit or branch defects than simple four-bar linkages. Mirth and Chase extended Waldron's solution rectification methods to rectify the circuit problem in Watt [20] and Stephenson [21] linkages. Watt circuit rectification was implemented in a late version of LINCAGES-6 [24]. Unfortunately, LINCAGES-6 was never migrated to computers running the WINDOWS operating system.

4 The Evolution of CAD Software for Linkage Design

The early linkage synthesis programs were typically written in FORTRAN. As the migration from minicomputers to personal computers occurred, FORTRAN became less and less used compared to C and C++. The early programs had to be rewritten to survive. During this transition, LINCAGES was maintained and enhanced, but RECSYN was not. While simplified aspects of RECSYN were reprogrammed in MATLAB for two and three positions [43], the original version of RECSYN was not reprogrammed to run on mouse driven, Windows-based platforms, and therefore the program ceased to be used.

As the personal computer became commonplace, the price of both computers and software tended to decrease significantly. In addition, both the graphics capabilities and the speed of computers increased dramatically. At the same time, equation solvers and constraint managers became more robust. This permitted the development of very sophisticated solid modeling software based on parametric design [29].

Two approaches to the development of linkage synthesis software evolved based on the increased computing and graphics capabilities available. The first utilized these capabilities directly to improve the user interface and to use search engines and knowledge bases to guide the user toward good solutions to complex design situations. Two programs which used this approach are LINCAGES and WATT [8]. LINCAGES in particular maintained the solid theoretical base discussed previously beneath a graphical user interface that guided even novice designers to good solutions to complex problems.

WATT was a suite of programs developed by Heron Technologies in the Netherlands. Not much has been published on the technical details for the WATT Suite; however, it appeared to have had a parameter reduction routine to limit the number of design parameters which must be considered. It then appeared to create a large number of trial solutions based on the most important design parameters and perform an efficient pattern-matching search of the data base to come up with viable solutions. These solutions then appeared to be refined using a genetic optimization algorithm, and a list of the best solutions were presented to the user. The user could quickly sift through the solutions by analyzing each for the full cycle of interest. The program was applicable to both path and motion synthesis, and the user could select from eight possible mechanisms[13]. The user interface was carefully designed to present the most important results on a single screen.

The WATT program was well suited to industry and was upgraded to run on Windows XP. Unfortunately, development seemed to have stopped around 2005 and the program was unavailable by 2012. Heron Technologies has removed the information for the program suite from the company web page.

During the last 10 years, solid-modeling programs such as ProEngineer, SolidEdge, and SolidWorks have incorporated kinematic analysis capabilities which provide the designer with visually realistic linkage animations along with analytical results for velocity, acceleration, forces, mechanical advantage, and interference. The analyses can be conducted quickly, making trial and error iterations possible for relatively simple problems. In addition, the very nature of the parametric design programs gives the designer access to the constraint manager. Constraints like perpendicularity, parallelism, concentricity, coincidence, etc., are integral to the function of solid modeling programs. Because these are also some of the same geometric operations required for kinematic synthesis, solid modeling programs provide a natural environment for direct kinematic synthesis.

The SyMech Design Modules [7, 47] utilize the ProEngineer platform for the synthesis of four-bar and multi-bar mechanisms. SyMech operates within the ProEngineer environment, so the designer must already be using ProEngineer. The program uses the equations from basic kinematic theory for four-bar linkages together with mathematical optimization and an interactive graphical user interface to guide the designer toward optimum solutions. The results are displayed and analyzed by ProEngineer directly. The four-bar module (SyMech-4) incorporates the basic equations (templates) for synthesis for motion generation, path generation, crank-rocker design, and function generation. The special cases for straight-line mechanisms and parallel motion mechanisms are also included. The user can check for circuit, branch, order, and interference defects by animating the solution within the ProEngineer environment, and she can adjust the design parameters to attempt to correct for these defects.

The multi-link version of the program is called SyMech-n. Technical details on the kinematic theory for the program do not seem to have been published. However, the program appears to be suited to problems which can be solved by a series of four-bar linkages which can be connected using function generation. Again, once a basic type of linkage is identified, it can be optimized by the user by analyzing and animating the linkage in the ProEngineer environment.

Because solid modeling programs already incorporate the graphic constructions required for kinematic synthesis as preprogrammed constraints, a novel approach to synthesis has been proposed by Kinzel, Schmiedeler, and Pennock [17]. This approach does not require a separate program for kinematic synthesis because all of the operations are accomplished within the parametric design program. The kinematic constructions are set up for a generic problem in the parametric design

---

[13] The available mechanisms were the four-bar, slider-crank, geared five-bar, Watt 1 six-bar, Watt 2 six-bar, Stephenson 1 six-bar, Stephenson 3 six-bar, and eight-bar for parallel motion.

environment. Design parameters are adjusted to produce a solution in real time by using either the drag function on the graphics display or by typing in numerical values to force the design to conform to predefined specifications. In essence, the designer produces a "program" in the graphics environment to solve an entire class of problems simply by changing the parametric variables directly. The construction constraints are maintained by the program's constraint manager. This approach is called "graphical constraint programming" or GCP. This procedure can be used directly for crank-rocker design, function generation and motion generation. Even the 5-position problem in motion generation can be solved using this procedure. The procedure also works for path generation if the solid modeling program can store the coordinates of points along the path to be followed. Mirth [19] applied the GCP approach to the synthesis of six-bar linkages.

5 A Possible Future for CAD Programs for Linkage Synthesis

Linkages permeate the design environment, but the number of individuals who solve problems complex enough to require programs like KINSYN, LINCAGES, REC-SYN, SyMech, and WATT is relatively small. The number is further reduced by the fact that modern solid modeling programs are so easy to use that even moderately complex problems can be solved iteratively in the graphics environment. The number of tractable design problems is further expanded by GCP, which does not require an extensive knowledge of kinematic theory or programming. Therefore, it is unlikely that any future company can survive if that company's only revenue stream is based on the sales of kinematic synthesis software. On the other hand, it is possible to develop and maintain sophisticated kinematic design programs within universities or in businesses which use the programs as a tool for product design or consulting. Such programs can be sold for supplementary revenue, which can justify making them available outside the home institution. Therefore, hopefully programs like LINCAGES and SyMech will survive well into the future.

In addition, it is recommended that solid modeling vendors include simple synthesis modules as part of their basic environment in much the same way that they provide analysis modules now. They should also promote using the basic environment for sophisticated designs using GCP. While a learning curve is associated with the process, it is well within the capabilities of technical school graduates or experienced designers.

6 Conclusions

The market for kinematic software is too small to expect a large number of linkage synthesis packages to be available in the future. Nevertheless, the impact of the computer assisted linkage synthesis programs that have been developed should not be underestimated. KINSYN may have constituted the first truly interactive computer aided engineering system, leveraging both the distinct computational capabilities of the machine and the intuitive capabilities of the human in the loop. LINCAGES and RECSYN extended that strategy. These groundbreaking programs may have inspired later interactive computer aided engineering applications.

The time savings of utilizing interactive linkage synthesis tools is dramatic. An optimal solution might be found in the course of an hour rather than days or even weeks. In some cases, the computer based tools were equally useful for bringing to light quickly that no practical solutions were available using a certain set of precision positions. Previously, the designer may have invested a great amount of effort to reach the same conclusion. The interactive programs can also indicate that the problem might be correctable by relaxing a constraint on the original positions.

As graphical user interfaces have improved, the scope of problems that can be solved quickly in the computer aided drafting environment has increased. Therefore, the need for standalone kinematic synthesis programs has decreased. However, a class of problems is always likely to exist that is sufficiently complex that they cannot be solved easily by manual iterative methods alone, even by experienced designers. Therefore, the development of special kinematic synthesis programs, especially in universities and research departments, continues to be justified.

References

1. Bawab, S., Sabada, S., Srinivasan, U., Kinzel, G.L., Waldron, K.J.: Automatic Synthesis of Crank Driven Four-Bar Mechanisms for Two, Three, or Four-Position Motion Generation. ASME Journal of Mechanical Design 119(2), 225–231 (1997)
2. Chase, T.R., Erdman, A.G., Riley, D.R.: Synthesis of Six-Bar Linkages Using and Interactive Package. In: Proceedings of the 7th OSU Applied Mechanisms Conference, Kansas City, MO, Paper #LI (December 1981)
3. Chase, T.R., Erdman, A.G., Riley, D.R.: Improved Centerpoint Curve Generation Techniques for Four Precision Position Synthesis Using the Complex Number Approach. ASME Journal of Mechanisms, Transmissions, and Automation in Design 107(3), 370–376 (1985)
4. Chase, T.R., Erdman, A.G., Riley, D.R.: Triad Synthesis for up to Five Design Positions With Application to the Design of Arbitrary Planar Mechanisms. ASME Journal of Mechanisms, Transmissions, and Automation in Design 109(4), 426–434 (1987)
5. Chase, T.R., Mirth, J.A.: Circuits and Branches of Single-Degree-of-Freedom Planar Linkages. ASME Journal of Mechanical Design 115(2), 223–230 (1993)
6. Chuang, J.C., Strong, R.T., Waldron, K.J.: Implementation of Solution Rectification Techniques in an Interactive Linkage Synthesis Program. ASME Journal of Mechanical Design 103(3), 657–664 (1981)
7. Cook, J.B., Olson, D.G.: The Design of a 10-Bar Linkage for Four Functions using SyMech. In: Proceedings of the 2002 ASME Design Technology Conference, Montreal, Quebec, Canada, September 29-October 2, Paper No. DETC2002/MECH-34369 (2002)
8. Draijer, H., Kokkeler, F.: Heron's Synthesis Engine Applied to Linkage Design. In: Proceedings of the 2002 ASME Design Technology Conference, Montreal, Quebec, Canada, September 29-October 2, Paper No. DETC2002/MECH-34373 (2002)
9. Erdman, A.G.: Three and Four Precision Point Kinematic Synthesis of Planar Linkages. Mechanism and Machine Theory 16(3), 227–245 (1981)

10. Erdman, A.G., Chase, T.R.: New Software Synthesizes Complex Mechanisms. Machine Design 57(19), 107–113 (1985)
11. Erdman, A.G., Gustafson, J.E.: LINCAGES: Linkage Interactive Computer Analysis and Graphically Enhanced Synthesis Package, Presented at the Design Engineering Technical Conference, Chicago, IL, USA, ASME Paper No. 77-DET-5 (September 1977)
12. Erdman, A.G., Lonn, D.: A Unified Synthesis of Planar Six-Bar Mechanisms Using Burmester Theory. In: Proceedings of the Fourth World Congress on the Theory of Machines and Mechanisms, Newcastle Upon Tyne, England (September 1975)
13. Filemon, E.: Useful Ranges of Centerpoint Curves for Design of Crank-and-Rocker Linkages. Mechanism and Machine Theory 7(1), 47–53 (1972)
14. Freudenstein, F., Sandor, G.N.: Synthesis of Path-Generating Mechanisms by Means of a Programmed Digital Computer. ASME Journal of Engineering for Industry Series B 81(2), 159–167 (1959)
15. Freudenstein, F., Sandor, G.N.: On the Burmester Points of a Plane. ASME Journal of Applied Mechanics Series E 28(1), 41–49 (1961)
16. Gupta, K.C.: Synthesis of Position, Path and Function Generating 4-Bar Mechanisms with Completely Rotatable Driving Cranks. Mechanism and Machine Theory 15(2), 93–101 (1980)
17. Kinzel, E.C., Schmiedeler, J.P., Pennock, G.R.: Kinematic Synthesis for Finitely Separated Positions Using Geometric Constraint Programming. ASME Journal of Mechanical Design 128(5), 1070–1079 (2006)
18. Krause, J.K.: Designing Mechanisms on a Personal Computer. Machine Design 55(6), 94–99 (1983)
19. Mirth, J.A.: The Application of Geometric Constraint Programming to the Design of Motion Generating Six-Bar Linkages. In: Proceedings of the 2012 ASME International Design Engineering Technical Conferences, Chicago, IL (August 2012)
20. Mirth, J.A., Chase, T.R.: Circuit Rectification for Four Precision Position Synthesis of Four-Bar and Watt Six Bar Linkages. ASME Journal of Mechanical Design 117(4), 612–619 (1995)
21. Mirth, J.A., Chase, T.R.: Circuit Rectification for Four Precision Position Synthesis of Stephenson Six-Bar Linkages. ASME Journal of Mechanical Design 117(4), 644–646 (1995)
22. Myklebust, A., Tesar, D.: The Analytical Synthesis of Complex Mechanisms for Combinations of Specified Geometric or Time Derivatives up to the Fourth Order. ASME Journal of Engineering for Industry Series B 97(2), 714–722 (1975)
23. Myklebust, A., Keil, M.J., Reinholtz, C.F.: MECSYN-IMP-ANIMEC: Foundation for a New Computer-Aided Spatial Mechanism Design System. Mechanism and Machine Theory 20(4), 257–269 (1985)
24. Nelson, L., Erdman, A.G.: Recent Enhancements to the LINCAGES-6 Synthesis Package, Including Circuit Rectification. In: Mechanism Synthesis and Analysis (ASME DE-Vol. 70), The 1994 ASME Design Technical Conferences - 23rd Biennial Mechanisms Conference, Minneapolis, MN, pp. 263–271 (1994)
25. Root, R.R., Ragsdell, K.M.: A Survey of Optimization Methods Applied to the Design of Mechanisms. ASME Journal of Engineering for Industry 98(3), 1036–1041 (1976)
26. Rubel, A.J., Kaufman, R.E.: KINSYN III: A New Human-Engineered System for Interactive Computer-Aided Design of Planar Linkages. ASME Journal of Engineering for Industry 99(2), 440–448 (1977)

27. Sabada, S.: Computer-Aided Design of Four-Bar Linkages for Four-Position Rigid Body Guidance Using Solution Rectification and Optimization Techniques. MSME Thesis, The Ohio State University (1988)
28. Sabada, S., Srinivasan, U., Kinzel, G., Waldron, K.: Automatic Synthesis of Four-Bar Mechanisms for Four-Position Motion Generation. In: Proceedings of the 20th Biennial ASME Mechanisms Conference. Orlando, FL, September 25-28. Trends and Developments in Mechanisms, Machines, and Robotics, vol. 2, pp. 121–128 (1988)
29. Shah, J.J., Mantylia, M.: Parametric and Feature-Based CAD/CAM. Wiley Interscience (1995)
30. Sivertsen, O., Myklebust, A.: MECSYN: An Interactive Computer Graphics System for Mechanism Synthesis by Algebraic Means. Presented at the Design Engineering Technical Conference, Beverly Hills, CA, USA, September 28-October 1, ASME Paper No. 80-DET-68 (1980)
31. Song, S.M., Waldron, K.J.: Theoretical and Numerical Improvements to an Interactive Linkage Design Program - RECSYN. In: Proceedings of the 7th OSU Applied Mechanisms Conference, Kansas City, MO, Paper #VIII (December 1981)
32. Spitznagel, K.L., Tesar, D.: Multiparametric Optimization of Four-Bar Linkages. ASME Journal of Mechanical Design 101(3), 386–391 (1979)
33. Srinivasan, U., Sabada, S., Kinzel, G., Waldron, K.: Automatic Synthesis of Four-Bar Mechanisms for Two and Three Position Motion Generation. In: Proceedings of the 20th Biennial ASME Mechanisms Conference. Trends and Developments in Mechanisms, Machines, and Robotics, September 25-28, vol. 2, pp. 113–120 (1988)
34. Strong, R.T., Waldron, K.J.: Joint Displacements in Linkage Synthesis Solutions. ASME Journal of Mechanical Design 101(3), 477–487 (1979)
35. Sun, J.W.H., Waldron, K.J.: Graphical Transmission Angle Control in Planar Linkage Synthesis. Mechanism and Machine Theory 16(4), 385–397 (1981)
36. Venkataraman, S.C., Kinzel, G.L., Waldron, K.J.: Optimal Synthesis of Four-Bar Linkages for Four-Position Rigid Body Guidance With Selective Tolerance Specifications. In: Proceedings of the 1992 ASME Mechanisms Conference. Mechanical Design and Synthesis, Scottsdale, AZ, September 13-16, DE-vol. 46, pp. 651–659 (1992)
37. Waldron, K.J.: Range of Joint Rotation in Planar Four-Bar Synthesis for Finitely Separated Positions: Part I - The Multiple Branch Problem. Presented at the Design Engineering Technical Conference, New York, NY, ASME Paper No. 74-DET-108 (October 1974)
38. Waldron, K.J.: Range of Joint Rotation in Planar Four-Bar Synthesis for Finitely Separated Positions: Part II - Elimination of Unwanted Grashof Configurations. Presented at the Design Engineering Technical Conference, New York, NY, ASME Paper No. 74-DET-109 (October 1974)
39. Waldron, K.J.: The Order Problem of Burmester Linkage Synthesis. ASME Journal of Engineering for Industry 97(4), 1405–1406 (1975)
40. Waldron, K.J.: Elimination of the Branch Problem in Graphical Burmester Mechanism Synthesis for Four Finitely Separated Positions. ASME Journal of Engineering for Industry 98(1), 176–182 (1976)
41. Waldron, K.J.: Graphical Solution of the Branch and Order Problems of Linkage Synthesis for Multiply Separated Positions. ASME Journal of Engineering for Industry 99(3), 591–597 (1977)
42. Waldron, K.J.: Location of Burmester Synthesis Solutions with Fully Rotatable Cranks. Mechanism and Machine Theory 13(2), 125–137 (1978)

43. Waldron, K.J., Kinzel, G.L.: Kinematics, Dynamics, and Design of Machinery. Wiley, New York (2003)
44. Waldron, K.J., Stevensen Jr., E.N.: Elimination of Branch, Grashof, and Order Defects in Path-Angle Generation and Function Generation Synthesis. ASME Journal of Mechanical Design 101(3), 428–437 (1979)
45. Waldron, K.J., Strong, R.T.: Improved Solutions of the Branch and Order Problems of Burmester Linkage Synthesis. Mechanism and Machine Theory 13(2), 199–207 (1978)
46. Waldron, K.J., Sun, J.W.-H.: Graphical Transmission Angle Control in Planar Linkages Synthesis. In: Proceedings of the 6th OSU Applied Mechanisms Conference, Denver, CO, Paper #XXXIV (October 1979)
47. http://www.symech.com (2013)

3 An Excellent Adventure

Vincent J. Vohnout

1 Introduction

This paper concerns aspects of the conception, design, building and testing of the Adaptive Suspension Vehicle (ASV) for which Ken Waldron was a Principal Investigator. The ASV was a hexapod vehicle of 7500 lb gross weight that was completely self-contained in terms of motive power and motion control. I (the author) was involved in the project, in turns, as Ken's graduate student, a university research associate, the chief mechanical engineer on the ASV, and an engineering contractor. This project was conducted at The Ohio State University from 1982 to about 1990 and funded, in the most part, by the Defense Advanced Research Projects Administration (DARPA). It is not my intention to provide much technical detail about the ASV; the interested reader is referred to the publications listed in the reference section of this paper and /or to Ken's own extensive bibliography on the topic. What I intend is to recount my experience with Ken Waldron as a director of the greatest engineering adventure of my career. Also, I will not be too rigorous as to exact dates or to the acknowledgement of the many who made contributions to the project. Space in this paper and my memory will not permit it. What is hoped for is to describe how such a complex system as the ASV could successfully be developed at a university principally by a group of graduate students and freshly minted engineers. Since this paper is ostensibly a technical document, I can (must) provide the answer here in the introduction. It was the team structure of the program that made the ASV one of the most successful hardware projects that our DARPA contract manager had known.

Teams, in any arena, are strongly influenced by their leaders/coaches, and the ASV team was no exception. Ken and Dr. Robert McGee from the Department of Electrical Engineering were the co-Principal Investigators (PI's) and thus had full

Vincent J. Vohnout
Director and Principal Engineer
Geo-Core , Columbus, Ohio U.S.A.
e-mail: vjvohnout@hotmail.com

V. Kumar et al. (Eds.): *Adv. in Mech., Rob. & Des. Educ. & Res.*, MMS 14, pp. 35–47.
DOI: 10.1007/978-3-319-00398-6_3 © Springer International Publishing Switzerland 2013

Fig. 1 Author driving the ASV during advanced field trials at the OSU agriculture fields, 1987. The posture shown is the first part of a large obstacle negotiation maneuver call the "Praying Mantis". In this stage the ASV balances and rears back on the rear four legs in order to place the front feet on top of an obstacle that is taller than the normal foot lift of the front legs.

executive power over the project. Rather than being dictators, though, they took the coach's approach. (Also, they worked well together as co-PI's.) This I believe was a fundamental enabling factor in the success of the ASV.

2   The Adventure Begins

In 1980 I applied to and was accepted into the Masters graduate program in the Mechanical Engineering Department of The Ohio State University. During my entrance interview with the then department chairman, I turned down a teaching assistant position (I don't like children much) but thankfully accepted a research assistant position. At the end of the interview the chairman provided a list of research projects in the department that were in the market for new research assistants. I started that afternoon working my way down the list in order to find a project for which I might be a good fit and have some genuine interest. I had interviewed with a few projects that included one on gear rattle, one on coal combustion and a very short meeting with E.O. Doebelin who told me as I stood in his office doorway, "I don't do research, I write textbooks. I don't know why they keep putting me on those lists" (or something like that). Ken's name was next on the list. I found his office after some wandering in the old Robinson Lab building. Ken was in and was willing to talk with me about his current need of a grad assistant. He explained how the program involved the development of walking machines. My response was … whatever is a walking machine?? He gave me a brief background on the current state of ambulating machines that used leg-like mechanisms in place of wheels or track loops as the locomotion elements. I was definitely interested in such a far-out program as I have always gravitated to the unconventional. He also made it explicit that this was a hardware-based program, not just a paper study. I in turn, described my major undergraduate course of study in Machine Design and my earlier experience as a Journeyman Tool and Die Maker. I stated that I would be very comfortable in a hardware-based research project, especially one as unconventional as involving machines with legs. Ken offered me the assistantship, and I accepted immediately. Leaving Ken's office, I deposited the unfinished list of my interviews in the first wastebasket I saw in the hallway.

After the short time to allow new assistants to settle into first quarter classes, Ken began the weekly (more or less) program meetings. These were small gatherings at this time since this was the pre-ASV period. Work was focused on laboratory-scale machines intended to demonstrate to our sponsors (current and potential) that we knew what we were doing and had the True Vision. Ken had an open approach in these early program meetings that I appreciated a great deal. I had quite enough strict autocratic direction during my Tool and Die apprenticeship. I recognized that some of his students were uncomfortable with his lack of specific, line-item direction for their research, but it is my opinion that Ken's approach promoted creative, independent thinking that generated the kind of results that such programs require (and is also a true goal of graduate-level education).

When I joined Ken's group, the OSU Hexapod was still the only walking machine at OSU and chiefly the toy of the electrical-controls guys. The OSU Hexapod had been, until then, a laboratory machine developed by Robert McGee as a test bed for digital control of multi-legged machines. It was quite successful for its time, and the resulting publications generated significant interest in the then new Digital Control community. However, non-academic onlookers (in the potential sponsor community) were critical of the very poor energy efficiency of the OSU Hexapod. They reasoned that such machines could never cut their power and data cables and leave the lab since they would never be structurally able to carry their own computers and power supplies. This was the conclusion of a simplistic linear extrapolation by the onlookers, but represented a real issue to be addressed if funding was to be secured from the deep pockets of DARPA. I believe it was this issue that led the electrical-controls guys to decide to talk with someone in Mechanical Engineering, but I am not certain since I was not there yet. More importantly, it turned out to be Ken they contacted, and as a consequence, the legged locomotion energy efficiency issue became the topic of my Master's work with Ken. As the prime directive for my thesis work, Ken was brief and clear. I needed to generate a leg design and actuation method of comparable scale to the OSU Hexapod that would be, as a minimum, an order of magnitude more efficient in a justifiably equivalent comparison. Hence, the title of the thesis was "Mechanical design of an energy efficient robotic leg for use on a multi-legged walking vehicle" [5]. Luckily, the program team included others working on other aspects of the ambulating machine problem which, with Ken's fairly open management style, provided for a synergistic atmosphere, and new ideas could be easily presented (excepting violations of the 2nd Law of Thermodynamics).

I recall the early program meetings with Ken concerned discussions on the optimum kinematics for an efficient walking machine leg. Nearly all experimental machines up to then used an insect-like design where the main foot motion was generated by a rotation about a vertical hip axis. Consequently, the plane of a leg was generally perpendicular to the machine plane of symmetry and principle

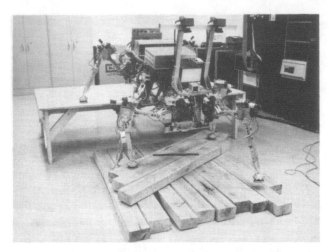

Fig. 2 The OSU Hexapod during laboratory testing of obstacle traversing. This 18 DOF machine weighed 220 lbs (100Kg) and was one of the first walking machines that could operate under full digital control. Not shown is the large power and control cables attached at the rear of the machine. (~1975)

direction of motion. Such leg kinematics generated large moments, as a function of machine weight, on the hip actuators and required continuous coordination of hip and knee rotations in order to keep the foot path straight (relative to the body) and the body height constant. For the PDP 1170 that was used (which was state of the art at the time), this represented a rather strenuous computational load. Ken suggested a more mammalian inspired leg design that would have its principle structural and motion plane parallel to the machine plane of symmetry and principle direction of motion. We termed this a planar-type leg. The advantages and disadvantages of a planar leg design were the subject of many meetings, with different mechanisms being presented for consideration. I have always been a proponent of simplicity, especially in machine design. As a consequence, I gravitated toward a simple four-bar leg mechanism solution. I no longer remember if I, Ken or someone else originally suggested a four–bar, but I recognized the advantages that included the potential elimination of two control DOF required by the Hexapod (abduction-adduction and body height) for straight motion. Additionally, a properly designed four-bar leg could recycle some of the main actuator energy during a stride [5]. After a lot of iteration and consultations with Ken, I found a four-bar solution with a coupler point curve that was straight over the hexapod stride length and remained so for a significant range of driven link lengths (foot heights). I presented my design at a program meeting and defended it. Ken did not give an easy pass to program elements that he considered important. It was not a good sign if he didn't have at least an opinion about your idea.

With the leg kinematics defined, there was still a high-level program decision to be made before commencing with the mechanical design phase. Specifically, was a complete six-legged machine required or could we get away with some smaller number of legs in some appropriate stabilizing apparatus? With my graduation timeline in mind and the amount of work I would be required to do, I proposed that a single leg was sufficient to establish the energy requirements of a machine having six such. Ken agreed with my argument, perhaps in part because it would cost only a fraction of an OSU-hexapod-size six-legged machine. The subsequent single-leg test bed was named the Monopod.

I designed the four-bar leg, employed a disk-type servo motor and a custom high efficiency 10:1 gearbox for the main drive, and custom designed ball screw actuators for the abduction-adduction and leg lift actuators, all of which provided big energy savings over the serial wound ac drill motors used on the Hexapod. The complete leg mechanism was mounted in the inner frame of a three-wheel cart. The inner frame had a guided vertical DOF that allowed the leg to rise in order to assume the body weight load of the virtual complete machine. A joystick was used to drive the monopod around the lab while collecting power and velocity data - sometimes just for fun. The Monopod cart provided a means to generate the required data, eliminating the need for an expensive stationary treadmill-type test stand. Also, lab space was easier to obtain if you promised not to bolt anything to the floor. After fixing the various and inevitable glitches, the energy efficiency tests were conducted and the data reduced in various ways in view of establishing a valid comparison. For the final assessment, I employed a dimensionless metric termed Specific Resistance (to motion) published by Gabrelli and Von Karmen in the 1950's for comparing all manner of locomotion systems including animals [1]. The metric was beautifully simple, consisting of the power consumed divided by the speed and the weight in motion. Using this metric it was shown that a virtual Hexapod-size walking machine employing the Monopod-style four-bar leg would be 25 times more efficient than the OSU Hexapod. Results from my thesis were included in a report and new program proposal to DARPA. When Ken returned from Washington, I was informed that DARPA had agreed to fund a follow-on program. I thought this was great and that I would be involved as a postgrad research associate in making the virtual machine used in my thesis the new generation laboratory walking machine to replace the old OSU Hexapod. I correctly predicted most of the goals of the new program except the laboratory part. DARPA did not want an incremental approach to legged vehicle technology; they insisted on advancing directly to a vehicle that would operate in real field conditions under fully autonomous control and energy means. Basically we were thrown into the deep end of the pool to learn to swim as designers of "useful" legged vehicles.

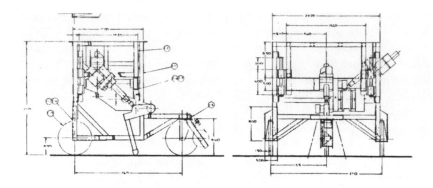

Fig. 3 Mechanical assembly drawing of the OSU Monopod

3   Diving into the Deep End

Ken's organization and executive skills were now to be tested at a higher level since our DARPA program manager informed us that if we failed, it would not be due to lack of funding. While my counterparts on the electronics/controls side of the team and I took up positions as full-time research associates in the new program, Ken was backfilling with new graduate students. Program meetings in the rather small conference room in the old Mechanical Engineering offices were getting crowded. A first order of business for the team was to thoroughly understand the program goals as DARPA envisioned them and renegotiate those that were not considered realistic even under very optimistic views. I recall that Ken listened to the views and opinions of everyone at the meetings, adding his own technical views of what could be accomplished and perhaps more importantly, what absolutely needed to be accomplished in order to convince DARPA to continue the funding. The final project goals that DARPA agreed upon were translated into design parameters for a "useful" legged vehicle. DARPA's definition of useful was a machine that could provide close logistic support to ground troops in the field without requiring the use of roads. In addition to being completely self-contained in terms of power supply and control, while operating in a field environment that included 50% slopes, loose soils, mud and forest, the machine had to be capable of carrying a 500 lb payload. Ultimately DARPA desired autonomous control under an artificial intelligence operating system. However, GPS was still in early development, artificial intelligence programming was still far from competent and the IBM 386 computer board was still a few years away. As a consequence, DARPA agreed to the use of Simulated Artificial Intelligence, i.e. there would be a driver. Today allowing for a driver in place of autonomous control may not be seen as a simplifying compromise but a complication due to the driver support required. Current drone aircraft development as replacement of piloted aircraft is a useful example of how an on-board driver/pilot complicates the design of the aircraft, thus increasing cost while, in certain circumstances, reducing performance.

Soon after the start of the "big machine" program, Ken's team meetings were largely concerned with discussions of leg kinematics. Due, at least in part, to the success of the Monopod, four-bar mechanisms were the preferred design. Many variations that approached the working volume of the foot needed to attain the mobility requirements were considered. None fully exhibited the characteristics of what was thought to be a workable design. As a new direction, a planar pantograph mechanism was suggested as an alternative to the four-bar. The more we investigated the pantograph, the better it looked. After my own study of the pantograph as a basis for the leg design, I was fully convinced of its superiority over the four-bar variations we had investigated. I presented my rational at the meetings. Among these reasons were that the foot working volume was a direct amplification of the path of the input points; therefore, orthogonal rectilinear inputs produced amplified rectilinear foot paths. Two of the three motions, foot lift and thrust, were completely decoupled, simplifying control tasks for the (ridiculously slow by today's standards) IBM 286 computers that were the best available to us at the time. Compactness of the pantograph mechanism was another advantage. A major disadvantage was the heavily loaded linear rectilinear joints required at the input points. Design of this type of joint is significantly more complex than the revolute joints of a four-bar mechanism. Ultimately, a consensus formed for the use of the pantograph leg, and further investigation into four-bar solutions was terminated.

The next major design decision addressed by the program team was leg actuator methods. By this time we knew that the machine needed to be much larger than the scale of the lab machines simply based on the terrain and obstacle negotiation requirements, the energy autonomy and the driver and payload requirements. Ken allowed that no concept was rejected out of hand, but the estimated leg power requirements made the use of hydraulics an almost obvious choice. With the leg kinematics and actuators decided, the program team started on the hardware design of a prototype full-size leg (PTleg) to be built and lab tested. The Prototype Leg was also often referred to as the Breadboard Leg, but the term was always misleading in my opinion.

The size of the prototype leg precluded a Monopod-style cart system, so we needed a lab space in which we could actually bolt a test stand to the floor. Nearly every square inch of space in the old Robinson lab building, that housed the Mechanical Engineering department, was spoken for. However, I found a storage room filled with odds and ends of metal stock and other stuff that had apparently not been needed or touched for years. I took Ken and others of the team to see it. It was just barely large enough to house the planned leg prototype test system, but it could work if Ken could wrestle it away from the faculty or staff who currently owned the rights to it. He did, and for the next year or so, it was the main home of Ken's part of the "big walking machine" program. It was about this time that the term Adaptive Suspension Vehicle (ASV) became the official designation for the machine and program; most people involved, however, referred to it simply as "the Walker."

After many months of intense activity by other team members and me, the prototype leg and its test stand were assembled. The hydraulic drive system was

Fig. 4 Ken in the Prototype Leg lab giving a "Dog and Pony" to some group. He did a lot of them over the course of the program. (The number balloons are for some forgotten publication and were not really there.)

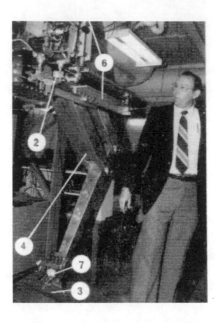

the main responsibility of grad students on the team as its design and control were the subjects of their individual theses. I was involved in the hardware integration of the drive system with the prototype leg. Testing of the completed PTleg involved initially simple unloaded foot motion path control. Loaded motion tests were accomplished by having the leg push against the test stand structure in a quasi-isometric exercise method. The thrust loads of walking were simulated by driving the foot along the test stand base platform under the appropriate vertical loading. Different shoes having various coefficients of friction were envisioned as the solution for attaining the required thrust loads, but a hardwood block shoe was the only one I recall being required. Once serious testing of the prototype leg system commenced, the choice of a remote location away from classrooms and offices was recognized as fortuitous. Operating at design walking speeds and loads, the test system generated great thumps at each footfall against the test frame base platform that could be heard well down the hallway from the lab. Had the lab been located on the second floor, as had been initially suggested and where most classrooms were, we would have had a lynch mob on our hands in no time.

The prototype leg proved to be a valuable learning exercise (as prototypes should be) for avoiding major pitfalls in the ASV design. Primarily we learned that the we had 1) basically a good kinematic and structural leg design, 2) a good solution to the heavily loaded rectilinear input joints, 3) to find an alternative to direct valve control of the hydraulic actuators due to poor efficiency, and 4) an ineffective ankle design in the form of a passive-spring centered DOF that needed to change. All these results and many others less primary were presented and discussed in Ken's program meetings so that even those team members engaged in other allied but indirect research efforts were kept involved. This was helpful to the cohesion of the team and promoted cross-disciplinary cooperation among team members without direct requests from Ken.

4   The Big Splash

DARPA apparently liked the results of the prototype leg phase enough to continue support. I know this since I continued to be paid as a research associate. I was not involved in the funding end - Ken and Robert McGee handled that. I concentrated on program technical advancements that they could sell to the sponsors. They had done a great job in that respect, and the team had the resources to tackle the design and construction of the field-operational machine.

The Prototype leg control system was largely developed and implemented by Mechanical Engineering team members, but the scope of the ASV control problem with 18 active DOF's and custom digital signal processing required more active participation of the Electronic and Controls group headed by Robert McGee. Two fellow Master's students, Denis Pugh and Eric Ribble, completed their Master's degrees on various control and computational improvements to the OSU Hexapod and also, like me, became postgrad research associates on the ASV program. Denis was into advanced state-space controls, and Eric designed and built specialized digital signal processors. We all held similar positions in the ASV program, and we worked well together, perhaps because none of our areas of expertise overlapped. Years later at some meeting concerning the ASV, Ken referred to himself and Robert McGee as the "Instigators" and Denis, Eric and myself as the "Perpetrators".

The ASV team had become significantly more technically diverse and larger. The potential for contentious meetings and turf wars are always a danger to large program teams with members of diverse backgrounds, but Ken's approach apparently dissipated or resolved such issues before they could become truly disruptive. Robert McGee had very similar talents in the team-building area. In the many years of the program, I recall only one turf conflict, and it was due to overlapping areas and responsibilities. Ken rectified the dispute in one short meeting.

The ASV mechanical design work had become much more than I and a graduate student or two could handle and meet the timeline that DARPA wanted. With the requirement for more manpower, Ken allowed me to hire select undergraduate students to work as draftsmen and technicians. We politely informed the other graduate students cohabitating my office but not involved with the ASV program that they would need to find another place to work. Drafting tables (this was before affordable desktop CAD) were moved in, and a series of 3 to 4 student drafters rotated in and out, working between classes. Ken never tried to micro-manage my student worker group so that I had mostly full control. I was able to keep the workers and let go of those that just wanted a place to rest between classes. With a staff of drafters, the detail design work advanced at a good pace to the point that some leg and frame components could be sent for fabrication even as hydraulic system design was still in the configuration stage. From the Prototype leg testing we knew that conventional valve control was a non-starter. Fortunately, a new postdoc, A. Pery, who had considerable hydraulic system design experience, joined the ASV team. He suggested a two-stage hydraulic-hydrostatic system that circumvented the large throttling losses (and waste heat gain) inherent in valve control of the large actuator flow rates required at full walking speeds. I saw the proposed hybrid system as the best of the actuator concepts yet considered. Arie and I made our case to Ken and the team and won approval for the unconventional drive system.

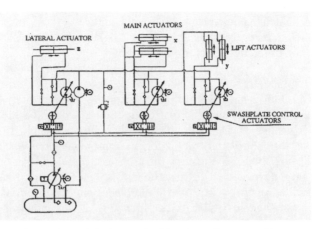

Fig. 5 Schematic of the ASV hybrid hydraulic drive for a single leg. There were six of these systems, one inside each leg. See reference [2] for a complete description.

After the aluminum ASV frame was fabricated in the main lab of the Welding Engineering building across the street, it was moved (with difficulty) into the largest laboratory room in Robinson lab for load-deflection tests. It could never fit in the prototype leg lab. Clearly a larger facility was needed for the construction and indoor testing of the ASV. We were totally unsuccessful in obtaining any suitable lab space on the main campus. However, I found an empty prefabricated building belonging to some OSU staff support office (maybe computer systems) one to two miles away on the west campus. I acquired access to the building and took Ken out to see it. The building had 3600 sq.ft. of high bay lab space with plenty of electric power and 1200 sq.ft. of finished office space. It wasn't much to look at, but it suited the ASV program perfectly. After walking though the building, Ken asked how much of it did I think we would need. I replied, *"All of it."* I recall that it was a tussle with the administration, but between Ken and Robert McGee and whomever else they recruited, the building was assigned to the ASV program for the duration. The building, which became known as the Walker Lab, was the new home to the ASV mechanical design group of student drafters, technicians and myself. The detail design task of creating the fully toleranced working drawings was accomplished almost entirely by undergraduate students working in a big central drafting room in the Walker Lab. At the height of the program there were about 15 students working. They generated nearly a thousand manually drafted drawings, explicitly defining every custom part required for the ASV.

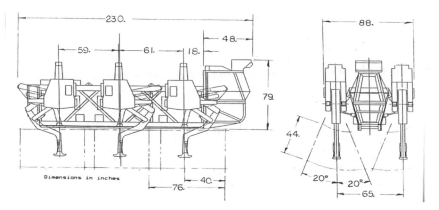

Fig. 6 Assembly drawing of the ASV with leg working volumes shown. Dimensions are in inches and degrees.

Ken kept well abreast of the program from the weekly meeting briefings and regular visits to the Walker Lab. As the detail design work was winding down, the fabrication and assembly work was ramping up. Most of the machine parts were sent to machine shops in the region, but the aluminum welding of the leg structures was kept in-house, principally for program schedule reasons, but also for quality control. The program was extremely lucky to find a welding engineering student who was also a qualified airframe welder to fabricate the frame. However, he just up and graduated soon after. The ASV fabrication required highly skilled welders and machinists. OSU had such talent on staff, but they were kept more than booked by other research programs. Due to the complex Civil Servant and other OSU institutional requirements, it was virtually impossible to hire the talent required. The work-around that Ken and Robert McGee came up with was innovative, perfectly legal and sent the ASV adventure to a new level for the author. They suggested that I form a start-up company with Denis and Eric, the two other program research associates, and submit a bid for a subcontract for the fabrication and testing of the ASV. The rationale was simple - an independent subcontractor could hire anyone deemed fit for the work and fire them if it proved otherwise. Moreover, a subcontractor could attract the required talent by offering wages beyond the rigid pay scale levels enforced for technicians at OSU. Denis, Eric and I formed a company called Adaptive Machine Technologies in 1985, submitted a bid and won the contract (it was not sole-sourced as far as I recall). I continued attending Ken's program meetings but as a subcontractor instead of an OSU research associate. In truth, it was a transparent change of team structure, so almost no one involved really noticed.

The ASV stood up under its own power and onboard computer control in early in 1986. Tethered indoor walking was demonstrated shortly afterward. In 1987 the ASV was operating outdoors. Field testing continued until 1990, culminating in a simulated timber harvesting trial conducted at Virginia Tech in Blacksburg [6]. I was the first driver of the ASV and operated it for many of the trials. It was operated through an aircraft-type joystick. The joystick generated the input in terms of a desired velocity vector, and the control computers did all the leg and body posture control. Also, the control program would not allow the machine to violate the stability margins. It was great fun to drive since the control computers did all the work and would not allow you to dump it over. The ASV was so easy to operate that even Ken and Robert McGee could do it with only 5 minutes of instruction.

The ASV program was a great adventure in engineering and a good lesson in the power of effective teams. The interested reader can see the ASV in action on YouTube.

(a)   (b)

Fig. 7 a) Asv frame with leg hip pivot mounts in place. b) Two student technicians mounting a leg onto the frame.

a)   b)

Fig. 8 a) A much younger author in front of the completed ASV. The indoor testing had just begun (1986). b) A much older author visiting the ASV in deep storage (2005).

References

[1] Gabrielli, G., Von Karmen, T.H.: What price speed? Specific power required for propulsion of vehicles. Mechanical Engineering 72(10), 775–781 (1950)
[2] Nair, S., Singh, R., Waldron, K.J., Vohnout, V.J.: Power system of a multi-legged walking robot. Robotics and Autonomous Systems 9, 149–163 (1992)

[3] Pugh, D.R., Ribble, E.A., Vohnout, V.J., Bihari, T.E., Walliser, T.M., Patterson, M.R., Waldron, K.J.: A technical description of the Adaptive Suspension Vehicle. International Journal of Robotics Research 9(2), 24–42 (1990)

[4] Song, S.M., Waldron, K.J., Vohnout, V.J., Kinzel, G.L.: Computer-aided design of a leg for an energy efficient walking machine. Mechanism and Machine Theory 19(1), 17–24 (1984)

[5] Vohnout, V.J.: Mechanical design of an energy efficient robotic leg for use on a multi-legged walking vehicle. M.S. Thesis, The Ohio State University, Columbus, Ohio (1982)

[6] Vohnout, V.J., Pugh, D.R.: Walking machines: A solution to mobility problems of the forestry industry. In: Proc. Robotics in the Forestry Symposium of the Eastern Division of the Forestry Engineering Inst. of Canada (1990)

[7] Waldron, K.J., Klein, C.A., Pugh, D., Vohnout, V.J., Ribble, M., Patterson, M., McGhee, R.B.: Operational experience with the Adaptive Suspension Vehicle. In: Proceedings of 7th World Congress on Theory of Machines and Mechanisms, vol. 3, pp. 1495–1498 (1987)

[8] Waldron, K.J., Pugh, D.R., Vohnout, V.J., Ribble, E.A., Walliser, T.M.: Walking machines for the forestry industry. In: Proceedings of IARP Workshop on Robotics in Agriculture and the Food Industry, pp. 219–228 (1990)

[9] Waldron, K.J., Vohnout, V.J.: The Adaptive Suspension Vehicle, Videotape. MIT Press (1988)

[10] Waldron, K.J., Vohnout, V.J., Brown, T.F., Kinzel, G.L., Srinivasan, K.: Two experiments in legged locomotion. In: Proceedings of 9th Applied Mechanisms Conference, vol. 1(II.B) (1985)

[11] Waldron, K.J., Vohnout, V.J., Pery, A., McGhee, R.B.: Configuration design of the Adaptive Suspension Vehicle. International Journal of Robotics Research 3(2), 37–48 (1984)

[12] Waldron, K.J., Vohnout, V., Pery, A., Song, S.M., Wang, S.L.: Mechanical and geometric design of the Adaptive Suspension Vehicle. In: Proceedings of Fifth CISMiFToMM Symposium on Theory and Practice of Robots and Manipulators, pp. 240–249 (1984)

[13] Waldron, M.B., Vohnout, V.J.: Formalizing knowledge in design for CAD/CAM integration. In: Proceedings of the International Workshop on Engineering Design and Manufacturing Management (1988)

4 Mechanisms and Robotics at the ASME Design Engineering Technical Conferences – The Waldron Years 1986-2012

Steven A. Velinsky

Abstract. The American Society of Mechanical Engineers' mechanisms and robotics community has sponsored the dissemination of research results through the Design Engineering Technical Conferences for the 67-year history of the Design Engineering Division. Kenneth J. Waldron has been a tremendous contributor and leader in this field for over forty years. This paper presents a brief synopsis of the history of the mechanisms and robotics portion of this conference in honor of Prof. Waldron's 70th birthday.

1 Introduction

The Design Engineering Division (DED) of the American Society of Mechanical Engineers (ASME) was established in 1966 as it became independent from ASME's General Engineering Department during the evolution of the structure of the society. The DED established an annual Design Engineering Conference (DETC) in 1973 in Cincinnati, OH. While independent conferences sponsored by the Mechanisms Committee predated the first DETC by almost 20 years, the Mechanisms Committee soon became a fixture at the DETC and was the DETC sponsor in alternate years. This paper discusses the Mechanisms Conferences for the years 1986 through 2012 with particular attention to the contributions of Professor Kenneth J. Waldron in honor of his 70th birthday.

The Mechanism Committee, which changed its name to the Mechanisms and Robotics Committee in 2004, conducted its annual conference in even numbered years starting in 1964. Due to the continued tremendous interest, the Mechanisms and Robotics Conference became an every year event starting in 2006.

Steven A. Velinsky
Department of Mechanical & Aerospace Engineering, University of California, Davis
e-mail: savelinsky@ucdavis.edu

V. Kumar et al. (Eds.): *Adv. in Mech., Rob. & Des. Educ. & Res.*, MMS 14, pp. 49–54.
DOI: 10.1007/978-3-319-00398-6_4 © Springer International Publishing Switzerland 2013

2 The Years 1986-1999

While the Mechanisms Conference was held for the 19th time in 1986 (October 5-8), this paper begins with this conference since it is a banner year to start highlighting Prof. Waldron's contributions as he chaired both the Mechanisms Conference as well as the Design Engineering Technical Conferences, sponsoring these conferences in Columbus, Ohio. Moreover, it was the first Mechanisms Conference attended by this author.

Many of the attendees can remember the excitement of the IMSA GT championship 500 km race through the streets of Columbus, walking distance from the conference hotel (Hyatt Regency) on October 5, 1986. Many of us were comparably excited by the opportunity to see Prof. Waldron's hexapod Adaptive Suspension Vehicle (ASV) operate during a tour of his Robotic Mechanisms Laboratory as part of the conference. Fortunately, the ASV did not shoot 6-foot flames out of its exhaust during downshifts!

Dr. Jason Lemon, Chairman of International TechneGroup, Inc. and founder of the Structural Dynamics Research Corporation was the conference's keynote speaker and discussed the emergence of computer-aided engineering methods in machine design practice. The conference contained approximately 150 papers. Ken Waldron contributed 3 papers in the areas of expert systems [1], walking machines [2], and industrial robots [3] and participated in two panel sessions on Current Issues in Robotics and ASME-NSF Research Study in Machine Dynamics.

The Mechanisms Committee did not participate in the DETC in the odd numbered years, which were held are follows: Sept. 27-30, 1987, Boston, MA; Sept. 17-21, 1989, Montreal; Sept. 22-25, 1991, Miami, FL; Sept. 19-22, 1993, Albuquerque, NM; Sept. 17-21, 1995, Boston, MA; Sept. 14-17, 1997, Sacramento, CA; Sept. 12-16, 1999, Las Vegas, NV. The 1995 DETC is particularly noteworthy as it was the 50th anniversary of the DED with numerous associated special events.

The 1988 DETC was held in Orlando, FL from Sept. 25-28. In addition to contributing technical papers, Kenneth Waldron received the ASME Leonardo da Vinci Award for eminent achievement in the design or invention of a product, which is universally recognized as an important advance in machine design – his Adaptive Suspension Vehicle highlighted at the 1986 conference.

The Mechanisms Conference was held Chicago, IL on Sept. 16-19, 1990. The conference had over 210 papers in 41 sessions. Its keynote speaker was Prof. Bernard Roth of Stanford University. In addition to contributing 3 papers, two on parallel dual [4,5] and one on motion synthesis [6], Prof. Waldron participated in a panel session on Research Opportunities in Machine Dynamics and was honored with the 1990 Mechanisms Committee Award recognizing his cumulative contribution to the field of mechanism design.

The Mechanisms Conference was held in Scottsdale, AZ on Sept. 13-16, 1992 with over 240 papers in 47 sessions. Prof. Waldron was the Keynote Speaker and contributed 7 papers in the following areas: mechanism synthesis [7], mixed mechanism kinematics [8], stewart platform [9], analysis and synthesis of 4-bars [10,11], navigation of mobile robots [12], quadruped galloping [13].

The Mechanisms Conference was held in Minneapolis, MN on Sept. 11-14, 1994 with about 180 papers in 40 sessions. Marc Dulude, Vice-President Parametric Technology, presented the Keynote on concurrent product and process development for mechanical design automation. In addition to contributing 4 papers on active mechanisms [14], synthesis of 6-bar [15], multi-circuit mechanisms [16], and articulated wheeled vehicles [17], Prof. Waldron participated in a panel session on Research Opportunities in Mechanisms. Moreover, he was honored with the 1994 Machine Design Award recognizing both his eminent achievement and his distinguished service in the field.

The 1996 conference was held in Irvine from Aug. 18-22 and included 152 mechanisms papers. The Keynote was delivered by Prof. Delbert Tesar, University of Texas-Austin, on The Future of Mechanical Design in the World of Computer Technology. Prof. Waldron contributed a paper on actively configurable wheeled vehicles [18]. The 1998 Mechanisms Conference was held in Atlanta, GA from Sept. 13-16. The Keynote was delivered by James Smith, Sandia National Laboratories, on Intelligent Micromachines. Prof. Waldron contributed a paper on actively kinematic geometry [19].

3   The Years 2000-2012

2001 was a devastating year for the country and the DETC, which was held Sept. 9-12 in Pittsburgh, PA. Not only was the conference interrupted by the terrible events of 9/11, but United flight 93 crashed into a field near Shanksville, Pennsylvania, approximately 65 miles southeast of Pittsburgh. The DETC was held in 2003 and 2005 as follows: Sept. 2-6, 2003, Chicago, IL; Sept. 24-28, 2005, Long Beach, CA.

The 2000 DETC was held in Baltimore, MD from Sept. 10-13 with 36 sessions sponsored by the Mechanisms Committee. Prof. Bernard Roth, Stanford University presented the Keynote Lecture on Mechanisms in the 20th Century. Prof. Waldron contributed a paper on discrete parallel arrays [20]. The 2002 conference was again held in Montreal from Sept. 29-Oct. 2 with 177 mechanisms papers and 8 invited talks. The DETC Keynote was delivered by Robert Ambrose, NASA Johnson Space Center, on Mechanism and Energetic Challenges in Humanoid Design. Prof. Waldron contributed a paper on robotic legs [21] and participated on a panel concerning New Directions for Research in Kinematics and Mechanisms. Prof. Waldron was honored for his educational contributions with the 2002 Ruth and Joel Spira Outstanding Design Educator Award.

The 2004 DETC was held in Salt Lake City, UT from Sept. 28-Oct. 2. The Mechanisms and Robotics Conference included 173 papers and 8 keynote talks. Prof. Waldron contributed a paper on galloping machines [22].

Starting in 2007, the Mechanisms and Robotics Committee held its conference every year. The conferences for the noted years are summarized in the following.

2006 Philadelphia Sep. 10-13 - 140 Mechanisms and Robotics papers.
   Plenary Lecture: Robert Kruse, General Motors Corp., Technical Research Challenges Facing the Automotive Industry
   Keynotes in Mechanisms and Robotics
   Delbert Tesar, Univ. of Texas-Austin - A New Wave of Technology: An Essential Step Built on Core Technologies in Mechanical Engineering
   Jorge Angeles, McGill University - The Kinematics and Dynamics of Parallel Schonflies-Motion Generators
   Pierre Larochelle, Florida Institute of Technology - The Computer Aided Synthesis of Linkages: Today and Tomorrow

2007 Las Vegas Sep. 4-7 - 140 Mechanisms and Robotics papers.
   Keynote – 3 NSF Program Directors – Shaking the Money Tree at NSF
   Mechanisms Plenary lectures
   Ashok Midha, University of Missouri-Rolla - Compliant Mechanisms: Memory Lane, the Journey, and the Exciting Road Ahead
   Mohan Budduluri, Founder Restoration Robotics - Image Guided Robotics: From Cancer Treatment to Hair Transplantation.

2008 New York Aug. 3-6 - 165 Mechanisms and Robotics papers.
   Mechanisms Keynote lectures
   Steven Velinsky, Univ. of California-Davis - On the Road from Cement to Silicon: Field Robotics for Highway Tasks
   Clement Gosselin, Laval University - Cable Driven Parallel Mechanisms: Force-Directed Design and Applications
   Ken Waldron received the Robert E. Abbott Award recognizing his devoted tireless and innovative leadership and service to the ASME DED, the international design engineering community, and the profession.

2009 San Diego Aug. 30-Sep. 2 - 137 Mechanisms and Robotics papers.
   Keynote – Francois Pierrot, Director of Research, Centre National de la Recherche Scientifique - Creating Novel Parallel Mechanisms with Industry in Mind

2010 Montreal Aug. 15-18 - 190 Mechanisms and Robotics papers.
   Keynotes
   Vijay Kumar, University of Pennsylvania - Cooperative Manipulation and Transport by Aerial and Ground Robots
   Manfred Husty, Leopold-Franzens-Universität Innsbruck - Mechanism Constraints - The Algebraic Formulation

2011 Washington DC Aug. 28-31 - 142 Mechanisms and Robotics papers.
   Keynote – Sridhar Kota, Univ. of Michigan - Innovation and US-based Manufacturing

2012 Chicago Aug. 12-15 - 178 Mechanisms and Robotics papers.
   Keynotes
   Paolo Dario, Director of The BioRobotics Institute of the Scuola Superiore Sant'Anna, Pisa, Italy - Problems and Challenges in the Design of Novel Mechanisms for Robotics Surgery
   Feng Gao, Shanghai Jiao Tong University - Development of Parallel Robotic Mechanisms for High-Payload Machines

4   Service to the Design Engineering Division

Kenneth Waldron has tirelessly served the ASME Design Engineering Division (DED) for over 40 years. Ken's active participation in ASME activities started at the committee level and progressed through division leadership up through national and international leadership. Ken has chaired several committees within the DED. He was chair of the Mechanisms Committee from 1984 to 1986, Honors and Awards Committee from 1995 to 1998; Member Interests Committee 1981-85; Government Relations Committee 1991-95; and USCToMM 1993-2002. He served as Technical Editor of one of the division's premier journals, *The ASME Transactions Journal of Mechanisms, Transmissions and Automation in Design* from 1988 through 1992. He joined the Design Engineering Executive Committee from 1996 to 2002 serving as Division Chair in 2000 and 2001. Furthermore, he completed a term as Technical Leader of the ASME Systems and Design Group (2005-08) after a term as Member-at-Large (2000-03). He was the President of the International Federation for Theory of Machines and Mechanisms (IFToMM), the premier international organization devoted entirely to this area from 2000 through 2007. ASME is the US affiliate organization of IFToMM. Ken has had countless other professional service activities both within and outside ASME. He has also served as an Associate Editor of Applied Mechanics Reviews and Chair of the Management Committee of the *IEEE/ASME Transactions on Mechatronics*. Simply, Ken is a colleague that can always be counted upon to serve the profession and his continuous enthusiasm is amazing.

5   Conclusions

It is an honor and a privilege to have the opportunity to contribute to the celebration of Kenneth J. Waldron's 70th birthday. Aside from being a prolific researcher and contributor to the field, Ken has always been modest and he has served as a mentor to large numbers of students and colleagues. On top of his valuable contributions to the profession, Ken is a true gentleman and scholar, and his friendship to so many of us is tremendously valued.

References

1. Waldron, K.J., Waldron, M.B., Wang, M.: An expert system for initial bearing selection. In: Proc. ASME Design Engineering Technical Conferences (1986)
2. Song, S.-M., Waldron, K.J.: Geometric design of a walking machine for optimal mobility. In: Proc. ASME Design Engineering Technical Conferences (1986)
3. Wang, S.-L., Waldron, K.J.: A study of the singular configurations of serial manipulators. In: Proc. ASME Design Engineering Technical Conferences (1986)
4. Murthy, V., Waldron, K.J.: The parallel dual of the Stanford arm. In: Proc. ASME Design Engineering Technical Conferences (1990)

5. Murthy, V., Waldron, K.J.: Position kinematics of the generalized lobster arm and its series-parallel dual. In: Proc. ASME Design Engineering Technical Conferences (1990)
6. Waldron, K.J.: Solution rectification in three position motion generation synthesis. In: Proc. ASME Design Engineering Technical Conferences (1990)
7. Bawab, S., Kinzel, G.L., Waldron, K.J.: Rectified synthesis of coupler-driven four-bar mechanisms for four-position motion generation. In: Proc. ASME Design Engineering Technical Conferences (1992)
8. Husain, M., Waldron, K.J.: Position kinematics of a mixed mechanism. In: Proc. ASME Design Engineering Technical Conferences (1992)
9. Husain, M., Waldron, K.J.: Direct position kinematics of the 3-1-1-1 Stewart platforms. In: Proc. ASME Design Engineering Technical Conferences (1992)
10. Srikrishna, P., Waldron, K.J.: An unified analysis and implementation of four mutually separated position synthesis of four-bar linkages. In: Proc. ASME Design Engineering Technical Conferences (1992)
11. Venkataraman, S., Kinzel, G.L., Waldron, K.J.: Optimal synthesis of four-bar linkages for four-position rigid-body guidance with selective tolerance specifications. In: Proc. ASME Design Engineering Technical Conferences (1992)
12. Sreenivasan, S.V., Waldron, K.J.: A drift free navigation system for a mobile robotic vehicle operating on unstructured terrain. In: Proc. ASME Design Engineering Technical Conferences (1992)
13. Nanua, P., Waldron, K.J.: Instability and chaos instability in quadruped gallop. In: Proc. ASME Design Engineering Technical Conferences (1992)
14. Sreenivasan, S.V., Waldron, K.J., Mukherjee, S.: Global optimal force allocation in active mechanisms with four frictional contacts. In: Proc. ASME Design Engineering Technical Conferences (1994)
15. Bawab, S., Kinzel, G.L., Waldron, K.J.: Rectified synthesis of six-bar mechanisms with well defined transmission angles for four-position motion generation. In: Proc. ASME Design Engineering Technical Conferences (1994)
16. Waldron, K.J., Sreenivasan, S.V.: A study of the position problem for multi-circuit mechanisms. In: Proc. ASME Design Engineering Technical Conferences (1994)
17. Sreenivasan, S.V., Waldron, K.J.: Displacement analysis of an articulated wheeled vehicle configuration on uneven terrain. In: Proc. ASME Design Engineering Technical Conferences (1994)
18. Waldron, K.J.: Steering algorithms for an actively configurable wheeled vehicle. In: Proc. ASME Design Engineering Technical Conferences (1996)
19. Waldron, K.J.: Lower pair joint geometry. In: Proc. ASME Design Engineering Technical Conferences (1998)
20. Waldron, K.J., Yang, P.H.: Force coordination of a discretely actuated parallel array. In: Proc. ASME Design Engineering Technical Conferences (2000)
21. Schmiedler, J., Waldron, K.J.: Leg stiffness and articulated leg design for dynamic locomotion. In: Proc. ASME Design Engineering Technical Conferences (2002)
22. Waldron, K.J., Kallem, V.: Control modes for a three-dimensional galloping machine. In: Proc. ASME Design Engineering Technical Conferences (2004)

5 Screws and Robotics for Metrology

Joseph K. Davidson, Samir Savaliya, and Jami J. Shah

Abstract. The pseudoinverse of a rectangular matrix of modified screws is used to compute the least-squares fit of a set of points that have been measured along a line-profile. Tolerances on line profiles are used to control cross-sectional shapes of parts, such as turbine blades. The specified profile is treated as a moving platform of a hypothetical, redundant, and planar in-parallel-actuated robot, and all the measured points are presumed to be fixed in it. The locations of the linear actuators are represented with screw coordinates, and these are arranged in a matrix equation that relates the three small displacements of the platform to the corresponding deviations (treated as small displacements) of the measured points.

1 Introduction

The objective of dimensional metrology is to check manufactured parts for conformance to tolerance specifications on drawings. The drawings define features, such as planes, cylinders, and profiles, and the associated tolerance specifications define tolerance-zones for each feature [1]. A tolerance-zone sets the limits for manufacturing variations that are permissible for a feature. Modern coordinate measurement machines (CMMs) provide an automated process of inspection that is replacing many traditional manual methods. The CMM measurements made on a part are represented as the coordinates for a large number of points in a 'cloud'. Although direct comparison of the coordinates for a set of points can determine whether or not all fit within a tolerance-zone, a feature representation is more meaningful when monitoring a manufacturing process. For instance, the boundaries of a minimum-zone, or a least-squares fit, of points is informative about the location of the feature in the tolerance-zone, and there is potential to use drift of the location as a monitoring tool. Considerable computation, arranged in computer software, is required to convert the points to a feature.

Joseph K. Davidson · Samir Savaliya · Jami J. Shah
Design Automation Lab, Department of Mechanical and Aerospace Engineering,
Arizona State University, Tempe, Arizona, USA
e-mail: j.davidson@asu.edu

V. Kumar et al. (Eds.): *Adv. in Mech., Rob. & Des. Educ. & Res.*, MMS 14, pp. 55–66.
DOI: 10.1007/978-3-319-00398-6_5   © Springer International Publishing Switzerland 2013

Existing methods for fitting a feature to a set of measured points are: a one-sided fit, a two-sided fit, or minimum-zone fit, and the least-squares fit. A recent article summarizing conversion algorithms in general is [2]; another is [3]. An example is [4,5] where measured points are converted to minimum-zone features for planes, lines, and cylinders. One measure of conformance is evaluating whether or not points lie in a specified tolerance-zone. Choi and Kurfess [3] present an algorithm that determines if a point-set, when displaced *en masse*, can fit into the intended zone, and in [6] extend this to determine minimum zones. Conversion algorithms that apply to profiles, including surface-profiles, are described in [3] and [7], although all examples presented are for objects having continuous curvature, such as cones, cylinders, and sculptured surfaces.

The focus of this paper is the use of screws to formulate a computational method to obtain the least-squares fit of a set of points that have been measured on a rectangular line-profile, one that has both C^1- and C^2-discontinuity. The results will be a location, similar to [6], and size for the line-profile.

The specifications for a raised boss are shown in Fig. 1. Its rectangular shape is controlled by the profile tolerance $t = 0.2$ mm relative to the Datums A, B, and C. This specification establishes two boundary rectangles at each cross-section of the raised profile. One is 0.1 mm larger along every line normal to the surface, and the other is 0.1 mm smaller, according to the Standard [1].

2   Regression Line in the Plane

There are two types of least-squares fit: the *linear* (vertical) fit and the *orthogonal* fit [8]. The distinction between the two is the reference system used for describing coordinates (x_i, y_i). To understand better this distinction, undertake a straight-line fit of n identified points in a plane. For both methods, the solution-line is of the form $y = mx + b$, and there are n linear equations that relate the x_i- and y_i-values. From the Gauss-Markov Theorem [9], the least-squares fit is obtained by minimizing the sum

$$\sum \{y_i - (mx_i + b)\}^2 \tag{1}$$

for $i = 1...n$. As one example, apply linear regression to the five points in Table 1 which are symmetrically disposed about the line $y = 1 + x/2$ in the $x_j y_j$-frame in Fig. 2. When standard software (e.g. MAPLE) for linear regression is applied to these five points, the result is $m = 24/53$ and $b = 69/53$ (line with long dashes in Fig. 2), values that represent the linear least-squares fit for the points.

Table 1 Coordinates of points for the example of least-squares fit for a line

Point	1	2	3	4	5
x	3	5	8	8	8
y	3	3	4	5	6

Fig. 1 Specification for a raised rectangular profile having sharp corners. Its shape is controlled by the profile tolerance t = 0.2 mm relative to Datums **A**, **B**, and **C**.

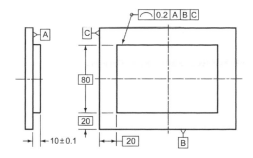

Fig. 2 Five points equally disposed about a line $y = 1 + x/2$ (solid line) in the $x_j y_j$-frame and the linear regression line (dashed line) for them

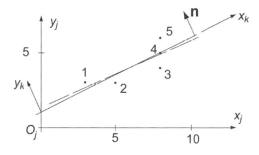

The set of n equations, which relate the n points to the linear regression line, may also be written

$$[\mathbf{y}_i] = [\mathbf{K}'][\$] = \begin{bmatrix} x_1 & 1 \\ x_2 & 1 \\ \vdots & \vdots \\ x_n & 1 \end{bmatrix}[\$], \qquad (2)$$

where $[\mathbf{y}_i] = [y_1 \ ... \ y_n]^\mathsf{T}$, $[\$] = [m \ b]^\mathsf{T}$, and $[\mathbf{K}']$ is an $n \times 2$ rectangular coefficient matrix. The n linear equations are, of course, inconsistent. However, one solution method is to use the pseudoinverse $[\mathbf{K}']^\#$ to obtain the unknowns m and b in $[\$]$ for the least-squares fit of the points [10, 11]. It ensures that the values m and b contained in $[\$]$ correspond to a minimization of the sum of the squares of all the differences $y_i - (mx_i + b)$. The set of y_i-values reside in matrix $[\mathbf{y}_i]$ and the corresponding set of directions for their measurement resides in one column of $[\mathbf{K}']$. For an overconstrained (and inconsistent) set of linear equations, $[\mathbf{K}']^\#$ is formed [10] as implied in the second of the equations

$$[\$] = [\mathbf{K}']^\# [\mathbf{y}_i] = \left\{([\mathbf{K}']^\mathsf{T}[\mathbf{K}'])^{-1}[\mathbf{K}']^\mathsf{T}\right\}[\mathbf{y}_i] \qquad (3)$$

Of course, an alternative method for solving Eqs (2) for $[\$]$ is to use singular value decomposition [12].

The computed values $m = 24/53$ and $b = 69/53$ for the least-squares line in Fig. 2 are not very close to the theoretical values of $1/2$ and 1 because the reference direction for error measurement was not made at right angles to the theoretical geometric shape. The orthogonal least-squares fit may be undertaken by measuring coordinates (x_i, y_i) for each point, and the coordinates m and b in [$], from the reference $x_k y_k$-frame that has its x_k-axis lying on the theoretical line $y_j = 1 + x_j/2$ (solid line in Fig. 2). When the matrix $[\mathbf{y}_i]$ in Eqs (2) and (3) is then formed from the y_k-values that are computed from this new reference direction, Eq (3) gives a solution that, when transformed from the $x_k y_k$-frame to the $x_j y_j$-frame, produces values $m = 1/2$ and $b = 1$.

For what follows in §4, it is helpful to note here that, when every y_i is increased (or decreased) by the same offset value ΔF, Eq. (2) produces an unchanged slope m and a value for b that is increased exactly by ΔF. This role for ΔF is the same for both the linear and orthogonal least-squares fits.

3   The Tolerance-Map for Rectangular Line-Profiles

The objective of this section is to describe briefly a geometric representation for the tolerance-zone for a line-profile, i.e. for the allowable limits of manufacturing variations for the profile; a more elaborate treatment may be found in [13]. We call this representation a Tolerance-Map (T-Map); it describes the freedom of a feature in its tolerance-zone. For line-profiles, the manufacturing variations will be represented with the true profile and profiles parallel to it. Each point in the T-Map corresponds to any one of these parallel profiles or to any one of them that is displaced, yet remains within the tolerance-zone. Four degrees of freedom are required to specify the manufacturing variations of a line-profile, such as any one cross-section of the rectangular boss in Fig. 1. Correspondingly, its T-Map will be four-dimensional (4-D). Therefore, it becomes necessary to choose five of the parallel and/or displaced profiles as basis profiles and to define the T-Map by placing five corresponding basis points $\psi_1 \ldots \psi_5$ to form the vertices of a basis simplex. Five barycentric coordinates $\lambda_1 \ldots \lambda_5$, each one at its basis point ψ_i, then identify any point ψ in the T-Map, and each such point corresponds to one manufacturing variation (one profile) in the tolerance-zone. When we set $\Sigma \lambda_i = 1$ ($i = 1 \ldots 5$), the coordinates $\lambda_1 \ldots \lambda_5$ become the *areal* coordinates of ψ [14].

Of the five basis-profiles required, two will be: ψ_1, the smallest-sized profile, and ψ_2, the largest-sized profile, i.e. the inner and outer boundaries to the tolerance-zone, respectively. These are both locked in place and cannot displace. The remaining basis-profiles are based on displacements of the *middle-sized* rectangular profile (MSP). Each MSP is represented by its components of translations, e_x and e_y, and its rotational displacement θ. The basis-profiles displaced to the limits $e_x = t/2$ and $e_y = t/2$ in the x- and y-directions are labeled ψ_3 and ψ_4, respectively, and the one rotated counterclockwise the maximum amount $\theta = t/2\,\overline{a}$ is ψ_5 (Fig. 3(a)). Note that the basis-profile ψ_5 corresponds to the limit to rotation determined by length $2\,\overline{a}$, the *shorter* side of the rectangle. Although this requires portions of this profile to lie outside the tolerance-zone, the T-Map boundary will still reflect the design intent by having its basis *point* ψ_5 lie beyond it. The result will be a T-Map consistent with choices made in [13].

The 3-D T-Map for all the middle-sized rectangular profiles is established with the four basis-points ψ_{12}, ψ_3, ψ_4, and ψ_5 shown in Fig. 3(b). Basis-points ψ_3, ψ_4, and ψ_5 are placed at the same distance $t/2$ from the origin along the three axes of a rectangular Cartesian frame of reference with axes e_x, and e_y, and θ'. Note that the angular limit $\theta = t/2\,\overline{a}$ is multiplied by the length $\overline{a}$, i.e. $\theta' = \overline{a}\,\theta$ so that the units along all axes are the same. The origin in Fig. 3(b) is labeled ψ_{12} because it represents the undisplaced MSP, i.e. the average of the limiting sizes ψ_1 and ψ_2.

Rectangular profiles that are larger or smaller than the MSP are more limited in their allowable displacements e_x, e_y, and θ, and the limits diminish linearly with change in size. Therefore, the full T-Map for the rectangular tolerance-zone in Fig. 3(a) is a double hyperpyramid in 4-D that is depicted in Fig. 4. The base for each single hyperpyramid is the 3-D solid shape from Fig. 3(b), and every other section (two are shown) at right angles to the direction of size is smaller and geometrically similar.

There now is another way to view the objective of this paper: reduce the measured points on one line-profile to a set of small-displacement coordinates that locate a single point within the T-Map of Fig. 4. The result is an *i*-Map, that displays the quality of manufacturing relative to tolerance specifications.

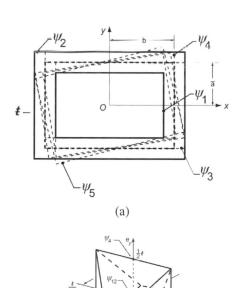

(a)

Fig. 3 (a) The middle-sized profile (dashed-lined rectangle) in the (exaggerated) tolerance-zone that is specified with the profile tolerance *t*; five variational possibilities are labeled, three with dotted lines. (b) The T-Map for all the middle sized rectangles in the sharp cornered tolerance zone of Fig. 3(a). Adapted from [13].

(b)

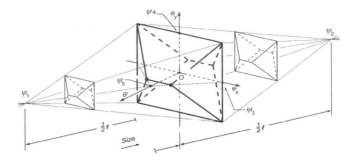

Fig. 4 The 4-D T-Map for the rectangular tolerance-zone in Fig. 3(a) and showing all five basis-points $\psi_1,\ldots,\psi_5$. For clarity of the graphics, the scale in the direction of size ($\psi_1\psi_2$) is exaggerated. Adapted from [13].

4   Least-Squares Fit of a Line-Profile to Measured Points

In Fig. 5(a), the MSP (dashed line) and the boundaries of its tolerance-zone are shown drawn on the platform of a planar in-parallel robot that is guided with three linear actuators that lie on the normalized screws $\$_1$, $\$_2$, and $\$_3$. The actuators are attached to the platform at three of the measured points, i.e. at A, B, and C, and the directions of the corresponding $\$_i$ are the same as for the inward unit normals $\mathbf{n}_i$ from the closest side of the rectangle to the profile that is represented as an envelope of four tangent lines. Assessing minimum distances at the corners of a profile can be problematic because the envelope tangent-lines are not *segments;* instead, each line (p, q, s) is of infinite extent. This matter will be resolved by assessing minimum distances from a parallel curve larger than the MSP. A larger parallel curve is generated easily from the envelope description of a MSP by increasing the value of coordinate s by the same amount for every tangent-line. For purposes of the profile and measured points shown in Fig. 5(a), the outer boundary to the tolerance-zone is an acceptable reference-envelope ($\Delta s = t/2 = 0.1$mm). Deviations d'_i may be obtained from $d = px + qy + s$, where (x, y) are the coordinates of a measured point and (p, q, s) are homogeneous coordinates of the nearest tangent to the MSP [11, 15].

Each of the three linear actuators in Fig. 5(a) exerts a force of magnitude F_i' and causes a velocity of *extension* v'_i of the actuator attached to the platform. Since each v'_i is an extension velocity, the following formulation leads to the orthogonal least-squares fit of the points. Speed and time are of no importance in measurement reduction, so each v'_i will be replaced with a differential displacement d'_i of the measured point in the direction of $\mathbf{n}_i$. The corresponding deviation screw for the platform body is represented by $[\$] \equiv (0, 0, \delta\theta; \delta x, \delta y, 0)$. Since displacements are confined to the xy-plane, the three zero-coordinates may be omitted. Each of the actuator forces in the xy-plane is represented with wrench coordinates, i.e. $F_i' \, \$_i \equiv (\mathbf{F}'_i; \mathbf{T}'_i) \equiv (\mathcal{L}'_i, \mathcal{M}'_i, 0; 0, 0, \mathcal{R}'_i)$, where $\mathcal{L}'_i$ and $\mathcal{M}'_i$ are the

x- and y-components of actuator-force $\mathbf{F}'_i$ and $\mathcal{R}'_i$ is the moment of $\mathbf{F}'_i$ about the origin, i.e. $\mathcal{R}'_i = -y_iL'_i + x_iM'_i$. Since all forces will lie in the xy-plane, the three zero-coordinates may be omitted, just as for [$]. Also, the geometry may be isolated from the statics by normalizing the wrench coordinates, i.e.

$$F'_i\$_i \equiv (L'_i, M'_i; \mathcal{R}'_i) \equiv F'_i(L'_i, M'_i; R'_i) \tag{5}$$

this making $(L'_i)^2 + (M'_i)^2 = 1$. The normalized coordinates L'_i, M'_i, and R'_i for each $\$_i$ are the scalar *screw* coordinates for the actuator-wrench $F'_i\$_i$; they contain only geometry, i.e. direction and location of $F'_i\$_i$.

Fig. 5 (a) The line-profile (dashed line) of Fig. 1, its tolerance-zone boundaries (with an exaggerated scale), and 19 measured points, all lying on the platform of a planar in-parallel robot which is guided by three linear actuators lying on the screws $\$'_1$, $\$'_2$ and $\$'_3$ at points A, B, and C. (b) The free-body diagram of the platform carrying the profile. The external loads are the force $\mathbf{F}'_1$ acting along the screw $\$'_1$ at point A and the equilibrium wrench $(\mathbf{F}_1; \mathbf{T}_1)$ exerted on the platform from the environment and represented with the coordinates $(F_x, F_y; T_z)$. Also shown is the differential displacement vector $\mathbf{d}'_1$ that is aligned with $\$'_1$ at A. The shape of the platform ABC, and the relative location of the xy-frame are together congruent to those same features in Fig. 5 (a).

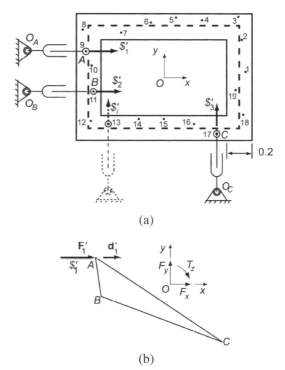

(a)

(b)

A free-body diagram of the platform in Fig. 5(a) contains the three forces $\mathbf{F}'_i$ ($i = 1, 2, 3$) and an equilibrium wrench, composed of a force and a couple, exerted on the platform from the environment. The force and couple are represented with the wrench $(\mathbf{F}; \mathbf{T})$. Consider now that all of the actuated joints have no force applied and are free to move except one, say $\$_1$, shown in Fig. 5(b). Then, the only additional loads on a free-body diagram of the platform are those *portions* of the equilibrium wrench reacting back on it from the environment which are required to equilibrate $F'_1\$'_1$, i.e. the force and couple $(\mathbf{F}_1; \mathbf{T}_1)$ shown in Fig. 5(b) with the components F_x, F_y, and T_z. Since the virtual work of all forces and moments on the free body must be zero for a kinematically admissible displacement of the platform arising from $\mathbf{d}'_i$, the system of forces and couples for the special case in Fig. 5(b) leads to

$$F_1'd'_1 + [T_z \ F_x \ F_y][\delta\theta \ \delta x \ \delta y]^{\mathrm{T}} = 0, \tag{6}$$

in which the order of the coordinates in $(\mathbf{F}_1 ; \mathbf{T}_1)$ has been changed to $(\mathbf{T}_1 ; \mathbf{F}_1)$ and the zero-coordinates again have been omitted. The term $F_1'd'_1$ represents the virtual work of force $\mathbf{F}'_1$ with virtual displacement $\mathbf{d}'_1$, both in the direction of $\$_1$, at point A on the platform, and the product $[T_z \ F_x \ F_y][\delta\theta \ \delta x \ \delta y]^{\mathrm{T}}$ represents the virtual work from the equilibrium-wrench acting on the platform whose deviation (twist multiplied by time) is $[\$] \equiv [\delta\theta \ \delta x \ \delta y]^{\mathrm{T}}$.

It is helpful to shift attention to the wrench $-(\mathbf{T}_1 ; \mathbf{F}_1)$ exerted on the environment and *produced* at the platform by the force $F_1'\$_1$ at A. Since the platform in Fig. 5(b) is a two-force (two-wrench) member, with each wrench intensity of equal magnitude, $-(\mathbf{T}_1 ; \mathbf{F}_1) \equiv -(T_z ; F_x, F_y) \equiv (\mathcal{R}_1 ; L'_1, \mathcal{M}') \equiv F_1' (\mathbf{R}_1 ; L'_1, \mathbf{M}_1')$. Making this substitution in Eq. (6) gives

$$F'_1d'_1 = F'_1[\mathbf{R}'_1; L'_1, \mathbf{M}'_1][\delta\theta \ \delta x \ \delta y]^{\mathrm{T}} \tag{7}$$

for the virtual work expression when force is exerted only at $\$_1$. Two more Eqs. (7), with subscripts $_2$ and $_3$, occur when force is applied only at $\$_2$ and only at $\$_3$. The force-amplitude at each actuated joint may be removed from each term, and all terms on the right come from the product of a row matrix and a column matrix of three elements each. When the three equations are ordered sequentially, then the rows of screw coordinates comprise a matrix $[\mathbf{K}']$ that is formed entirely from the (normalized) coordinates for $\$_1$, $\$_2$ and $\$_3$, and the three equations may be written

$$\begin{bmatrix} d'_1 \\ d'_2 \\ d'_3 \end{bmatrix} = \begin{bmatrix} R'_1 & L'_1 & M'_1 \\ R'_2 & L'_2 & M'_2 \\ R'_3 & L'_3 & M'_3 \end{bmatrix} \begin{bmatrix} \delta\theta \\ \delta x \\ \delta y \end{bmatrix} \quad \text{or} \quad [\mathbf{d}'_i] = [\mathbf{K}'][\$]. \tag{8}$$

From robotics we recognize $[\mathbf{K}']$ as a Jacobian for the actuators of the robot platform in which the normalized coordinates have been rearranged (see e.g. [11]).

So long as the screws $\$_1$, $\$_2$ and $\$_3$ are independent for the three measured deviations d'_1, d'_2, and d'_3 at locations A, B, and C around the profile, the solution to Eq. (8) for $[\$]$, i.e. $[\$] = [\mathbf{K}']^{-1}[\mathbf{d}'_i]$, is unique and all three scalar Eqs. (8) are satisfied exactly. This solution ensures that actuator extensions d'_1, d'_2, and d'_3 are kinematically consistent with the platform (profile) displacement $[\$]$. However, in practical situations, there are many more measured points around a line-profile than three. For instance, in Fig. 5(a) there are 19 points. For every additional point, there would be an added, and redundant, linear actuator with its normalized screw $\$_i$ exerting a force of amplitude F_i' on the platform. One example is shown with dashed lines at Point 13 in Fig. 5(a). Each of these additional points adds a row to the matrices $[\mathbf{d}'_i]$ and $[\mathbf{K}']$ in Eq. (8), so that, for all the measured points,

$$[\mathbf{d}'_i] = \begin{bmatrix} d'_1 \\ d'_2 \\ \vdots \\ d'_n \end{bmatrix} = [\mathbf{K}'] \ [\$] = \begin{bmatrix} R'_1 & L'_1 & M'_1 \\ R'_2 & L'_2 & M'_2 \\ \vdots & \vdots & \vdots \\ R'_n & L'_n & M'_n \end{bmatrix} [\$]. \tag{9}$$

The coordinates ($\delta\theta$, δx, δy) of [$\$$] appear only in a 3-D cross-section of the T-Map (Fig. 3(b)), e.g. in the base of the 4-D double hyperpyramid in Fig. 4; they do not represent the *size* of the least-squares envelope, i.e. the fourth dimension of the T-Map. The values for displacements d'_i, then, may all contain a constant value $-\Delta F$ that represents the change in feature size between that of the MSP and the least-squares profile, and they *must* contain a value Δs that was introduced artificially to establish the correct proximity of a measured point to the profile. For reduction of CMM data, then, each generic Eq (7) must be augmented to

$$\left. \begin{aligned} d'_i &= [R'_i \ L'_i \ M'_i][\delta\theta \ \delta x \ \delta y]^{\mathrm{T}} + (\Delta s - \Delta F) \\ &= [R'_i \ L'_i \ M'_i \ I][\delta\theta \ \delta x \ \delta y \ (\Delta s - \Delta F)]^{\mathrm{T}} \end{aligned} \right\} \tag{10}$$

(compare to $y_i = mx_i + b$ in §2). The size-change ΔF is introduced in Eq. (10) with a negative sign because all the d'_i-values are directed inward in Fig. 5, corresponding to a *reduction* in size. Yet the 4-D T-Map in Fig. 4 is arranged with the rightward sense, a more natural positive sense, corresponding to *increase* in size, i.e. from the smallest (ψ_1) to the largest (ψ_2) profile allowable in the tolerance-zone.

The scalar relation in Eq (10) forms the transition between the setting of in-parallel robotics and the setting of metrology where CMM data are reduced to geometric variables related to Tolerance-Maps. Now the least-squares fit is obtained by minimizing the sum

$$\sum [d'_i - \{ R'_i\delta\theta + L'_i\delta x + M'_i\delta y + (\Delta s - \Delta F) \}]^2 \tag{11}$$

for $i = 1\dots n$ [15]. Matrix [$\$$] in Eq (9) is augmented to contain the *four* components $\delta\theta$, δx, δy and ($\Delta s - \Delta F$), and the matrix [$\mathbf{K}'$] in Eq (9) is augmented on the right with a column of ones so that the n Eqs (10) (for the n measured points) produce the matrix equation

$$[\mathbf{d}'_i] = \begin{bmatrix} d'_1 \\ d'_2 \\ \vdots \\ d'_n \end{bmatrix} = [\mathbf{K}'][\$] = \begin{bmatrix} R'_1 & L'_1 & M'_1 & 1 \\ R'_2 & L'_2 & M'_2 & 1 \\ \vdots & \vdots & \vdots & \vdots \\ R'_n & L'_n & M'_n & 1 \end{bmatrix} [\$] \tag{12}$$

The Moore-Penrose solution to Eq. (9) for [$\$$], i.e. [$\$$] = $[\mathbf{K}']^{\#}[\mathbf{d}'_i]$ (see Eq. (3)), produces the least-squares location ($\delta\theta$, δx, δy) and size-adjustment ($\Delta s - \Delta F$) for the profile [10, 12], i.e. that location and size for a profile which minimizes the sum in Eq (11). (Compare the pair of Eqs (1) and (2) to the pair (11) and (12).)

Note that coordinates ($\delta\theta$, δx, δy) correspond to coordinates (θ, e_x, e_y) in the T-Map of Fig. 4.

5   Example

As one example, consider the measured points that are shown around the MSP in Fig. 5(a). The points represent an imperfectly manufactured rectangular profile. The coordinates (L'_1, M_1'; R'_1) for the actuator screws at each point, and the deviations d', are presented in Table 2 for each of the measured points; the deviations are all measured from the outer boundary of the tolerance-zone, so Δs = 0.1 mm (Fig. 3(a)). The values in Table 2 are used to build matrices [K'] and [d'_i] in Eq. (12). The solution of [K'] produces the least-squares solution

$$[\$] = [\delta\theta \ \delta x \ \delta y \, (\Delta s - \Delta F)]^\mathsf{T} = [0.000303 \ 0.011078 \ 0.033305 \ 0.095444]^\mathsf{T}.$$

Table 2 Coordinates of measured points around a manufactured rectangular profile

Points	1	2	3	4	5	6	7	8	9	10
L'_i	−1	−1	0	0	0	0	0	1	1	1
M_i'	0	0	−1	−1	−1	-1	-1	0	0	0
R_i, mm	5	30	−60	−30	-10	10	35	-38	-20	-10
d_i', mm	1	1	1	1	1	1	1	1	1	1

Points	11	12	13	14	15	16	17	18	19
L'_i	1	1	0	0	0	0	0	-1	-1
M_i'	0	0	1	1	1	1	1	0	0
R_i, mm	11	35	-45	-20	0	24	42	-30	-10
d_i', mm	1	1	1	1	1	1	1	1	1

The resultant least-squares profile of this solution is shown as the profile with the thin line in Fig 6. Note that the scale of the tolerance-zone is enlarged by a factor of 10 in Figs. 5(a) and 6, and the scale for the profile dimensions is diminished by a factor of 10. Consequently, the least-squares profile is drawn at $\delta\theta$ = 0.0303 rad = 1.73^0 in the counterclockwise direction. Further, to make the appearance of the displaced origin '+' in Fig. 6 be consistent with the displayed points, its coordinates δx = 0.011078 mm and δy = 0.033305 mm have been scaled up by a factor of 10 with respect to the MSP. The corresponding size adjustment from the MSP is ΔF = 0.1 − 0. 095444 = 0. 004556 mm, a small growth in size.

Fig. 6 The resultant least-squares profile shown with the thin line. Its displacement from origin O is shown with the '+' mark.

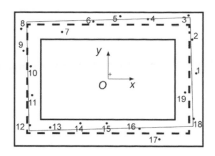

6   Conclusion

The fitting method in this paper is the same as that in [15]. It is an alternative to the one proposed in [6]: both techniques provide a rigid body transformation that locates a set of points that have been measured on a profile relative to a specified tolerance-zone. In [6] a minimum-zone capture of the points is computed, whereas here the orthogonal least-squares fit of the points is utilized. However, in this paper another variable is added to the computed results, the size of the profile, so identifying a corresponding point (i-Map) within the T-Map of tolerance specifications in Fig. 4. Although the least-squares fit is just one of several possible fits to measured points, it is an important one because it recognizes (a) the inter-penetration of mating surfaces (asperities), which violate computed minimum-zone boundaries, and (b) the potential existence of other points further from the intended feature than any of the measured ones. Any one such point could noticeably change a computed minimum zone, but it would have little effect on a least-squares computation that is based on a large number of measured points.

Acknowledgments. The authors are grateful for funding provided by National Science Foundation Grant #CMMI-0969821 and to Mr. Shyam Rao for constructing the figures.

References

1. American National Standard ASME Y14.5M, Dimensioning and Tolerancing. Amer. Soc. Mech. Engs, NY (2009)
2. Mani, N., Shah, J.J., Davidson, J.K.: Standardization of CMM fitting algorithms and development of inspection maps for use in statistical process control. In: ASME Int. Manuf. Sci. Eng. Conf. Corvallis, OR, USA, June 13-17, vol. 2, MSEC2011 50152 (2011)
3. Choi, W., Kurfess, T.R.: Dimensional measurement data analysis, part 1: A zone fitting algorithm. J. Manuf. Sci. Eng. 121, 238–245 (1999)
4. Carr, K., Ferreira, P.: Verification of form tolerances part I: Basic issues, flatness, and straightness. Prec. Eng. 17, 131–143 (1995)

5. Carr, K., Ferreira, P.: Verification of form tolerances part II: Cylindricity and straightness of a median line. Prec. Eng. 17, 144–156 (1995)
6. Choi, W., Kurfess, T.R.: Dimensional measurement data analysis, part 2: Minimum zone evaluation. J. Manuf. Sci. Eng. 121, 246–250 (1999)
7. Barari, A., ElMaraghy, H.A., Knopf, G.K.: Evaluation of Geometric Deviations in Sculptured Surfaces Using Probability Density Estimation. In: Davidson, J.K. (ed.) Models for Computer-Aided Tolerancing in Design and Manufacturing, Proc., 9th CIRP Int'l Seminar on CAT, Tempe, AZ, USA, April 10-12, 2005, pp. 45–54. Springer, Dordrecht (2007)
8. Dowling, M.M., Griffin, P.M., Tsui, K.L., et al.: A comparison of the orthogonal least squares and minimum enclosing zone methods for form error estimation. Mfg. Rev. 8, 120–138 (1995)
9. Choi, S.C.: Introductory applied statistics in science, pp. 23–35. Prentice-Hall, Englewood Cliffs (1978)
10. Ben-Israel, A., Greville, T.N.E.: Generalized inverse: Theory and application, 2nd edn. Springer, New York (2003)
11. Davidson, J.K., Hunt, K.H.: Robots and screw theory: Application of kinematics and statics to robotics, Oxford (2004)
12. Strang, G.: Linear Algebra and Its Applications, 4th edn. Thomson Brooks/Cole (2006)
13. Davidson, J.K., Shah, J.J.: Modeling of geometric variations for line-profiles. J. Comp. Inform. Sc. Eng. 12, 041004 (2012)
14. Coxeter, H.S.M.: Introduction to geometry, 2nd edn. Wiley, New York (1969)
15. Davidson, J.K., Savaliya, S.B., He, Y., et al.: Methods of robotics and the pseudoinverse to obtain the least-squares fit of measured points on line-profiles. In: ASME Des. for Mfg. & Life Cycle Conf., Chicago, IL, USA, August 12-15, DETC2012-70203 (2012)

6 Mobility Analysis and Type Synthesis with Screw Theory: From Rigid Body Linkages to Compliant Mechanisms

Hai-Jun Su, Lifeng Zhou, and Ying Zhang

Abstract. Mobility analysis is one of fundamental problems in kinematics and an important tool in type synthesis of linkages. In this paper, we will review screw theory as a mathematical tool for mobility analysis of overconstrained linkages and compliant mechanisms. Established by Ball in late 1800, screw theory has become one of the fundamental theories for characterizing instantaneous kinematics of spatial movements. In mid to late 1960, Waldron was one of the first modern kinematicians who systematically developed screw theory and its applications to the constraint analysis and synthesis of overconstrained linkages. Due to the screw theory, several overconstrained spatial linkages have been invented and designed, including the well known Waldron six-bar loop overconstrained linkage. In recent years, mobility analysis has been extended to compliant mechanisms which achieve motion through deflection of flexure joints. By the concept of relative compliance/stiffness, we can also define mobility of compliant mechanisms similar to their rigid body counterparts. This paper will summarize some recent work on applying screw theory to mobility analysis and synthesis of compliant mechanisms.

1 Introduction

Waldron (1966) [1] defined "mobility" or number of degrees of freedom of a mechanism as "the number of transformation parameters of joints of the mechanism which are required to determine the position of every point of every member with respect to a coordinate frame fixed to one of the members." Numerous authors have attempted to come up a general formula for calculating the mobility of general mechanisms. The most popular moblity formula is probably Kutzbach-Gruebler criterion, written as

Hai-Jun Su · Lifeng Zhou · Ying Zhang
Department of Mechanical and Aerospace Engineering, The Ohio State University, USA
e-mail: su.298@osu.edu

V. Kumar et al. (Eds.): *Adv. in Mech., Rob. & Des. Educ. & Res.*, MMS 14, pp. 67–81.
DOI: 10.1007/978-3-319-00398-6_6   © Springer International Publishing Switzerland 2013

$$M = d(n - j - 1) + \sum_{i=1}^{j} f_i \tag{1}$$

where M is the mobility, n is the number of links, j is the number of joints, $d = 6$ for the general spatial case and $d = 3$ for planar and spherical cases, and f_i is the connectivity of the ith joint.

A screw is the geometric entity that underlies the foundation of statics and instantaneous (first-order) kinematics. Ball [2] was the first to establish a systematical formulation for screw theory. In the era of modern kinematics, a number of authors [3, 4, 5, 6] have contributed the development of screw theory and its application to analysis and design of spatial linkages. The two fundamental concepts in screw theory are "*twist*" representing a general helical motion of a rigid body about an instantaneous axis in space, and "*wrench*" representing a system of force and moment acting on a rigid body. These two concepts are often called duality [7] in kinematics and statics.

Waldron was probably one of the first modern kinematicians who applied screw theory for mobility analysis and synthesis of spatial linkages in 1960. In particular, Waldron systematically investigated and invented overconstrained linkages [1, 8]. This includes the well known Waldron six-bar overconstrained linkage [9]. Since then, screw theory has been applied to various research topics ranging from robotics [6, 10], mobility analysis [11], assembly analysis [12, 13] and topology synthesis [14] of parallel mechanisms.

In recent years, compliant mechanisms [15, 16] have received increasingly attention from the community due to their applications to precision machinery, aerospace and space structures and so on. Compliant mechanisms gain their mobility at least partially from deformation of their flexible members. Compared with their rigid body counterparts, compliant mechanisms or flexures have many advantages, such as high precision and a simplified manufacturing and assembly process due to integration of joints with rigid links. However the design and analysis of compliant mechanisms is complex due to the nonlinearity of deformation of the flexible members.

Similar to rigid body mechanisms, one important task in design of compliant mechanisms is so called "type synthesis" whose goal is to find one or more compliant mechanisms for achieving a prescribed motion pattern. As an important task of type synthesis, mobility analysis is to characterize the motion pattern for a particular compliant mechanism. Recently screw theory has been applied to mobility analysis [17, 18, 19] and type synthesis [20, 21] of compliant mechanisms. The basic principle is to first characterize freedom and constraints of flexure elements using twists and wrenches in screw theory under the assumption of ideal geometries of compliant mechanisms. For instance, a circular notch is considered as an idealized rotational joint, hence can be characterized as a pure rotational twist. Then we consider a compliant mechanism as a system of rigid bodies interconnected by these flexure elements. By applying kinematic transformation of screws, we can analyze and synthesize mobility of compliant mechanisms in a similar manner of rigid body mechanisms. In this paper, we will summarize some recent advances in this area.

2 Screw Theory Overview

In this section, we first review basic concepts of screw theory as a background preparation for the following sections.

In screw theory, an instantaneous screw motion is represented by a twist $\hat{T}$. And a constraint or forbidden motion is represented by a wrench $\hat{W}$. Both twist $\hat{T}$ and wrench $\hat{W}$ are 6 by 1 column vectors, written as

$$\hat{T} = \left\{ \begin{matrix} \boldsymbol{\Omega} \\ \mathbf{V} \end{matrix} \right\} = \left\{ \begin{matrix} \boldsymbol{\Omega} \\ \mathbf{c} \times \boldsymbol{\Omega} + p\boldsymbol{\Omega} \end{matrix} \right\}, \tag{2}$$

$$\hat{W} = \left\{ \begin{matrix} \mathbf{F} \\ \mathbf{M} \end{matrix} \right\} = \left\{ \begin{matrix} \mathbf{F} \\ \mathbf{c} \times \mathbf{F} + q\mathbf{F} \end{matrix} \right\}, \tag{3}$$

where p and q are called pitches of twist and wrenches. And $\hat{T}$ and $\hat{W}$ satisfy the so called reciprocal condition:

$$\hat{T} \circ \hat{W} - \boldsymbol{\Omega} \cdot \mathbf{M} + \mathbf{V} \cdot \mathbf{F} = 0. \tag{4}$$

A general rotational or translational freedom respectively corresponds to a twist with zero or infinite pitch, written as

$$\hat{T}_R = \left\{ \begin{matrix} \boldsymbol{\Omega} \\ \mathbf{c} \times \boldsymbol{\Omega} \end{matrix} \right\}, \quad \hat{T}_P = \left\{ \begin{matrix} \mathbf{0} \\ \mathbf{V} \end{matrix} \right\}. \tag{5}$$

Similarly a general rotational or translation constraint removes a rotation or translation along a particular direction. They respectively correspond to a wrench with infinite or zero pitch, written as

$$\hat{W}_R = \left\{ \begin{matrix} \mathbf{0} \\ \mathbf{M} \end{matrix} \right\}, \quad \hat{W}_P = \left\{ \begin{matrix} \mathbf{F} \\ \mathbf{c} \times \mathbf{F} \end{matrix} \right\} \tag{6}$$

For convenience, we define six principal twists as the rotation and translations about all the three coordinate axes, written as

$$\hat{R}_x = \begin{pmatrix} 1\ 0\ 0\ 0\ 0\ 0 \end{pmatrix}^T \quad \hat{R}_y = \begin{pmatrix} 0\ 1\ 0\ 0\ 0\ 0 \end{pmatrix}^T \quad \hat{R}_z = \begin{pmatrix} 0\ 0\ 1\ 0\ 0\ 0 \end{pmatrix}^T$$
$$\hat{P}_x = \begin{pmatrix} 0\ 0\ 0\ 1\ 0\ 0 \end{pmatrix}^T \quad \hat{P}_y = \begin{pmatrix} 0\ 0\ 0\ 0\ 1\ 0 \end{pmatrix}^T \quad \hat{P}_z = \begin{pmatrix} 0\ 0\ 0\ 0\ 0\ 1 \end{pmatrix}^T \tag{7}$$

Similarly, we define six principal wrenches as the rotational and translational constraint about all the three coordinate axes, written as

$$\hat{F}_x = \begin{pmatrix} 1\ 0\ 0\ 0\ 0\ 0 \end{pmatrix}^T \quad \hat{F}_y = \begin{pmatrix} 0\ 1\ 0\ 0\ 0\ 0 \end{pmatrix}^T \quad \hat{F}_z = \begin{pmatrix} 0\ 0\ 1\ 0\ 0\ 0 \end{pmatrix}^T$$
$$\hat{M}_x = \begin{pmatrix} 0\ 0\ 0\ 1\ 0\ 0 \end{pmatrix}^T \quad \hat{M}_y = \begin{pmatrix} 0\ 0\ 0\ 0\ 1\ 0 \end{pmatrix}^T \quad \hat{M}_z = \begin{pmatrix} 0\ 0\ 0\ 0\ 0\ 1 \end{pmatrix}^T \tag{8}$$

The coordinate transformation of a twist or wrench is calculated as

$$\hat{T}' = [Ad]\hat{T}, \quad \hat{W}' = [Ad]\hat{W}, \tag{9}$$

where $\hat{T}, \hat{W}$ and $\hat{T}', \hat{W}'$ correspond to the twist and the wrench before and after the transformation. And $[Ad]$ is the so-called 6×6 adjoint matrix, written as

$$[Ad] = \begin{bmatrix} R & 0 \\ DR & R \end{bmatrix} \qquad (10)$$

where $[R]$ is a 3 by 3 rotation matrix and $[D]$ is the 3 by 3 skew-symmetric matrix defined by the translational vector $\mathbf{d} = (d_x, d_y, d_z)^T$. They have the form

$$[R] = \begin{bmatrix} \mathbf{x} & \mathbf{y} & \mathbf{z} \end{bmatrix}, \quad [D] = \begin{bmatrix} 0 & -d_z & d_y \\ d_z & 0 & -d_x \\ -d_y & d_x & 0 \end{bmatrix}$$

3   Mobility Analysis of Overconstrained Linkages

It is well known that the Kutzbach-Gruebler formula applies to mechanisms with general dimensions and may fail for overconstrained mechanisms which gain some extra mobility due to their special dimensions. Many modern kinematicians have already contributed to generalize this formula to include various kinds of overconstrained mechanisms. For instance, in 1966, Waldron [1] proposed a formula for single loop linkages, written as

$$M = (N + n) - (m + n - k) \qquad (11)$$

where N is the number of degrees of freedom of the serial chain by breaking the single loop linkage at any joint, n is the connectivity of the closing joint, and $m + n - k$ is the order of the equivalent screw system of the loop linkage.

3.1   Waldron Six-Bar Linkages

Waldron six-bar linkage also called hybrid six-bar linkage [9] is a kind of overconstrained linkage that has one mobility. It is formed by two Bennett four-bar linkages [22] sequentially connected. It is well known that a Bennett four-bar linkage is a spatial overconstrained linkage with one degree of freedom. And during the movement of the Waldron six-bar linkage, the two Bennett linkages will keep their geometrical constrains without change.

As shown in Fig. 1, the geometry of a Waldron six-bar linkage is described as the following. Let us denote the eight axes of the two Bennett linkages as $z_1 - z_8$ with the first four axes belong to the first Bennett linkage and the last four belong to the second Bennett linkage. When constructing the six-bar linkage with two four-bars linkages, z_1 coincides with z_5 while links 1, 8 are replaced by a single link 9 and links 4, 5 are replaced by link 10. As a result, links 2, 3, 10, 6, 7 and 9 are connected by six joints $z_2, z_3, z_4, z_6, z_7, z_8$ to form a single loop six-bar linkage.

The relative position of the links and joints is described by using Denavit and Hartenberg parameters. These parameters are presented as $\alpha_i, a_i, d_i, \theta_i$. α_i is the twist

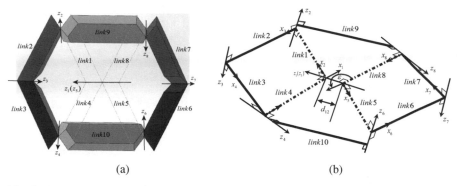

Fig. 1 The Waldron six-bar linakge is formed by two Bennett four-bar linkages

angle between the axes of z_i and z_{i+1}, a_i is the distance between z_i and z_{i+1}, d_i is the offset between links i and $i+1$ along z_i, θ_i is the angle from x_i to x_{i+1} measured about z_i. The values of those parameters are showed in Table 1. d_{12} is the distance of the two Bennett linkage along z_1, and ε is the angle between x_1 and x_8. And here in order to establish the position relationship of the two Bennett linkages clearly, we substituted the D-H parameters of z_4 to z_1 and z_1 to z_5 for z_4 to z_5. According to the geometric constraint of Bennett linkage, we have $\sin \alpha_1 / a_1 = \sin \alpha_2 / a_2$ and $\sin \alpha_{11} / a_{11} = \sin \alpha_{22} / a_{22}$.

Table 1 D-H parameters of the Waldron six bar linkage

i	joint i-joint j	α_i	a_i	d_i	θ_i
1	$z_1 - z_2$	α_1	a_1	0	θ_1
2	$z_2 - z_3$	α_2	a_2	0	θ_2
3	$z_3 - z_4$	α_1	a_1	0	$2\pi - \theta_1$
4	$z_4 - z_1$	α_2	a_2	0	$2\pi - \theta_2$
5	$z_1 - z_5$	0	0	d_{12}	ε
6	$z_5 - z_6$	α_{11}	a_{11}	0	θ_{11}
7	$z_6 - z_7$	α_{22}	a_{22}	0	θ_{22}
8	$z_7 - z_8$	α_{11}	a_{11}	0	$2\pi - \theta_{11}$
9	$z_8 - z_5$	α_{22}	a_{22}	0	$2\pi - \theta_{22}$

3.2 Mobility Analysis of Waldron Six-Bar Linkages

In this section, we show how to use screw theory to calculate the mobility of Waldron hybrid six-bar linkage.

Since the relative position between the first and second Bennett linkages is decided by d_{12} and ε, we only need to analyze one of two Bennett linkages and the second one could be calculated by using transformation of coordinates. Based on the Denavit and Hartenberg parameters, the transformation between the adjacent joints can be obtained easily by

$$T_i = X[\alpha_i, a_i]Z[\theta_i, d_i] \tag{12}$$

where $Z[\cdot]$ and $X[\cdot]$ represents the screw displacement along z and x axis respectively. The transformation from join z_i to the first joint can be calculated by

$${}_i^1T = T_1 T_2 \cdots T_{i-1}, \quad i = 2, 3, 4 \tag{13}$$

The transform matrix ${}_i^1T$ could be written as

$${}_i^1T = \begin{bmatrix} {}_i^1R & {}_i^1\mathbf{d} \\ 0 & 1 \end{bmatrix} \tag{14}$$

For the screw of z_1, we can choose it as $\$_1 = (0 \ \ 0 \ \ 1 \ \ 0 \ \ 0 \ \ 0)^T$. Then the other axis vector could be calculated as $\mathbf{s}_i = {}_i^1R(0,0,1)^T$. An arbitrary point on z_i could be chosen as $\mathbf{r}_i = {}_i^1\mathbf{d}$. Then the screw of z_i will be $\$_i = (\mathbf{s}_i; \mathbf{r}_i \times \mathbf{s}_i)$.

To obtain the four screws $(\$_1', \$_2', \$_3', \$_4')$ of the second Bennett linkage, we only need to replace the $\alpha_1, \alpha_2, a_1, a_2, \theta_1, \theta_2$ in $(\$_1, \$_2, \$_3, \$_4)$ by $\alpha_{11}, \alpha_{22}, a_{11}, a_{22}, \theta_{11}, \theta_{22}$ of the second Bennett linkage in the coordinate of joint 5 which are the D-H parameters of the second Bennett linkage given in Table 1. And the screws of the second Bennett linkage in the coordinate frame of joint 1 can be calculated as

$$\$_{i+4} = [Ad]\$_i', \quad i = 1, 2, 3, 4. \tag{15}$$

where $[Ad]$ is the six by six adjoint transformation matrix of screws by substituting the following matrices

$$[R] = \begin{bmatrix} \cos(\varepsilon) & -\sin(\varepsilon) & 0 \\ \sin(\varepsilon) & \cos(\varepsilon) & 0 \\ 0 & 0 & 1 \end{bmatrix}, \quad [D] = \begin{bmatrix} 0 & -d_{12} & 0 \\ d_{12} & 0 & 0 \\ 0 & 0 & 0 \end{bmatrix}$$

into formula (10).

Finally the screw system of the hybrid six-bar linkage is obtained as $\$ = (\$_2, \$_3, \$_4, \$_6, \$_7, \$_8)$. The order N of the loop linkage is calculated as the rank of the 6 by 6 matrix formed by these six screws. By using Mathematica program, it is easy to figure out that this order is five. Since the six-bar linkage is connected by six revolute joints, the mobility of its serial chain will be 5 and the connectivity of the closing joint will be 1. Therefore, the mobility of the six bar linkage is calculated as $M = 6 - 5 = 1$ using Waldron mobility formula (11).

4 Mobility Analysis and Type Synthesis of Compliant Mechanisms

Inspired by the above work on rigid body mechanisms, we recently extended this work to mobility analysis and type synthesis of compliant mechanisms. Compliant mechanisms can be considered as a collection of relative rigid members (links) connected with flexible members (flexure joints). Compliant mechanisms gain at least part of their mobility from deformation of flexible members.

4.1 Mobility and Compliance

The mobility of a compliant mechanism is a subtle concept as virtually any material deforms more or less, hence results in movement. As we know, compliance C is defined as the ratio of movement over loading exerted for any specific direction determined by a screw $\hat{T}$. There are two kinds of compliance: rotational and translational, which have the unit of rad/Nm and $1/N$ respectively. For any member of a compliant mechanism, there are three rotational compliances and three translational compliances along the axes of coordinate system attached to that member, denoted by $C_{Rx}, C_{Ry}, C_{Rz}, C_{Tx}, C_{Ty}, C_{Tz}$. To compare rotational compliance with a translational one, we multiple the rotational compliances by a chosen constant l, i.e.

$$C_{tx} = C_{Rx}l, \quad C_{ty} = C_{Ry}l, \quad C_{tz} = C_{Rz}l. \tag{16}$$

The constant l can be chosen as the overall dimension of the member of interest, typically the motion stage of a compliant mechanism. Compliances C_{tx}, C_{ty}, C_{tz} represents the translation of the tip of a bar with length l that is attached to the motion stage of the mechanism when a tangent force is applied at the tip.

Now we redefine compliances of a member of a compliant mechanism as

$$C_1 = C_{tx}, \quad C_2 = C_{ty}, \quad C_3 = C_{tz}, \quad C_4 = C_{Tx}, \quad C_5 = C_{Ty}, \quad C_6 = C_{Tz}. \tag{17}$$

To define the mobility of a compliant mechanism, we introduce the concept of "compliance ratio" which is essentially the ratio of the compliance of the mechanism in a particular direction over the maximum compliance in all directions, i.e.

$$CR_i = \frac{C_i}{\max(C_i)}, \quad i = 1,\ldots 6 \tag{18}$$

The range of CR_i is between 0 and 1. If CR_i is below a specified small threshold, e.g. 0.01, we consider the mechanism has no mobility in that direction. Note this represents that the movement of the mechanism in the direction $\hat{T}_i$ is two order smaller than that in the direction with the maximum compliance when the same force is exerted. And the mobility of a compliant mechanism is counted as the number of mobility in three rotational and three translational directions.

4.2   Commonly Used Flexure Primitives

Here we first study the mobility of a list of flexure primitives commonly used in compliant mechanisms. A flexure primitive is defined as an "atomic" flexure mechanism that consists of only one flexure element and zero intermediate body. They cannot be further divided into substructures. In this section we first categorize commonly used flexure primitives and derive their freedom and constraint spaces. Then we will discuss a general synthesis methodology for constructing serial and parallel kinematic chains of these flexure primitives.

According to the mobility or the rank of their twist system, we can categorize the most commonly used flexure primitives. For instance, notch hinges, short beams and split tubes have one rotational degree of freedom. A spherical notch or short wire/rod has three rotational degrees of freedom. A thin beam or blade flexure, rotational symmetric cylinder or a disc coupling has two rotational and one translational mobility. And a long wire or corner blade has three rotational and two translational mobility. These flexure primitives and their freedom space and twist and wrench matrices are summarized in Table 2.

These primitives are basic building blocks for constructing more complex flexure systems. In what follows, we show how to build more complex mechanisms with these flexure primitives using a serial, parallel or hybrid structure.

4.3   Serial Flexure Chains

A serial flexure mechanism is formed by connecting a functional body to a fixed reference body through a serial chain of flexure elements that are joined with intermediate bodies. Let us denote the motion space of the jth flexure element in a serial flexure mechanism by a twist matrix $[T_j]$. The motion space of the rigid body constrained by this flexure system is the superimposition of the motion of individual elements. Mathematically the motion space of a serial chain of m flexures is given by the range space of the following matrix

$$[T] = [Ad_1 T_1 \quad Ad_2 T_2 \quad \cdots \quad Ad_m T_m]. \tag{19}$$

which is the column-wise combination of each $[T_j]$ after an appropriate coordinate transformation $[Ad_j]$. The column rank of $[T]$ gives the mobility of the functional body, denoted by $f = rank(T)$. Since it is not uncommon that the column vectors of $[T_j]$ are dependent, the mobility f is typically less than or equal to the total number of columns of $[T]$. By column reducing the matrix $[T]$, we can obtain a basis of the motion space of the flexure system. And the complementary constraint space is obtained by the standard screw algebra, denoted by a 6 by $6 - f$ wrench matrix $[W]$.

Figure 2(a) shows a serial chain of two identical blade flexures. Blade flexure 2 is perpendicular to blade 1. We place the stage and its local coordinate system at the end of the second blade. The twist matrix for both blade flexures is $[T_b]$, already given in Table 2. The coordinate transformation from blade 1 to functional body is a pure translation along y axis for l units,

Table 2 The motion and constraint spaces of commonly used flexure primitives

Flexure	Freedom Symbol	$[T]$	$[W]$
Notch/Living Hinge Short Beam Split Tube	R	$[\hat{R}_z]$	$[\hat{F}_x\ \hat{F}_y\ \hat{F}_z\ \hat{M}_x\ \hat{M}_y]$
Spherical Notch Short Wire/Rod	S=3R	$[\hat{R}_x\ \hat{R}_y\ \hat{R}_z]$	$[\hat{F}_x\ \hat{F}_y\ \hat{F}_z]$
Blade/Sheet/ Rotational Disc Coupling Leaf Spring Symmetric Cylinder	B=2R-P	$[\hat{R}_x\ \hat{R}_z\ \hat{P}_y]$	$[\hat{F}_x\ \hat{F}_z\ \hat{M}_y]$
long wire/rod corner blade	W=3R-2P	$[\hat{R}_x\ \hat{R}_y\ \hat{R}_z\ \hat{P}_y\ \hat{P}_z]$	$[\hat{F}_x]$
Bellow Spring	B_s=2R-3P	$[\hat{R}_x\ \hat{R}_y\ \hat{P}_x\ \hat{P}_y\ \hat{P}_z]$	$[\ddot{M}_z]$

$$R_1 = \begin{bmatrix} 1 & 0 & 0 \\ 0 & 1 & 0 \\ 0 & 0 & 1 \end{bmatrix}, \quad D_1 = \begin{bmatrix} 0 & 0 & l \\ 0 & 0 & 0 \\ -l & 0 & 0 \end{bmatrix}.$$

And the transformation from the blade 2 to the functional body is pure rotation about z axis for $-\pi/2$,

$$R_2 = [Z(-\frac{\pi}{2})] = \begin{bmatrix} 0 & 1 & 0 \\ -1 & 0 & 0 \\ 0 & 0 & 1 \end{bmatrix}, \quad D_2 = \begin{bmatrix} 0 & 0 & 0 \\ 0 & 0 & 0 \\ 0 & 0 & 0 \end{bmatrix},$$

where $[Z(\alpha)]$ represents the rotation about z axis for an angle of α.

The twist matrix of a serial chain of two blades is obtained by substituting them into (19),

$$[T_{bb}] = [Ad_1T_b \quad Ad_2T_b] = \begin{bmatrix} 0 & 0 & 1 & 0 & 0 & 0 \\ 0 & 0 & 0 & 0 & 0 & 1 \\ 1 & 0 & 0 & 1 & 0 & 0 \\ l & 0 & 0 & 0 & -1 & 0 \\ 0 & 1 & 0 & 0 & 0 & 0 \\ 0 & 0 & -l & 0 & 0 & 0 \end{bmatrix}$$

$$\triangleq \begin{bmatrix} 1 & 0 & 0 & 0 & 0 & 0 \\ 0 & 1 & 0 & 0 & 0 & 0 \\ 0 & 0 & 1 & 0 & 0 & 0 \\ 0 & 0 & 0 & 1 & 0 & 0 \\ 0 & 0 & 0 & 0 & 1 & 0 \\ -l & 0 & 0 & 0 & 0 & 0 \end{bmatrix}, \tag{20}$$

where the last step is a column-wise reduction process. Obviously $f = rank(T_{bb}) = 5$ as the elements of the last column are all zeros. Therefore, the flexure system provides a mobility of five degrees-of-freedom.

The corresponding reciprocal wrench matrix is

$$[W_{bb}] = [0, \quad 0, \quad 1; \quad l, \quad 0, \quad 0]^T, \tag{21}$$

which represents a constraint along a line parallel to z axis at the point $\mathbf{r}$ shown as the blue line in Fig. 2(a).

4.4 Parallel Flexure Chains

A parallel flexure mechanism is formed by connecting a functional body to a reference body through two or more flexure elements in parallel. Let us denote the

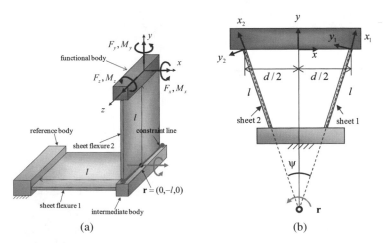

(a)   (b)

Fig. 2 (a) A serial chain compliant mechanism formed by two perpendicular blade flexures, (b) A parallel chain flexure mechanism formed by two parallel ideal blade flexures

constraint space of the jth flexure element by a wrench matrix $[W_j]$. The constraint space of the functional body is the superimposition of the constraint space of each element. Mathematically the constraint space of a parallel flexure mechanism with m flexures is given by the following wrench matrix

$$[W] = [Ad_1W_1 \quad Ad_2W_2 \quad \cdots \quad Ad_mW_m]. \tag{22}$$

Again matrices $[Ad_j]$ are coordinate transformation of jth flexures.

The column rank of $[W]$ gives the degree-of-constraint of the functional body, denoted by $c = rank(W)$. Similar to the case of serial chains, c is typically less than or equal to the total number of columns of $[W]$ as some column vectors are dependent. By column reducing the matrix $[W]$, we can obtain a basis of the constraint space of the flexure system. And the complementary motion space is obtained by the standard screw algebra, denoted by a 6 by $6 - c$ twist matrix $[T]$.

Figure 2(b) shows a trapezoidal leaf-type flexure pivot that is formed by two identical blade flexures assembled symmetrically at an angle of ψ and a distance of d. The coordinate transformations for blade 1 and 2 are respectively

$$R_1 = [Z(\frac{\pi - \psi}{2})], \quad \mathbf{d}_1 = (\frac{d}{2}, 0, 0),$$

$$R_2 = [Z(\frac{\pi + \psi}{2})], \quad \mathbf{d}_2 = (-\frac{d}{2}, 0, 0).$$

Substituting the above formula into (22) and applying a column-wise reduction, we obtain the following wrench matrix,

$$
\begin{aligned}
[W_t] &= [Ad_1W_b \quad Ad_2W_b] \\
&= \begin{bmatrix}
0 & \sin(\frac{\psi}{2}) & 0 & 0 & -\sin(\frac{\psi}{2}) & 0 \\
0 & \cos(\frac{\psi}{2}) & 0 & 0 & \cos(\frac{\psi}{2}) & 0 \\
1 & 0 & 0 & 1 & 0 & 0 \\
0 & 0 & -\cos(\frac{\psi}{2}) & 0 & 0 & -\cos(\frac{\psi}{2}) \\
-\frac{d}{2} & 0 & \sin(\frac{\psi}{2}) & \frac{d}{2} & 0 & -\sin(\frac{\psi}{2}) \\
0 & \frac{1}{2}d\cos(\frac{\psi}{2}) & 0 & 0 & -\frac{1}{2}d\cos(\frac{\psi}{2}) & 0
\end{bmatrix} \\
&\triangleq \begin{bmatrix}
\sin(\frac{\psi}{2}) & 0 & 0 & 0 & 0 \\
0 & 1 & 0 & 0 & 0 \\
0 & 0 & 1 & 0 & 0 \\
0 & 0 & 0 & 1 & 0 \\
0 & 0 & 0 & 0 & 1 \\
\frac{1}{2}d\cos(\frac{\psi}{2}) & 0 & 0 & 0 & 0
\end{bmatrix} = \begin{bmatrix} 0 \end{bmatrix},
\end{aligned} \tag{23}
$$

where W_b is the reciprocal wrench of T_b. Again the last step is obtained by a column-wise reduction. The corresponding reciprocal twist matrix of $[W_t]$ is

$$[T_t] = \begin{bmatrix} 0 \\ 0 \\ \sin\left(\frac{\psi}{2}\right) \\ -\frac{1}{2}d\cos\left(\frac{\psi}{2}\right) \\ 0 \\ 0 \end{bmatrix}, \tag{24}$$

which represents the rotation about the intersection line of the blades shown as $\mathbf{r} = (0, -d\cot(\psi/2)/2, 0)$ in Fig. 2(b).

4.5 Design of Freedom Elements

By using serial or parallel chains of flexure primitives as shown in Table 2, we have synthesized a catalogue of freedom and constraint elements which provide translational or rotational freedom or constraints. For convenience, we list all the possible freedom elements with one rotational (R) and translational (P) DOF motion, i.e. R-joints and P-joints, in Figs. 3 and 4.

These freedom elements are basic building blocks to construct hybrid structures of flexure mechanisms. For instance, if we would like to design a parallel structure with three rotations, we just need to use three translational constraint elements to remove all translations. As shown in Fig 5(a), We first design a serial chain of two blade flexures (denote as B-B) that functions as a single translational constraint. By combining three serial chains of B-B, we obtain a design. The functional body **A** can rotate about its center relative to the base body **B**, while its translations are constrained.

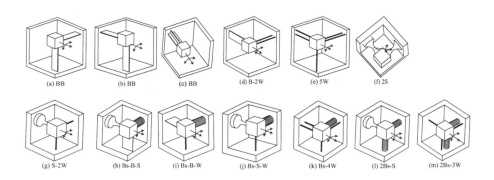

Fig. 3 Various designs of R-joints with flexure primitives: B, W, S and B$_s$. The double arrow arcs represent the rotation allowed by flexure R-joints. The box represents the functional body.

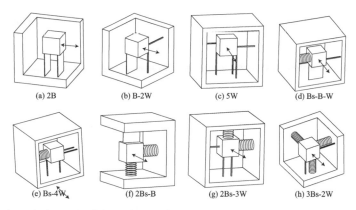

(a) 2B	(b) B-2W	(c) 5W	(d) Bs-B-W
(e) Bs-4W	(f) 2Bs-B	(g) 2Bs-3W	(h) 3Bs-2W

Fig. 4 Various designs of P-joints with parallel structures of flexure primitives: B, W, B_s. The arrowed lines indicate the direction of translation. The box represents the functional body.

4.6 Synthesis of Hybrid Structures

We can further build more complex flexure mechanisms with hybrid structures of flexure primitives together with the freedom and constraint elements synthesized in the previous sections. Here a hybrid structure is a structure with both serial and parallel connections.

Figure 5(a) shows a compliant parallel platform mechanism that has three rotational degrees of freedom. Each limb is a serial chain of two blade flexures. The functional body A can rotate about its center relative to the base body B while its translations are constrained.

As another example, we would like to design a parallel structure with three translational degrees of freedom. We just need to use three rotational constraint elements to remove all rotations. If we choose the BB design in for all three rotational

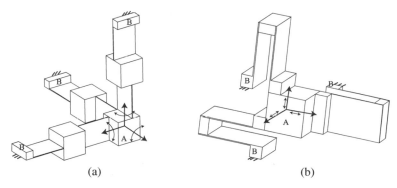

(a)	(b)

Fig. 5 (a) A parallel platform with three rotational degrees of freedom and (b) a parallel compliant platform with three translational degrees of freedom. The body B are fixed. The body A is the functional body.

constraints, we obtain the design shown in Fig 5(b). The functional body A can translate in all directions while its rotations are constrained.

5   Conclusions

In this paper, we first reviewed screw theory and its applications to mobility analysis and synthesis of rigid body linkages. In particular, we highlighted contributions of Waldron to mobility formula for general loop linkages with screw theory and synthesis of overconstrained linkages in 1960. As an example, we studied the mobility of the Waldron hybrid six-bar linkages using screw theory. Inspired by these work, we then reviewed some recent advances in applying screw theory to mobility analysis of compliant mechanisms. This screw theory based mobility formula is the foundation of mobility analysis and type synthesis of compliant mechanisms. We presented a screw theory representation of freedom and constraint spaces of commonly used flexure primitives, synthesis of R and P joints with flexure primitives and synthesis of hybrid structures such as compliant parallel platform mechanisms.

References

1. Waldron, K.: The constraint analysis of mechanisms. Journal of Mechanisms 1(2), 101–114 (1966)
2. Ball, R.S.: The Theory of Screws. Cambridge University Press, Cambridge (1998) (Originally published in 1876 and revised by the author in 1900, now reprinted with an introduction by H. Lipkin and J. Duffy)
3. Hunt, K.H.: Kinematic Geometry of Mechanisms. Oxford University Press, New York (1978)
4. Phillips, J.: Freedom in Machinery. Introducing Screw Theory, vol. 1. Cambridge University Press, Cambridge (1984)
5. Phillips, J.: Freedom in Machinery. Screw Theory Exemplified, vol. 2. Cambridge University Press, Cambridge (1990)
6. Davidson, J.K., Hunt, K.H.: Robots and Screw Theory: Applications of Kinematics and Statics to Robotics. Oxford University Press, New York (2004)
7. Shai, O., Pennock, G.R.: A study of the duality between planar kinematics and statics. ASME Journal of Mechanical Design 128(3), 587–598 (2006)
8. Waldron, K.J.: Symmetric overconstrained linkages. Journal of Engineering for Industry 91(1), 158–162 (1969)
9. Waldron, K.: Hybrid overconstrained linkages. Journal of Mechanisms 3(2), 73–78 (1968)
10. McCarthy, J.M.: Geometric Design of Linkages. Springer, New York (2000)
11. Dai, J.S., Huang, Z., Lipkin, H.: Mobility of overconstrained parallel mechanisms. ASME Journal of Mechanical Design 128(1), 220–229 (2006)
12. Adams, J.D., Whitney, D.E.: Application of screw theory to constraint analysis of mechanical assemblies joined by features. ASME Journal of Mechanical Design 123(1), 26–32 (2001)
13. Smith, D.: Constraint Analysis of Assemblies Using Screw Theory and Tolerance Sensitivities. MS Thesis, Brigham Young University, Provo, UT (2003)

14. Kong, X., Gosselin, C.M.: Type synthesis of 3-dof translational parallel manipulators based on screw theory. ASME Journal of Mechanical Design 126(1), 83–92 (2004)
15. Howell, L.L., Midha, A.: Parametric deflection approximations for end-loaded, large-deflection beams in compliant mechanisms. ASME Journal of Mechanical Design 117(1), 156–165 (1995)
16. Howell, L.L.: Compliant Mechanisms. Wiley-Interscience, New York (2001)
17. Su, H.-J., Dorozhkin, D.V., Vance, J.M.: A screw theory approach for the conceptual design of flexible joints for compliant mechanisms. ASME Journal of Mechanisms and Robotics 1(4), 041009 (2009)
18. Su, H.-J.: Mobility analysis of flexure mechanisms via screw algebra. ASME Journal of Mechanisms and Robotics 3(4), 041010 (2011)
19. Su, H.-J., Tari, H.: On line screw systems and their application to flexure synthesis. ASME Journal of Mechanisms and Robotics 3(1), 011009 (2011)
20. Su, H.-J., Tari, H.: Realizing orthogonal motions with wire flexures connected in parallel. ASME Journal of Mechanical Design 132(12), 121002 (2010)
21. Yu, J., Li, S., Su, H.J., Culpepper, M.L.: Screw theory based methodology for the deterministic type synthesis of flexure mechanisms. ASME Journal of Mechanisms and Robotics 3(3), 031008 (2011)
22. Bennett, G.: A new mechanism. Engineering 76 (1903)

7 How Far are Compliant Mechanisms from Rigid-body Mechanisms and Stiff Structures?

G.K. Ananthasuresh

Abstract. Stiff structures made of elastic bodies and linkages consisting of rigid bodies are studied for centuries. Compliant mechanisms that straddle these two are extensively researched only in the last two decades, barring a few insightful results before that. The question addressed in this note is simple: Are compliant mechanisms really different from stiff structures and rigid-body linkages? If so, how much and in what ways? By discussing the fundamental concepts in structures and linkages, it is argued here that compliant mechanisms are as much similar to stiff structures and rigid-body linkages as they are different from them. Similarities and differences among the three categories of engineering solid entities are delineated from the viewpoints of function, mobility, analysis, synthesis, materials, fabrication, scaling, and balancing. It is noted here that the contrast among the three mainly arises due to viewing them from the continuum or discrete perspectives.

1 Introduction

Systems made of solids studied by engineers are primarily of two kinds: structures and mechanisms. Structures are designed to be stiff so that they do not move or deform; they simply bear the loads. Mechanisms, on the other hand, are supposed to move to transmit motion and forces. They are traditionally made of jointed assemblies of rigid bodies. Viewed in this manner, there is a clear demarcation between stiff structures and movable mechanisms. They are therefore treated differently in terms of analysis and synthesis. But in reality all structures do move and/or deform slightly due to unintended but inevitable clearances and non-rigid displacements. Likewise, mechanisms possess some characteristics of stiff

G.K. Ananthasuresh
Mechanical Engineering, Indian Institute of Science, Bangalore
e-mail: suresh@mecheng.iisc.ernet.in

V. Kumar et al. (Eds.): *Adv. in Mech., Rob. & Des. Educ. & Res.*, MMS 14, pp. 83–94.
DOI: 10.1007/978-3-319-00398-6_7   © Springer International Publishing Switzerland 2013

structures because they are indeed stiff in portions between the kinematic joints but they also undergo elastic deformations, sometimes substantial, under large loads and/or at high speeds. The matter of distinguishing one from the other becomes complicated when we also bring in compliant mechanisms [1] into discussion. Compliant mechanisms utilize elastic deformation to perform the functions of mechanisms but are similar to structures in their physical form when they do not contain any joints. Thus, compliant mechanisms straddle stiff structures and rigid-body mechanisms.

While stiff structures and rigid-body mechanisms have been studied extensively for centuries, the studies on compliant mechanisms are sporadic and are intensified only in the last two decades. Although the treatment of compliant mechanisms looks different prima facie from that of stiff structures and rigid-body mechanisms, hindsight shows that it is really not so different. In this short note, we examine the similarities and differences among the three (or two?) types of solid systems. The criteria for comparison include function, mobility and degrees of freedom, analysis and simulation methods, design (i.e., synthesis) approaches, materials used, manufacturing techniques employed, static balancing, and size (because in today's world micro and nano technologies are pursued with vigor).

The next section considers the aforementioned criteria one by one. Some remarks are included in the closing section of this note, which is by no means the final word on the topic.

2   Comparing Stiff Structures, Compliant Mechanisms, and Rigid-body Mechanisms

2.1   Function

Structures such as buildings and bridges are meant to be stiff to bear the loads without excessive deformation. Mechanism transmit force and motion from one point to another. It is fair to say that they also transmit energy. They also bear the loads but need actuation to do it. Compliant mechanisms are not different from rigid-body mechanisms in terms of function. An additional feature of compliant mechanisms is that they can transform energy from one form to another form across different energy domains. That is, it is much more than transforming mechanical energy from potential energy to kinetic energy. An electro-thermal-compliant microactuator [2] receives electrical energy and outputs mechanical strain energy via thermal energy en route. All mechanisms, whether rigid-body or compliance based, need a fixed frame to attach them. Machine frames with moving parts need to be stiff to contain the vibrations.

The physical forms of structures and rigid-body mechanisms are easily distinguishable. A structure is monolithic (i.e., it has single-piece construction) or is an assembly of rigid components fastened rigidly to one another. A rigid-body mechanism is an assemblage of rigid components (called "links" in the parlance of kinematics—a source of confusion to people outside the field) connected together with kinematic joints. Kinematic joints allow one or more relative motions out of the possible six between the two bodies, and restrict the rest of the relative motions. Joints do this by means of specific geometries. Certain assemblies of rigid bodies with kinematic joints can also lead to structures that cannot move but can only deform. Trusses are good examples. They deform appreciably only under large loads. If they are designed to be stiff, they can be safely considered to be rigid for all practical purposes.

The physical form of a compliant mechanism with no kinematic joints at all cannot be distinguished from that of a structure. Early on, Midha and co-workers [3] gave a simple example to make this point. A cantilever beam can be a compliant mechanism or a structure. A cantilever of a diving board is a mechanism as it stores the strain energy when the swimmer jumps on it and releases it as kinetic energy needed for diving. On the other hand, a cantilever beam holding a brush against the commutator in an electric motor is simply a structure. Figure 1 shows another example where the same physical form can serve as a structure or a compliant mechanism depending on the intended function and applied loads.

Fig. 1 (a) a compliant gripper, (b) symmetric half of (a), (c) a slight modification of the mechanism in (b) could serve as a structure, say, as a frame for a bus shelter

Tensegrity structures [4] and cable-driven robots [5] are interesting examples. In a tensegrity structure significant stiffness can be achieved with wires that undergo tension and rigid struts that undergo compression. But a tensegrity structure can deform under internal or external loads. Cytoskeleton of a biological cell, which some say is a tensegrity structure [6], endows locomotion and shape-changing abilities to the cell. Similarly, a cable-drive robot blurs the distinction between a structure and a mechanism. Tensegrity structures and cable-driven mechanisms are in fact compliant mechanisms.

Based on the foregoing, it is safe to call something a structure or a mechanism based on the function it serves. Therefore, a compliant mechanism can shift its role from a structure to a mechanism or vice versa. Mobility analysis makes this point more clearly than this, as explained in the next section.

2.2 Mobility

Mobility, i.e., the ability of a solid entity to be mobile, is well developed for rigid-body mechanisms. Grübler's formula [7] gives the number of degrees of freedom (dof), M, which gives the number of independent actuations needed to completely specify the configuration of the mechanism.

$$M = 6(n-1) - \sum_{i=1}^{5}(6-i)f_i \text{ in 3D} \tag{1a}$$

$$M = 3(n-1) - \sum_{i=1}^{2}(3-i)f_i \text{ in 2D} \tag{1b}$$

where n is the number of rigid bodies and f_i the number of kinematic joints that allow i relative motions between the bodies that they connect. For a rigid-body assembly to be a mechanism, $M \geq 1$. There is an equivalent formula for pin-jointed truss structures, which is known as the Maxwell's rule [8]. It gives the number of *states of self-stress*, s.

$$s = b - 3j + 6 \text{ in 3D} \tag{2a}$$

$$s = b - 2j + 3 \text{ in 2D} \tag{2b}$$

where b is the number of bars (equal to n in the dof formula due to Grübler) and j is the number of vertices where pin-joints are located. By state of self-stress, it is meant that the truss cannot be assembled in general without extending/contracting as many number of bars, and hence causing stress even in the absence of external forces.

One may check that the number of dof given by Grübler's formula will be equal but opposite in sign to the number of states of self-stress given by Maxwell's rule. We can use either formula as long as we know how to interpret dof and states of self-stress. It is well known in kinematics that Grübler's formula fails when there are special geometric conditions. So does Maxwell's rule! Both are based purely on topology (i.e., connectivity) information and do not account for symmetries. Fowler and Guest [9] have given symmetry-extension to Maxwell's rule. One can use that, with some interpretation, to compute the number of dof. A fool-proof method of computing the number of dof is treating the pin-jointed rigid-body assembly as a truss and computing the rank of the stiffness matrix. Deficiency in rank is equal to the number of dof. Thus, we see close connection between structures and mechanisms.

It is sometimes said that compliant mechanisms have infinitely many dof. That is not true if we fix the forces applied on it. By considering a general compliant mechanism that has rigid and flexible bodies as well as kinematic and flexible joints, Murphy et al. [10] presented a dof formula for compliant mechanisms. It was applied to a number of case-studies and extended slightly to an easily interpretable form in [11].

$$M = 6(n_{seg}-1) - \sum_{i=0}^{5}(6-i)f_i - \sum_{j=1}^{5}(6-j)c_j \sum_{k=1}^{q} k(sc)_k \text{ in 3D} \tag{3a}$$

$$M = 3(n_{seg}-1) - \sum_{i=0}^{2}(3-i)f_i - \sum_{j=1}^{2}(3-j)c_j \sum_{k=1}^{q} k(sc)_k \text{ in 2D} \tag{3b}$$

where n_{seg} is the number of segments in a compliant mechanism; f_i is the same as in Eqs. (1a-b) with the additional implication that f_o implies a "fixed" connection where no relative motions are allowed between the connected segments; c_j is the number of compliant joints that allow j relative motions; nc is the segment compliance; and q is the maximum segment compliance present in the mechanism. The new term here is the segment compliance, which simply means the number of flexible modes (not be confused with normal modes of vibration) of a compliant segment. For example, a rod that can only extend/contract but cannot bend has unit segment compliance. See [10, 11] for details.

   Thus, we see that from the mobility viewpoint, compliant mechanisms share much with stiff structures and rigid-body mechanisms. The connection between the latter two is also quite intriguing in light of relationship between Grübler's formula and Maxwell's rule.

2.3 Materials

There is absolutely no restriction on associating different classes of materials (viz. ceramics, elastomers, glasses, natural materials, and polymers and hybrids (composites, foams, lattice networks, etc.) [12]) with stiff and compliant bodies. This is because stiffness and flexibility depend not on the material alone but also on the geometric form. In fact, it is the geometry that influences stiffness/flexibility more than the material. A steel spring can be more flexible than a rubber ball of the same volume of material. A brittle material such as silicon has been extensively used to make compliant mechanisms.

   A question is often raised about the vulnerability of compliant mechanisms due to large displacements. First, let us answer that by noting that stiff structures and rigid-body mechanism also fail by breakage of material under excessive loads as compared to what they are designed for. So do compliant mechanisms. But it must be noted that large displacement does not mean large strain (and hence stress). Flexibility and strength are not complementary. Figure 2a shows a narrow spiral-shaped slit cut through the thickness of an acrylic sheet using laser machining. Figure 2b shows that it results in highly flexible spring. Even though its displacement is quite large under its own weight, it has low level of stress.

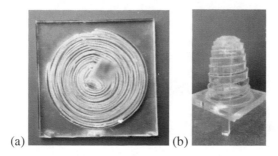

(a)   (b)

Fig. 2 (a) An acrylic plate with a spiral cut; (b) its deformation under self-weight. A rod is attached vertical to the plane of the cut-out plate for support.

Thus, compliant mechanisms can be made of any of the aforementioned classes of materials just as structures and rigid-body mechanisms can be constructed out of any material suitable for an application hand. Let us consider manufacturing techniques next.

2.4   Manufacturing

Until now, the compliant mechanisms reported in the literature and the ones found in the market are made using molding, casting, extrusion, machining, lithography, water-jet cutting, laser machining, punching, electro-discharge machining, etc. In fact, no technique needs to be excluded to make compliant mechanisms just as is the case with stiff structures and rigid-body mechanisms. As noted in the preceding sub-sections, geometric forms of the three classes of the solid systems cannot be distinguished from one another either from the viewpoint of materials used or from the perspective of manufacturing.

2.5   Analysis and Simulation

Analysis of rigid-body mechanisms entails solving algebraic equations (for statics) and ordinary differential equations (for dynamics). Structural analysis of arbitrarily shaped structures needs numerical techniques such as finite element analysis or the boundary element analysis, the former being popular, for solving the partial differential equations governing elastic deformation. Compliant mechanisms too often need finite element analysis if their actual physical form needs to be analyzed. In this respect they are closer to structures. However, some compliant mechanisms can be analyzed using rigid-body mechanism models.

Figure 3a shows a compliant slider-crank mechanism with three flexural pivots replacing the pin joints. Such a mechanism can be modeled as a rigid-body slider-crank mechanism with torsional spring constants to model the flexures. This is shown in Fig. 3b. A simple but effective model for flexural hinges was developed by Paros and Weisbord [13].

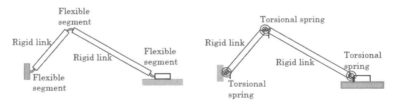

Fig. 3 (a) A compliant slider-crank mechanism; (b) its rigid-body model with torsional springs

Another significant advance for modeling compliant mechanism using rigid-body models was developed by Burns and Crossley [14] and Howell and Midha [15]. It was observed in [14] that the tip-deflection of a cantilever beam under transverse tip-load can be well approximated with a pin-jointed rigid crank of 5/6 the length of the cantilever. In [15], this was extended by noting that the elastic resistance of the cantilever can be represented with a torsional spring of linear spring constant. See Figs. 4a-b. This is called a pseudo rigid-body model and it has been used extensively in the last two decades.

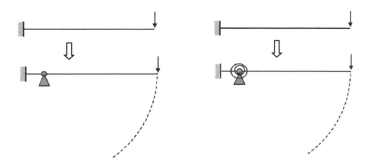

Fig. 4 Left: A cantilever beam approximated with a pin-jointed crank [14] and with an added torsional spring of linear spring constant [15]

More recently, in [16] another rigid-body model shown in Fig. 5 is developed. This, called a spring-lever (SL) model, captures the essential terminal characteristics of a compliant mechanism. An extension of this for dynamic analysis is reported in [17] and is called the spring-mass-lever (SML) model. These models too are amenable for large displacement analysis. More importantly, they are useful for synthesis.

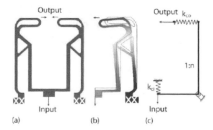

Fig. 5 (a) a compliant gripper, (b) symmetric half showing its deformation, and (c) a spring-lever model [16]

Here again, we see that compliant mechanism utilize structural and rigid-body analyses almost equally. In practice, even though compliant mechanisms can be accurately simulated by numerically solving the partial differential equations, several rigid-body models are used for their analysis. The aim of rigid-body models is not so much as to increase the computational efficiency but is to gain insight into the way compliant mechanisms work. They also help in synthesis.

2.6 Synthesis

Synthesis of rigid-body mechanisms is a rich field of research. Beginning with graphical techniques and analytical techniques, today we have sophisticated computational techniques for the synthesis of rigid-body mechanisms. Synthesis of structures too is a widely researched topic. Optimization is a versatile tool for synthesizing both mechanisms and structures. But structures got an edge over rigid-body mechanisms when shape and topology optimization techniques [18] came into existence. Compliant mechanism synthesis has benefitted from both classes of techniques—rigid-body mechanism and structural synthesis methods.

Howell and Midha [19] used their pseudo rigid-body models of compliant mechanisms in loop-closure equations and brought compliant mechanisms into the ambit of rigid-body mechanism synthesis. In this method, one can begin with an assumed linkage and synthesize it for desired kinetoelastic response. The solution obtained is transformed into a compliant mechanism by replacing the optimized torsional spring constants with flexural hinges or flexible beams of uniform cross-section. In [16, 17], which also use another kind of pseudo rigid-body models, a suitable compliant mechanism is chosen from a database using selection maps and then re-designed interactively, if necessary.

Alternatively, [20-22] adapted structural topology optimization techniques (originally developed for stiff structures) to design compliant mechanisms. In this technique, the designer is relieved from the task of choosing even the topology. Only the nominal specifications are enough in this method. Here, structural design is transformed into a material distribution problem. Much has been accomplished in this technique [18, 23].

Topology optimization of compliant mechanisms is sometimes thought of as distinctly different from that of stiff structures. It appears not so, some differences notwithstanding. The difference is that the points of interest in stiff structures are those where forces are acting. But in compliant mechanisms, there is always an output point of interest where there may or may not be a force. However, the problems faced in the topology optimization of stiff structures also arise, albeit in a slightly different form, in the topology optimization of compliant mechanisms.

See Fig. 6 that shows a topology-optimized image of a compliant mechanism. The point-flexures [24], which form one-node hinges, are conspicuous. The reason for their appearance was explained in [24] and a strategy to avoid these was offered for *distributed compliance*. This is important because the *lumped compliance* of the kind seen in Fig. 3a or Fig. 6 should be avoided as the flexural hinge regions are heavily stressed and they become the 'weakest link'. A similar problem arises stiff structures in the form checker-board pattern. This is attributed to numerical instability but one can argue on physical basis just as the occurrence of point-flexures in the optimal compliant topologies is explained.

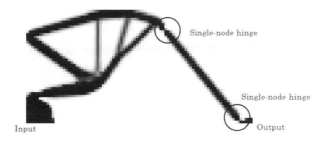

Fig. 6 A topology-optimized compliant mechanism with single-node hinges

A more subtle similarity between the topology optimization of compliant mechanisms and stiff structures becomes evident when body-force are considered. (e.g., self-weight). Figure 7 shows an image of a topology-optimized design under its self-weight. It is an ill-converged solution. The issue here is that whenever material is placed at a point, additional load ensues. So, there is no motivation for placing material anywhere at all. So, usually a small dummy force is used [25]. This is artificial. A similar requirement exists in compliant mechanisms where a small output load or a spring needs to be used to coax the algorithm to make a connection to the output point.

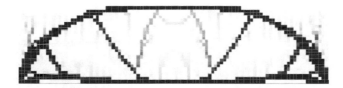

Fig. 7 An ill-converged solution of topology optimization of a structure with self-weight

Another intriguing issue is the first application of optimization technique for designing structures for desired deflection (as in compliant mechanisms) led to a rigid-body mechanism. Barnett [26] noticed this first. See [20] for details. Thus, we see that the line between stiff structures, rigid-body linkages, and compliant mechanism blurs when their synthesis (i.e., design) is considered. Wise counsel seems to be that one should use what is best for an application as per the available resources.

2.7   Size

Stiff structures exist across several orders of magnitude from mountains to the tips of atomic force microscopes. Making stiff structures is relatively easy at all sizes as compared to rigid-body mechanisms. Jointed mechanisms require assembly, which is a problem at all sizes. At macro scale, even though we are used to it, it is often said that more than 50% of cost of making a product goes into assembly.

At small scales—micro to nano—assembly is very difficult and uneconomical [27]. Compliant mechanisms help because they do not need assembly. The problems of wear and friction become dominant as the surface to volume ratio increases as miniaturization continues. In Nature too, we see more compliant designs at the small scales as compared to jointed skeletal design as we go down the size scale [20]. The matter of size is best left to the practitioner who knows the constraints that go beyond design into economy of material and manufacturing.

2.8   Static Balancing

With regard to energy efficiency, there is big disadvantage with compliant mechanisms as compared to rigid-body mechanisms. In the latter, barring frictional losses at joints, all of the input energy is delivered to the output. But in compliant mechanisms, significant part of input energy goes into deforming their elastic bodies. A remedy for this is static balancing. That is, by preloading a compliant mechanism, the energy stored a priori in it can be used to deform the mechanism to perform its intended function. Here, once again, it helps to take the detour to rigid-body linkages. A technique developed in [28] to statically balance a spring-loaded four-bar linkage (see Fig. 8a) was used to balance a spring-steel compliant mechanism shown in Fig. 8b. This route offers useful insight (e.g., what is the lowest level of preload and what are the options available to the designer?) than using structural optimization directly.

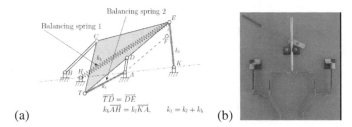

(a)   (b)

Fig. 8 (a) A spring-loaded four-bar linkage statically balanced with two more springs [28] and (b) a spring-steel compliant mechanism statically balanced with a rigid-body linkage and a spring

3   Closure

We considered a brief overview of compliant mechanisms, not exhaustive by any means, and compared and contrasted compliant mechanisms with stiff structures and rigid-body mechanisms. Several criteria that we considered show that similarities are many. Some differences do exist. While it is surprising to see why compliant designs were ignored by engineers for such a long time, it is intriguing to see that compliant mechanisms are after all not that much different from the other two kinds of solid systems. The difference comes about only when we consider compliant mechanisms in their "discretized" or "lumped" or "pseudo rigid-body" models as opposed to "continuum" structural models. Any insights gathered from studies on compliant mechanism do enrich the fields of rigid-body mechanisms and stiff structures.

Acknowledgments. The author is indebted to all his research students who worked with him on topics related to compliant mechanisms at University of Pennsylvania, Philadelphia, and Indian Institute of Science, Bangalore.

References

1. Howell, L.L.: Compliant Mechanisms. John Wiley, New York (2001)
2. Moulton, T., Ananthasuresh, G.K.: Design and Manufacture of Electro-Thermal-Compliant Micro Devices. Sensors and Actuators A 90, 38–48 (2001)
3. Midha, A., Norton, T.W., Howell, L.L.: On the Nomenclature, Classification, and Abstractions of Compliant Mechanisms. ASME J. Mech. Des. 116, 270–279 (1994)
4. Motro, R.: Tensegrity: Structural Systems for the Future. Elsevier, New York (2003)
5. Bosscher, P., Riechel, A.T., Ebert-Uphoff, I.: Wrench-Feasible Workspace Generation for cable-Driven Robots. IEEE Trans. Robotics 22(5), 890–902 (2006)
6. Ingber, D.E.: Tensegrity-based Mechanosensing from Macro and Micro. Progress in BioPhy. and Mol. Bio. 97, 163 179 (2008)
7. Erdman, A.G., Sandor, G.N., Kota, S.: Mechanism Design: Analysis and Synthesis, vol. 1. Prentice-Hall, Upper Saddle River (2001)
8. Calladine, C.R.: Buckminster Fuller's 'Tensegrity' Structures and Clerk Maxwell's Rules for the Construction of Stiff Frames. Int. J. Solids and Struc. 14, 161–172 (1978)
9. Fowler, P.W., Guest, S.D.: A Symmetry Extension of Maxwell's Rule for Rigidity of Frames. Int. J. Solids and Struc. 37, 1793–1804 (2000)
10. Murphy, M.D., Midha, A., Howell, L.L.: On the Mobility of Compliant Mechanisms. In: Proc. 1994 ASME Mechanisms Conference, Minneapolis, USA, September 11-14. DE-vol. 71, pp. 475–479 (1994)
11. Ananthasuresh, G.K., Howell, L.L.: Case Studies and a Note on the Degrees of Freedom in Compliant Mechanisms. In: ASME Mechanisms Conference, Irvine, USA, August 18-22, Paper No. 96-DETC/MECH-1217 (1996)
12. Ashby, M.F., Hugh, S., Cebon, D.: Materials: Engineering, Science, processing, and Design. Elsevier, Amsterdam (2007)
13. Paros, J.M., Weisbord, L.: How to Design Flexural hinges? Machine Design, 151–156 (November 25, 1965)
14. Bruns, R.H., Crossley, F.R.E.: Kinetostatic Synthesis of Flexible Link Mechanisms. ASME Paper no. 68-Mech-36 (1968)
15. Howell, L.L., Midha, A.: Parametric Deflection Approximations for End-loaded, Large-deflection Beams in Compliant Mechanisms. ASME J. Mech. Des. 117(1), 156–165 (1995)
16. Hegde, S., Ananthasuresh, G.K.: Design of Single-Input-Single-Output Compliant Mechanisms for Practical Applications Using Selection Maps. ASME J. Mech. Des. 132, 08107-1–08107-8 (2010)

17. Hegde, S., Ananthasuresh, G.K.: A Spring-Mass-Lever Model, Stiffness, and Inertia Maps for Single-Input-Single-Output Compliant Mechanisms. Mech. Mach. Theory (2012) (in press)
18. Bendsoe, M.P., Sigmund, O.: Topology Optimization: Theory, Methods, and Applications. Springer, Berlin (2003)
19. Howell, L.L., Midha, A.: A Loop-Closure Theory for the Analysis and Synthesis of Compliant Mechanisms. ASME J. Mech. Des. 118(1), 121–125 (1994)
20. Ananthasuresh, G.K.: A New Design Paradigm for Micro-Electro-Mechanical Systems and Investigations on Compliant Mechanism Synthesis. PhD Thesis, University of Michigan, Ann Arbor, USA (1994)
21. Frecker, M., Ananthasuresh, G.K., Nishiwaki, S., et al.: Topological Synthesis of Compliant Mechanisms using Multicriteria Optimization. ASME J. Mech. Des. 119(2), 238–245 (1997)
22. Sigmund, O.: On the Design of Compliant Mechanisms using Topology Optimization. Mech. of Struc. and Mach. 25(4), 495–526 (1997)
23. Deepak, S.R., Dinesh, M., Sahu, D., Ananthasuresh, G.K.: A Comprative Study of the Formulations and Benchmark Problems for the Topology Optimization of Compliant Mechanisms. ASME J. Mech. Rob. 1(1), 20–27 (2008)
24. Yin, L., Ananthasuresh, G.K.: Design of Distributed Compliant Mechanisms. Mech. based Des. of Struc. 31(2), 151–179 (2003)
25. Bruyneel, M., Duysinx, P.: Note on Topology Optimization of Continuum Structures Including Self Weight. Struc. Multidisc Optim. 29, 245–256 (2005)
26. Barnett, R.L.: Minimum-weight Design of Beams for Deflection. J. Eng. Mech. Div. EM 1, 75–109 (1961)
27. Ananthasuresh, G.K., Howell, L.L.: Mechanical Design of Compliant Microsystems: A perspective and Prospects. ASME J. Mech. Des. 127(4), 736–738 (2005)
28. Deepak, S.R., Ananthasuresh, G.K.: Perfect Static Balance of Linkages by Addition of Springs but not Auxiliary Bodies. ASME J. Mech. and Rob. 4, 021104-1–021104-12 (2012)

8 On a Compliant Mechanism Design Methodology Using the Synthesis with Compliance Approach for Coupled and Uncoupled Systems

Ashok Midha, Yuvaraj Annamalai, Sharath K. Kolachalam,
Sushrut G. Bapat, and Ashish B. Koli

Abstract. Compliant mechanisms are defined as those that gain some or all of their mobility from the flexibility of their members. Suitable use of pseudo-rigid-body models for compliant segments, and state-of-the-art knowledge of rigid-body mechanism synthesis types, greatly simplifies the design of compliant mechanisms. Starting with a pseudo-rigid-body four-bar mechanism, with one to four torsional springs located at the revolute joints to represent mechanism characteristic compliance, a simple, heuristic approach is provided to develop various compliant mechanism types. The *synthesis with compliance* method is used for three, four and five precision positions, with consideration of one to four torsional springs, to develop design tables for standard mechanism synthesis types. These tables reflect the mechanism compliance by specification of either energy or torque. The approach, while providing credible solutions, experiences some limitations.

Ashok Midha
Missouri University of Science and Technology, Rolla, MO
e-mail: midha@mst.edu

Yuvaraj Annamalai
Bombardier Aerospace, Wichita, KS
e-mail: yuvaraj.annamalai@gmail.com

Sharath K. Kolachalam
Foro Energy Inc., Littleton, CO
email: skolachalam@gmail.com

Sushrut G. Bapat · Ashish B. Koli
Missouri University of Science and Technology, Rolla, MO
e-mail: sgb8cc@mail.mst.edu, abkqm2@mst.edu

V. Kumar et al. (Eds.): *Adv. in Mech., Rob. & Des. Educ. & Res.*, MMS 14, pp. 95–116.
DOI: 10.1007/978-3-319-00398-6_8 © Springer International Publishing Switzerland 2013

The method is not yet robust, and research is continuing to further improve it. Examples are presented to demonstrate the use of weakly or strongly coupled sets of kinematic and energy/torque equations, as well as different compliant mechanism types in obtaining solutions.

1   Nomenclature

$\mathbf{Z}_n$ Vector notation of link n
R_n Magnitude of $\mathbf{Z}_n$
θ_n Angle of $\mathbf{Z}_n$, measured ccw from x-axis
P_j j^{th} precision point
δ_j Vector from first to j^{th} precision point
ϕ_j Rotation of the input link from first to j^{th} position
γ_j Rotation of the coupler from first to j^{th} position
ψ_j Rotation of the output link from first to j^{th} position
E_j Energy of the mechanism at j^{th} precision position
k_i Spring constant of the ith torsional spring
β_{ij} j^{th} angular position of the ith torsional spring
T_j Torque specified at j^{th} precision position
h_{ij} First-order kinematic coefficient of the ith link at j^{th} position
K_θ Characteristic stiffness coefficient
γ Characteristic radius factor
Θ Pseudo-rigid-body angle

2   Introduction

Methods developed in recent times for synthesizing compliant mechanisms comprise numerical synthesis [1, 2], a systematic application of structural optimization [3-6], graphical synthesis [7], loop closure [8, 9], and homotopy technique [10]. In using the pseudo-rigid-body model concept, loop closure offers invaluable benefits, such as use of existing knowledge base in rigid-body mechanism synthesis [11], generation of multiple solutions, and expediency of solution with accuracy. This paper describes a methodology to synthesize compliant mechanisms using the pseudo-rigid-body model concept and the loop-closure technique, while taking into account the mechanism's non-prescribed energy-free state.

The pseudo-rigid-body model concept [12, 13], which readily accommodates large deflection of flexible members, naturally bridges rigid-body synthesis to compliant mechanism design, providing the greatest benefit of all. For each flexible member (segment), a derived equivalent pseudo-rigid-body model predicts its deflection path and force-deflection characteristics. These segments are modeled by two or more rigid links attached at pin joints. A torsional spring, located at a pin joint, is used to model the force-deflection relationships of a compliant segment accurately.

A large-deflection cantilevered compliant (fixed-free) beam of length l and its equivalent pseudo-rigid-body model are shown in Figs. 1(a) and 1(b), respectively. It is assumed that the nearly circular path of the beam end can be modeled by two rigid links joined at a "characteristic" pivot along the beam [14]. A torsional spring at the pivot represents the beam's resistance to deflection. The stiffness coefficient K_θ is related to the torsional spring constant, k, of the beam. The location of this characteristic pivot is measured from the beam end as a fraction of the beam's length, γl, where γ is the characteristic radius factor [15, 16]. This distance γl is also known as the characteristic radius.

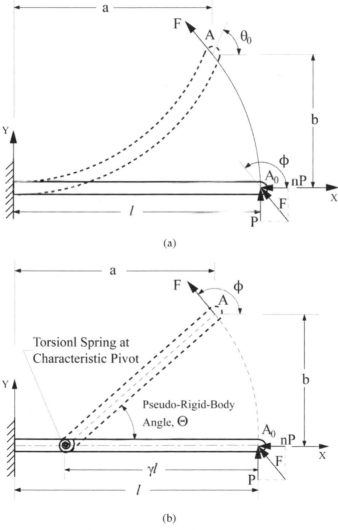

(a)

(b)

Fig. 1 (a) A cantilever beam with a force at the beam end, and (b) its pseudo-rigid-body model

The average value of the characteristic radius factor is found to be 0.85 [15], and may be used as a preliminary estimate in problem solving. The angle by which the characteristic radius is rotated is referred to as the pseudo-rigid-body angle, Θ. This concept, along with existing rigid-body mechanism theories for function, path, and motion generation, and path generation with prescribed timing [11], can be used advantageously to synthesize compliant mechanisms.

A loop-closure technique was developed to synthesize compliant mechanisms by combining loop-closure equations with energy/torque relations, which reflect the mechanism compliance [9]. This technique, termed as *synthesis with compliance*, thus relates the energy storage characteristics of compliant segments to the kinematic mobility of the mechanism. Therefore, for these two sets of equations, there are two sets of unknowns: 1) the kinematic variables, consisting of link lengths and angles of the pseudo-rigid-body model, corresponding to some select positions known as precision positions; and 2) the energy variables, composed of the undeflected spring position β_{i0}, related to the initial pseudo-rigid-body angle Θ_{n0}, and the spring stiffness k_i, related to the characteristic stiffness coefficient K_Θ for each compliant segment (Fig. 2).

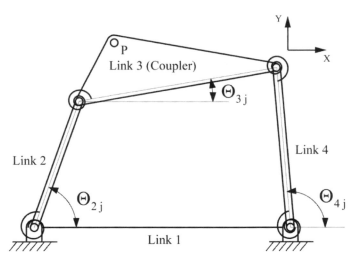

Fig. 2 Schematic of a pseudo-rigid-body four-bar mechanism with torsional springs

Using the above reduction technique, a pseudo-rigid-body kinematic chain with discrete compliances at the characteristic pivots may be obtained. With this in mind, the basic kinematic four-bar chain is selected, with its revolute joints representing the aforementioned characteristic pivots, and the springs the segment compliances.

Depending on the compliances (or the springs) introduced *synthesis with compliance* yields a set of weakly coupled or strongly coupled equations [15]. In weakly coupled set of equations, the kinematic equations are solved independently of energy/torque equations, whereas in the latter case, both the kinematic and energy/torque equations are simultaneously solved for all the unknowns. The possible compliant mechanism configurations, with fully compliant, small-length flexural pivot [15], and rigid segments, are heuristically derived [17] and improved upon in Fig. 3. In using equivalent pseudo-rigid-body model representations for various compliant segment types [15], and assuming four torsional springs in the rigid-body four-bar mechanism (Fig. 2), three possible compliant mechanisms (Figs. 3A-C) may be conceptualized. Similarly, five mechanism types result (Figs. 3D-H) from use of three springs, eight (Figs. 3I-P) from two springs, and two (Figs. 3Q, R) from a single spring, giving a total of 18 possible configuration types for solution. It is from these types that we shall draw upon for later examples.

3   Synthesis with Compliance for Energy Specifications

3.1   Kinematic Equations

In Function generation [11], the vector loop closure $Z_2 \rightarrow Z_3 \rightarrow Z_4 \rightarrow Z_{4j} \rightarrow Z_{3j} \rightarrow Z_{2j}$ (Fig. 4) gives the following equation:

$$Z_2(1 - e^{i\phi_j}) + Z_3(1 - e^{i\gamma_j}) + Z_4(e^{i\psi_j} - 1) = 0 \tag{1}$$

where, j is the mechanism position.

For path generation, motion generation (rigid-body guidance), and path generation with prescribed timing [11], the loop-closure equations (2) and (3) are obtained for loops $Z_2 \rightarrow Z_5 \rightarrow \delta_j \rightarrow Z_{5j} \rightarrow Z_{2j}$ and $Z_4 \rightarrow Z_6 \rightarrow \delta_j \rightarrow Z_{4j} \rightarrow Z_{6j}$ (Fig. 5), formed by dyads $Z_2 \rightarrow Z_5$ and $Z_4 \rightarrow Z_6$, respectively.

$$Z_2(e^{i\phi_j} - 1) + Z_5(e^{i\gamma_j} - 1) = \delta_j \tag{2}$$

$$Z_4(e^{i\psi_j} - 1) + Z_6(e^{i\gamma_j} - 1) = \delta_j \tag{3}$$

3.2   Energy Equations

The stored energy of the compliant mechanism in the jth precision position is estimated [8, 18] by the potential energy stored in the torsional springs of the pseudo-rigid-body model (Fig. 2) as:

$$E_j = \frac{1}{2}\sum_{i=1}^{m} k_i\left(\beta_{ij} - \beta_{i0}\right)^2; \qquad 1 \leq m \leq 4 \tag{4}$$

where, k_i is the spring constant, β_{ij} the jth angular position of the ith torsional spring, β_{i0} the angular position of the ith spring in its undeflected position, and m the number of torsional springs.

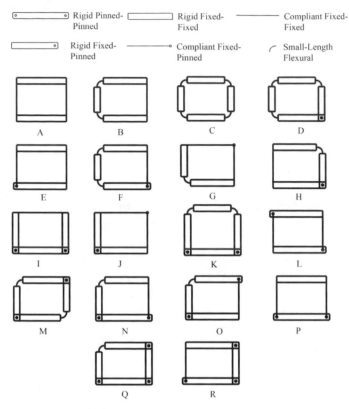

Fig. 3 Schematic representation of compliant mechanism types from a pseudo-rigid-body four-bar mechanism

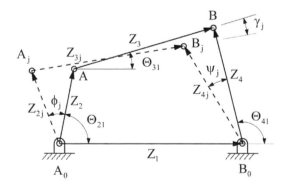

Fig. 4 Vector schematic of a four-bar function generation mechanism in its 1st and j^{th} positions

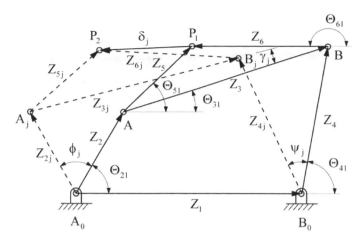

Fig. 5 Vector schematic of a four-bar mechanism showing vector dyads in the 1st and jth positions

The angle β_{ij} is related to the pseudo-rigid-body mechanism angles, Θ, [7, 8] as follows:

$$\beta_{1j} = \Theta_{2j} \tag{5a}$$

$$\beta_{2j} = 180° - (\Theta_{2j} - \Theta_{3j}) \tag{5h}$$

$$\beta_{3j} = \Theta_{4j} - \Theta_{3j} \tag{5c}$$

$$\beta_{4j} = \Theta_{4j} \tag{5d}$$

where, Θ_{nj} represents the angles of the nth link at the jth position. From equations (4) and (5), therefore, the mechanism potential energy in the jth position is given as:

$$E_j = \frac{1}{2}\left\{ \begin{array}{l} k_1\left(\Theta_{2j} - \Theta_{20}\right)^2 \\ + k_2\left[\left(\Theta_{3j} - \Theta_{30}\right) - \left(\Theta_{2j} - \Theta_{20}\right)\right]^2 \\ + k_3\left[\left(\Theta_{4j} - \Theta_{40}\right) - \left(\Theta_{3j} - \Theta_{30}\right)\right]^2 \\ + k_4\left(\Theta_{4j} - \Theta_{40}\right)^2 \end{array} \right\} \tag{6}$$

where, Θ_{n0} represents the angular position of the nth link in the initial energy/torque free spring state.

Based on equations (1)-(3) and (6), Tables 1-4 outline the number of equations, unknowns, and free choices for a given number of torsional springs (m), for function, path, and motion generation, and path generation with prescribed timing, respectively. These tables encapsulate a methodology for the synthesis of an appropriate pseudo-rigid-body four-bar mechanism for the given criteria. As an example, in Table 1, for a single torsional spring (m=1) and the three-precision-position case, there are 7 equations and 10 unknowns, and hence 3 free choices, theoretically yielding solutions in the order of $(\infty)^3$ [11]. In the last column of the table, the notations s.c. and w.c. signify a strongly and weakly coupled system, respectively.

For a function generation five-precision-position synthesis, and m=1, the system is over-constrained with more equations than unknowns, and hence is excluded from Table 1. In Tables 3 and 4, the cases of five-precision-position synthesis for m=1 & 2 are not included for a similar reason. The numbers in brackets in Tables 2-4 refer to additional equations or unknowns arising from the case wherein torque (instead of energy) is specified, as explained below.

Table 1 Design choices based on number of torsional springs for *function generation* synthesis with compliance

Number of Torsional Springs	Number of Equations	Number of Unknowns		Number of Free Choices
		Three Precision Positions		
1	7	$Z_2, Z_3, Z_4, \gamma_2, \gamma_3, k_1, \beta_{10}$	(10)	3 (s.c.[‡])
2	9[†]	$Z_2, Z_3, Z_4, \gamma_2, \gamma_3, k_1, k_2, \Theta_{20}, \Theta_{30},$		4 (w.c.[‡])
		Θ_{40}	(13)	
3	9[†]	"+ k_3	(14)	5 (w.c.)
4	9[†]	" + k_4	(15)	6 (w.c.)
		Four Precision Positions		
1	10	$Z_2, Z_3, Z_4, \gamma_2, \gamma_3, \gamma_4, k_1, \beta_{10}$	(11)	1 (s.c.)
2	14[†]	$Z_1, Z_2, Z_3, Z_4, \gamma_2, \gamma_3, \gamma_4, k_1, k_2, \Theta_{20},$		2 (s.c.)
		Θ_{30}, Θ_{40}	(16)	
3	12	$Z_2, Z_3, Z_4, \gamma_2, \gamma_3, \gamma_4, k_1, k_2, k_3, \Theta_{20},$		3 (w.c.)
		Θ_{30}, Θ_{40}	(15)	
4	12	" + k_4	(16)	4 (w.c.)
		Five Precision Positions		
2	17[†]	$Z_1, Z_2, Z_3, Z_4, \gamma_2,$		0 (s.c.)
		$\gamma_3, \gamma_4, \gamma_5, k_1, k_2, \Theta_{20}, \Theta_{30}, \Theta_{40}$	(17)	
3	17[†]	$Z_1, Z_2, Z_3, Z_4, \gamma_2, \gamma_3, \gamma_4, \gamma_5, k_1, k_2,$		1 (s.c.)
		$k_3, \Theta_{20}, \Theta_{30}, \Theta_{40}$	(18)	
4	15	" + k_4	(17)	2 (w.c.)

[†] Equation (14) gives two more scalar equations; [‡]s.c.$\equiv$ strongly coupled system; w.c.$\equiv$ weakly coupled system.

4 Synthesis with Compliance for Torque Specifications

4.1 Kinematic Equations

These equations remain the same as in the case of "Energy Specifications" above.

4.2 Torque Equations

The general torque equation [8, 18] is given by

$$T_{2j} = \sum_{i=1}^{m} k_i \left(\beta_{ij} - \beta_{i0} \right) \frac{d\beta_{ij}}{dS}; \qquad 1 \leq m \leq 4 \tag{7}$$

where, S represents the input variable for the mechanism, and all other variables are as defined in equation (4). If Θ_2 is the input, then $d\beta_{ij}/dS$ may be expressed as:

$$\left(\frac{d\beta_1}{d\Theta_2} \right)_j = 1 \tag{8a}$$

$$\left(\frac{d\beta_2}{d\Theta_2} \right)_j = \left(\frac{d\Theta_3}{d\Theta_2} \right)_j - 1 = h_{3j} - 1 \tag{8b}$$

$$\left(\frac{d\beta_3}{d\Theta_2} \right)_j = \left(\frac{d\Theta_4}{d\Theta_2} \right)_j - \left(\frac{d\Theta_3}{d\Theta_2} \right)_j = h_{4j} - h_{3j} \tag{8c}$$

$$\left(\frac{d\beta_4}{d\Theta_2} \right)_j = \left(\frac{d\Theta_4}{d\Theta_2} \right)_j = h_{4j} \tag{8d}$$

where, h_{ij} represents the first-order kinematic coefficient of the ith link at the jth position, and is defined [19] as follows:

$$h_{3j} = \frac{R_2 \sin(\Theta_{4j} - \Theta_{2j})}{R_3 \sin(\Theta_{3j} - \Theta_{4j})} \tag{9}$$

$$h_{4j} = \frac{R_2 \sin(\Theta_{3j} - \Theta_{2j})}{R_4 \sin(\Theta_{3j} - \Theta_{4j})} \tag{10}$$

Substituting the values of β_{ij} in equation (7),

$$
\begin{aligned}
T_{2j} = &\ k_1\left(\Theta_{2j}-\Theta_{20}\right) \\
&+k_2\left[\left(\Theta_{3j}-\Theta_{30}\right)-\left(\Theta_{2j}-\Theta_{20}\right)\right]\left(h_{3j}-1\right) \\
&+k_3\left[\left(\Theta_{4j}-\Theta_{40}\right)-\left(\Theta_{3j}-\Theta_{30}\right)\right]\left(h_{4j}-h_{3j}\right) \\
&+k_4\left(\Theta_{4j}-\Theta_{40}\right)h_{4j}
\end{aligned}
\tag{11}
$$

and expanding with the help of equations (9) and (10), we have

$$
\begin{aligned}
T_{2j} = &\ k_1\left(\Theta_{2j}-\Theta_{20}\right)+k_2\begin{bmatrix}\left(\Theta_{3j}-\Theta_{30}\right)- \\ \left(\Theta_{2j}-\Theta_{20}\right)\end{bmatrix}\left(\frac{R_2\sin(\Theta_{4j}-\Theta_{2j})}{R_3\sin(\Theta_{3j}-\Theta_{4j})}-1\right) \\
&+k_3\begin{bmatrix}\left(\Theta_{4j}-\Theta_{40}\right)- \\ \left(\Theta_{3j}-\Theta_{30}\right)\end{bmatrix}\left(\frac{R_2\sin(\Theta_{3j}-\Theta_{2j})}{R_4\sin(\Theta_{3j}-\Theta_{4j})}-\frac{R_2\sin(\Theta_{4j}-\Theta_{2j})}{R_3\sin(\Theta_{3j}-\Theta_{4j})}\right) \\
&+k_4\left(\Theta_{4j}-\Theta_{40}\right)\frac{R_2\sin(\Theta_{3j}-\Theta_{2j})}{R_4\sin(\Theta_{3j}-\Theta_{4j})}
\end{aligned}
\tag{12}
$$

This equation, involving first-order kinematic coefficients, requires Θ_{3j} (yet an unknown), where j represent the jth precision position of the mechanism. When j> 1, Θ_{3j} is given by $\Theta_{31}+\gamma_j$, where γ_j is the coupler position relative to the first precision position. Hence, if Θ_{31} is determined, then Θ_{3j} may be calculated. In function generation, this may either be a free choice, or be solved for explicitly from the kinematic equations. In all other synthesis methods, requiring the use of dyads, Θ_{31} is not readily available. However, it is easily obtained from the following equation (Fig. 4):

$$
\mathbf{Z}_3 - \mathbf{Z}_5 + \mathbf{Z}_6 = 0
\tag{13}
$$

Accordingly, for a strongly coupled, torque specification case, except for function generation synthesis, the number of unknowns is increased by consideration of Θ_{31} and R_3, as indicated in Tables 2-4 within brackets [21].

Table 2 Design choices based on number of torsional springs for *path generation* synthesis with compliance

Number of Torsional Springs	Number of Equations	Number of Unknowns		Number of Free Choices
		Three Precision Positions		
1	$11\,[+2^*]$	$Z_2, Z_4, Z_5, Z_6, \phi_2, \phi_3, \gamma_2, \gamma_3, \psi_2, \psi_3, k_1,$ β_{10}	$(16)[+2^{**}]$	5 (s.c.‡)
2	13	$Z_2, Z_4, Z_5, Z_6, \phi_2, \phi_3, \gamma_2, \gamma_3, \psi_2, \psi_3, k_1,$ $k_2, \Theta_{20}, \Theta_{30}, \Theta_{40}$	(19)	6(w.c.‡)
3	13	$" + k_3$	(20)	7(w.c.)
4	13	$" + k_4$	(21)	8(w.c.)
		Four Precision Positions		
1	$16\,[+2^*]$	$Z_2, Z_4, Z_5, Z_6, \phi_2, \phi_3, \phi_4, \gamma_2, \gamma_3, \gamma_4, \psi_2,$ $\psi_3, \psi_4, k_1, \beta_{10}$	$(19)[+2^{**}]$	3(s.c.)
2	$22^{\dagger}$	$Z_1, Z_2, Z_3, Z_4, Z_5, Z_6, \phi_2, \phi_3, \phi_4, \gamma_2, \gamma_3,$ $\gamma_4, \psi_2, \psi_3, \psi_4, k_1, k_2, \Theta_{20}, \Theta_{30}, \Theta_{40}$	(26)	4 (s.c.)
3	18	$Z_2, Z_4, Z_5, Z_6, \phi_2, \phi_3, \phi_4, \gamma_2, \gamma_3, \gamma_4, \psi_2, \psi_3,$ $\psi_4, k_1, k_2, k_3, \Theta_{20}, \Theta_{30}, \Theta_{40}$	(23)	5 (w.c.)
4	18	$" + k_4$	(24)	6 (w.c.)
		Five Precision Positions		
1	$21\,[+2^*]$	$Z_2, Z_4, Z_5, Z_6, \phi_2, \phi_3, \phi_4, \phi_5, \gamma_2, \gamma_3, \gamma_4,$ $\gamma_5, \psi_2, \psi_3, \psi_4, \psi_5, k_1, \beta_{10}$	$(22)[+2^{**}]$	1 (s.c.)
2	$27^{\dagger}$	$Z_1, Z_2, Z_3, Z_4, Z_5, Z_6, \phi_2, \phi_3, \phi_4, \phi_5, \gamma_2,$ $\gamma_3, \gamma_4, \gamma_5, \psi_2, \psi_3, \psi_4, \psi_5, k_1, k_2, \Theta_{20},$ Θ_{30}, Θ_{40}	(29)	2 (s.c.)
3	$27^{\dagger}$	$" + k_3$	(30)	3 (s.c.)
4	23	$Z_2, Z_4, Z_5, Z_6, \phi_2, \phi_3, \phi_4, \phi_5, \gamma_2, \gamma_3, \gamma_4,$ $\gamma_5, \psi_2, \psi_3, \psi_4, \psi_5, k_1, k_2, k_3, k_4, \Theta_{20},$ Θ_{30}, Θ_{40}	(27)	4 (w.c.)

*Equation (13) contributes two more scalar equations. **Z_3 introduces two additional unknowns. † Equation (14) gives two additional scalar equations.
‡s.c.$\equiv$ strongly coupled system; w.c.$\equiv$ weakly coupled system

5 General Synthesis Case with a Non-prescribed Energy-Free State

Let us consider a general case, where the energy-free position of the compliant mechanism is different from the prescribed positions. Currently, for this case, in a pseudo-rigid-body four-bar mechanism with more than one torsional spring, the deflection-free state of one spring does not govern the deflection-free states of the remaining springs. However, in a monolithic (one-piece) compliant mechanism, the energy-free state of one flexural segment implies that all other compliant segments are also in their energy-free states corresponding to that position. Thus, to model the compliant mechanism in an optimal way, all deflection-free angular positions of the torsional springs in a pseudo-rigid-body four-bar mechanism should be related to one another, as should be expected from the energy-free position of the compliant mechanism.

Without using additional equations, the torsional springs in their undeflected states are not yet constrained, and even though the resulting mechanism solution may be a valid pseudo-rigid-body four-bar mechanism with independent springs, it is not an acceptable one-piece compliant mechanism solution. These additional constraints will need to relate the deflection-free state angles (β_{i0}) in the energy/torque equations to one another and to the link angles of the pseudo-rigid-body four-bar mechanism.

At the energy-free position of the mechanism or its zero[th] position, equation (5) relates β_{i0} of the ith torsional spring to the pseudo-rigid-body angles. Additionally, as Θ_{n0} are part of the designed pseudo-rigid-body four-bar mechanism, they need to satisfy the four-bar loop-closure equation in the energy-free state. Hence, the vector equation shown below will need to be enforced:

$$\mathbf{Z}_{20} + \mathbf{Z}_{30} = \mathbf{Z}_1 + \mathbf{Z}_{40} \tag{14}$$

where, the subscript '0' represents the energy-free position. This provides additional constraints with no further unknowns appended to the system.

This additional equation (14) would suffice to satisfactorily synthesize a weakly coupled system. In a strongly coupled system, however, few more equations need to be included to ensure a satisfactory solution of the system. For a strongly coupled function generation synthesis case, with equations (1), (6) or (12), and (14) included in the system, $\mathbf{Z}_1$ is additionally an unknown. To accommodate this, the first precision position four-bar loop-closure equation, i.e.

$$\mathbf{Z}_2 + \mathbf{Z}_3 = \mathbf{Z}_1 + \mathbf{Z}_4 \tag{15}$$

is used, resulting in two more scalar equations added to the system. For the remaining three cases of strongly coupled system synthesis, $\mathbf{Z}_1$ and $\mathbf{Z}_3$ become additional unknowns. In order to solve them, the coupler loop-closure equation (13) is used in addition to equation (15). Consequently, the system accumulates four more scalar equations. The above discussion is applicable only for a pseudo-rigid-body four-bar mechanism with two or more torsional springs.

In a pseudo-rigid-body four-bar mechanism with a single spring ($m = 1$), a torsional spring deflection-free angle, β_{i0}, that identifies the energy-free state of the mechanism, does not impose any conditions on the other pseudo-rigid-body links that are without torsional springs. Hence, in this case, no additional constraints are required.

Table 3 Design choices based on number of torsional springs for *motion generation* synthesis with compliance

Number of Torsional Springs	Number of Equations	Number of Unknowns		Number of Free Choices
		Three Precision Positions		
1	$11 [+2^*]$	$Z_2, Z_4, Z_5, Z_6, \phi_2, \phi_3, \psi_2, \psi_3,$ k_1, β_{10}	$(14)[+2^{**}]$	3 (s.c.[‡])
2	13	$Z_2, Z_4, Z_5, Z_6, \phi_2, \phi_3, \psi_2, \psi_3, k_1, k_2,$ $\Theta_{20}, \Theta_{30}, \Theta_{40}$	(17)	4 (w.c.[‡])
3	13	" $+ k_3$	(18)	5 (w.c.)
4	13	" $+ k_4$	(19)	6 (w.c.)
		Four Precision Positions		
1	$16 [+2^*]$	$Z_2, Z_4, Z_5, Z_6, \phi_2, \phi_3, \phi_4, \psi_2, \psi_3,$ ψ_4, k_1, β_{10}	$(16)[+2^{**}]$	0 (s.c.)
2	$22^{†}$	$Z_1, Z_2, Z_3, Z_4, Z_5, Z_6, \phi_2, \phi_3, \phi_4,$ $\psi_2, \psi_3, \psi_4, k_1, k_2, \Theta_{20}, \Theta_{30}, \Theta_{40}$	(23)	1 (s.c.)
3	18	$Z_2, Z_4, Z_5, Z_6, \phi_2, \phi_3, \phi_4, \psi_2, \psi_3,$ $\psi_4, k_1, k_2, k_3, \Theta_{20}, \Theta_{30}, \Theta_{40}$	(20)	2 (w.c.)
4	18	" $+ k_4$	(21)	3 (w.c.)
		Five Precision Positions		
4	23	$Z_2, Z_4, Z_5, Z_6, \phi_2, \phi_3, \phi_4, \phi_5, \psi_2, \psi_3, \psi_4,$ $\psi_5, k_1, k_2, k_3, k_4, \Theta_{20}, \Theta_{30}, \Theta_{40}$	(23)	0 (w.c.)

*Equation (13) contributes two more scalar equations. **Z_3 introduces two additional unknowns. [†] Equation (14) gives two additional scalar equations.
[‡]s.c.≡ strongly coupled system; w.c.≡ weakly coupled system.

Accordingly, Tables 1-4 have been updated to reflect these changes in the required number of equations, unknowns, and free choices, for different number of torsional springs and various precision position requirements [21]. Example 1 shows the application of this technique in designing a compliant mechanism with one fixed-fixed compliant segment, where three precision positions and the corresponding energies are specified.

As mentioned earlier, the above discussion assumes that the energy-free position of the compliant mechanism is different from the prescribed positions. If the energy-free position of the mechanism happens to be one of the prescribed positions, a reduced system of equations can be used to synthesize a compliant mechanism and is shown in example 2.

Table 4 Design choices based on number of torsional springs for *path generation with prescribed timing* synthesis with compliance

Number of Torsional Springs	Number of Equations	Number of Unknowns		Number of Free Choices
Three Precision Positions				
1	11 [+2*]	$Z_2, Z_4, Z_5, Z_6, \phi_2, \phi_3, \gamma_2, \gamma_3, k_1,$ β_{10}	(14) [+2**]	3 (s.c.‡)
2	13	$Z_2, Z_4, Z_5, Z_6, \phi_2, \phi_3, \gamma_2, \gamma_3, k_1, k_2,$ $\Theta_{20}, \Theta_{30}, \Theta_{40}$	(17)	4 (w.c.‡)
3	13	" + k_3	(18)	5 (w.c.)
4	13	" + k_4	(19)	6 (w.c.)
Four Precision Positions				
1	16 [+2*]	$Z_2, Z_4, Z_5, Z_6, \phi_2, \phi_3, \phi_4, \gamma_2, \gamma_3,$ $\gamma_4, k_1, \beta_{10}$	(16)[+2**]	0 (s.c.)
2	22†	$Z_1, Z_2, Z_3, Z_4, Z_5, Z_6, \phi_2, \phi_3, \phi_4,$ $\gamma_2, \gamma_3, \gamma_4, k_1, k_2, \Theta_{20}, \Theta_{30}, \Theta_{40}$	(23)	1 (s.c.)
3	18	$Z_2, Z_4, Z_5, Z_6, \phi_2, \phi_3, \phi_4, \gamma_2, \gamma_3,$ $\gamma_4, k_1, k_2, k_3, \Theta_{20}, \Theta_{30}, \Theta_{40},$	(20)	2 (w.c.)
4	18	" + k_4	(21)	3 (w.c.)
Five Precision Positions				
4	23	$Z_2, Z_4, Z_5, Z_6, \phi_2, \phi_3, \phi_4, \phi_5, \gamma_2, \gamma_3,$ $\gamma_4, \gamma_5, k_1, k_2, k_3, k_4, \Theta_{20}, \Theta_{30}, \Theta_{40}$	(23)	0 (w.c.)

*Equation (13) contributes two more scalar equations. **Z_3 introduces two additional unknowns.
† Equation (14) gives two additional scalar equations.
‡s.c.≡ strongly coupled system; w.c.≡ weakly coupled system.

6 Examples

6.1 Example 1

It is desired to design a compliant mechanism for three-precision-position path generation with prescribed timing, with energy specified at the precision positions as follows:

$\delta_2 = -3 + 0.5i$; $\delta_3 = -5 + 0.25i$; $\phi_2 = 20°$; $\phi_3 = 35°$; $E_1 = 6.3$ in-lb; $E_2 = 28$ in-lb; $E_3 = 51.6$ in-lb

Assuming two torsional springs are used in the pseudo-rigid-body four-bar mechanism, Table 4 shows there are 13 equations, 17 unknowns and 4 free choices, resulting in a weakly coupled system. Hence, the kinematic and energy variables can be solved for independently. A compliant mechanism configuration with one fixed-fixed segment, as shown in Figure 3(I), is chosen for synthesis. Four free choices are expended on R_2, R_4, θ_{21}, and θ_{41}. Using equations (2), (3), (6) and (14), the following solution is obtained:

$\mathbf{Z}_1 = 2.875 + 3.019i$; $\mathbf{Z}_2 = 0.479 + 5.479i$
$\mathbf{Z}_3 = 5.355 + 3.884i$; $\mathbf{Z}_4 = 2.958 + 6.344i$
$\mathbf{Z}_5 = 4.652 + 6.397i$; $\mathbf{Z}_6 = -0.703 + 2.537i$
$\gamma_2 = 9.286°$; $\gamma_3 = 14.613°$
$\psi_2 = 22.064°$; $\psi_3 = 36.74°$
$k_3 = 78.27$ in-lb/rad; $k_4 = 78.27$ in-lb/rad
$\Theta_{20} = 70°$ $\Theta_{30} = 24.152°$
$\Theta_{40} = 43.921°$

The length of the fixed-fixed compliant segment is determined using the equation:

$$\gamma l = |\mathbf{Z}| \tag{16}$$

where, γ is the characteristic radius factor, l the length of the compliant segment, and $|\mathbf{Z}|$ the magnitude of pseudo-rigid-body link length. The moment of inertia is obtained using the equation:

$$k = 2\gamma K_\Theta \frac{EI}{l} \tag{17}$$

where k is the torsional spring stiffness, K_Θ the stiffness coefficient, with an average value of 2.65, E the modulus of elasticity, and I the moment of inertia. Considering a rectangular cross section of width, w, and thickness, t, using Polypropylene, a thermoplastic material, and assuming the width, w to be 0.5 in., the value of the thickness, t, is obtained as 0.258 in. The resulting compliant mechanism is shown in Fig. 6.

The synthesis results obtained using the pseudo-rigid-body model (PRBM) are compared with the finite element analysis (FEA) software ABAQUS®. The coupler curve is obtained using the PRBM and precision position locations from both PRBM and FEA are shown plotted on the coupler curve in Fig. 7.

6.2   Example 2

A fully-compliant mechanism is to be designed for three-precision-position path generation with prescribed timing and energy specifications:

$\delta_2 = -3 + 0.5i$; $\delta_3 = -5 + 0.25i$; $\phi_2 = 20°$; $\phi_3 = 35°$; $E_1 = 0$ in-lb; $E_2 = 15$ in-lb; $E_3 = 44.8$ in-lb.

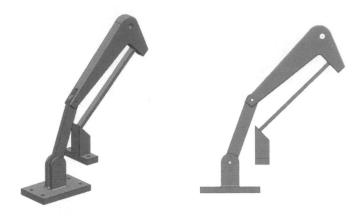

Fig. 6 Solid model of the compliant mechanism (with one fixed-fixed segment) designed in example 1

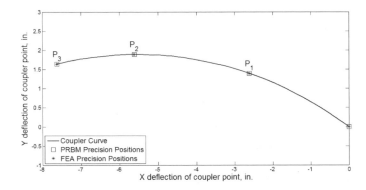

Fig. 7 Coupler curve of the mechanism (with precision positions) for example 1

In this example, the energy-free state of the mechanism is assumed to be the first precision position, and therefore the reduced system of kinematic and energy equations is used. Assuming four torsional springs are used in the pseudo-rigid-body four-bar mechanism, Table 4 shows that the resulting system is weakly coupled. A compliant mechanism configuration with two fixed-fixed compliant segments, as shown in Figure 3(A), is chosen for synthesis, and results in 10 equations, 14 unknowns, and hence, 4 free choices. Selecting R_2, Θ_{21}, R_4, and Θ_{41} as free choices, the following solution is obtained:

$Z_1 = 2.876 + 3.019i;$ $Z_2 = 0.4794 + 5.479i$

$Z_3 = 5.355 + 3.885i;$ $Z_4 = 2.958 + 4.110i$

$Z_5 = 4.652 + 6.422i;$ $Z_6 = -0.703 + 2.537i$

$\gamma_2 = 9.286°;$ $\gamma_3 = 14.613°$

$\psi_2 = 22.064°;$ $\psi_3 = 36.74°$

$k_1 = 80.88$ in-lb/rad; $k_2 = 80.88$ in-lb/rad

$k_3 = 87.73$ in-lb/rad; $k_4 = 87.73$ in-lb/rad

The lengths and moment of inertias of the two compliant segments are determined using equations (16) and (17), respectively. The lengths of the input and output compliant segments are obtained as 6.4705 in. and 8.2350 in., respectively. The width, w, is assumed to be 0.5 in. for both segments, resulting in the thickness, t, as 0.2406 in. and 0.2679 in., for input and output segments, respectively. The resulting compliant mechanism is shown in Fig. 8.

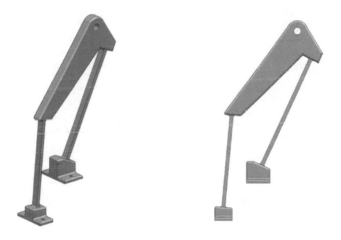

Fig. 8 Solid model of the compliant mechanism (with two fixed-fixed segments) designed in example 2

Fig. 9 shows the resulting coupler curve for this mechanism obtained using the PRBM. The precision position locations are shown plotted as obtained using the PRBM and FEA.

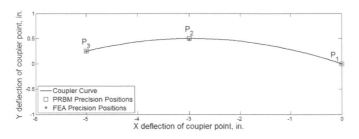

Fig. 9 Coupler curve of the mechanism (with precision positions) for example 2

6.3   *Example 3*

A compliant mechanism with small-length flexural pivots (SLFPs) is to be designed for three-precision-position path generation with specified energy:

$\delta_2 = -3 + 0.5i$; $\delta_3 = -5 + 0.2i$; $E_1 = 11.75$ in-lb; $E_2 = 37.25$ in-lb; $E_3 = 60$ in-lb.

Assuming four torsional springs are used in the pseudo-rigid-body four-bar mechanism; from Table 2, it can be seen that the resulting system is weakly coupled. A compliant mechanism configuration with four small-length flexural pivots, as shown in Figure 3(C), is chosen for synthesis, resulting in 13 equations, 21 unknowns, and hence, 8 free choices. Selecting R_2, Θ_{21}, R_4, Θ_{41}, γ_2, γ_2, Θ_{20}, and Θ_{30} as free choices, the following solution is obtained:

$Z_1 = 3.479 - 1.537i$; $Z_2 = 3 + 5.196i$
$Z_3 = 5.841 - 2.233i$; $Z_4 = 5.362 + 4.499i$
$Z_5 = -1.464 + 3.809i$; $Z_6 = -7.305 + 6.042i$
$\phi_2 = 24.473°$; $\phi_3 = 40.214°$
$\psi_2 = 22.635°$; $\psi_3 = 36.136°$
$\Theta_{40} = 8.7545°$
$k_1 = 30.688$ in-lb/rad; $k_2 = 29.0019$ in-lb/rad
$k_3 = 19.975$ in-lb/rad; $k_4 = 35.1406$ in-lb/rad

The length of SLFP is assumed to be 5% of the longer rigid segment. Thus, for the input and output pseudo-rigid-body link, the lengths of the corresponding SLFP are obtained as 0.2857 in. and 0.3333 in., respectively. The moment of inertia is calculated using the equation:

$$k = \frac{EI}{l} \tag{18}$$

Using the moment of inertias obtained, and assuming the width, w, to be 2.0 in. for all SLFPs, the thickness, t, is obtained for the four SLFPs, from bottom left to bottom right, as 0.06402 in., 0.06287 in., 0.05845 in. and 0.07056 in., respectively.

The resulting compliant mechanism is shown in Fig. 10, and Fig. 11 shows the coupler curve obtained using the PRBM for the mechanism, as well as the precision position locations.

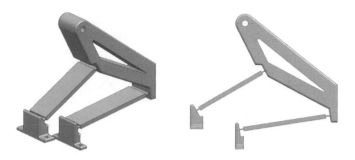

Fig. 10 Solid model of the compliant mechanism (with four small-length flexural pivots) designed in example 3

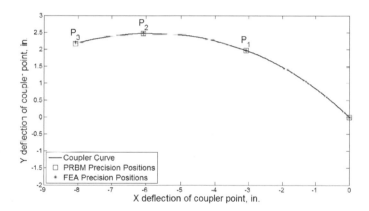

Fig. 11 Coupler curve of the mechanism (with precision positions) for example 3

7   Some Limitations Experienced in the Use of the Synthesis with Compliance Method

Though the *synthesis with compliance* technique is very useful for compliant mechanism design with energy/torque specifications, it suffers from some limitations in its current form, and are mentioned below:

This method solves the sets of kinematic and energy/torque equations as either a strongly coupled or a weakly coupled system depending on the number of unknowns shared between them. Generally, solving these coupled nonlinear kinematic and energy/torque equations presents increased complexity.

In a weakly coupled system, the kinematic and energy/torque equations are solved separately, and the kinematic configuration is solved for before solving the energy/torque equations. As a result, the latter system of equations frequently yields negative solutions for spring stiffness values. Although good solution were obtained, following a cumbersome process of iterations.

The number of variables involved in the sets of kinematic and energy equations are typically greater than number of equations available. In order to solve the equations, the user is required to assign reasonable values for the free choices and initial estimates. This process is highly cumbersome, and no guidelines currently exist to alleviate the situation.

Due to the nonlinearity of the sets of kinematic and energy/torque equations, the solutions obtained are rather sensitive to the values assigned for the free choices and initial estimates. Even the slightest changes in their values result in dramatic changes in the outcomes, which are frequently unrealistic.

In order to overcome the aforementioned limitations associated with the current state of the *synthesis with compliance* method, a more robust approach is currently being researched.

8   Conclusions

This paper is based on the use of existing concepts of equivalent pseudo-rigid-body models for compliant segments, and rigid-body mechanism synthesis for function, path and motion generation, and path generation with prescribed timing. Starting with a pseudo-rigid-body four-bar mechanism, with use of one to four torsional springs at the revolute joints to represent segment compliance, a heuristic approach is employed to develop a variety of compliant mechanism types. The *synthesis with compliance* technique has been used for a variable number of springs, for three, four and five precision positions of the mechanism. Exhaustive design tables have been systematically developed which enumerate the number of equations, unknowns and free choices for the above-mentioned synthesis types. The tables appropriately reflect these differences which result from the specification of either energy or torque. Examples have been presented to demonstrate the use of the *synthesis with compliance* method using different compliant segment types in obtaining the solutions. The results obtained are favorably compared with finite element analysis. Some insight is provided as to the limitations encountered in the method presented. Currently, the development of a more robust design methodology is underway, and will be reported in the near future.

References

1. Midha, A., Her, I., Salamon, B.A.: A Methodology for Compliant Mechanism Design: Part I – Introduction and Large-Deflection Analysis. In: Advances in Design Automation, 18th ASME Design Automation Conference, DE-vol. 44(2), pp. 29–38 (1992)
2. Her, I., Midha, A., Salamon, B.A.: A Methodology for Compliant Mechanism Design: Part II – Shooting Method and Application. In: Advances in Design Automation, 18th ASME Design Automation Conference, DE-vol. 44(2), pp. 39–45 (1992)
3. Ananthasuresh, G.K.: A New Design Paradigm for Micro-Electro-Mechanical Systems and Investigations on Compliant Mechanisms Synthesis. Ph.D. Thesis University of Michigan Ann Arbor (1994)
4. Frecker, M.I., Ananthasuresh, G.K., Nishiwaki, N., Kikuchi, N., Kota, S.: Topological Synthesis of Compliant Mechanisms using Multi-Criteria Optimization. ASME Journal of Mechanical Design 119, 238–245 (1997)
5. Sigmund, O.: On the Design of Compliant Mechanisms using Topology Optimization. Mechanics of Structures and Machines 25(4), 495–526 (1997)
6. Saggere, L., Kota, S.: Synthesis of Planar, Compliant Four-Bar Mechanisms for Compliant-Segment Motion Generation. ASME Journal of Mechanical Design 123(4), 535–541 (2001)
7. Mettlach, G.A., Midha, A.: Graphical Synthesis Techniques toward Designing Compliant Mechanisms. In: Proceedings of the 4th National Applied Mechanisms & Robotics Conference, vol. II, pp. 61-01 – 61-10 (1995)
8. Howell, L.L.: A Generalized Loop-Closure Theory for the Analysis and Synthesis of Compliant Mechanisms. Ph.D. Dissertation Purdue University (1993)
9. Howell, L.L., Midha, A.: A Generalized Loop-Closure Theory for the Analysis and Synthesis of Compliant Mechanisms. In: Pennock, G.R., et al. (eds.) Machine Elements and Machine Dynamics, 23rd Biennial Mechanisms Conference, vol. 71, pp. 491–500 (1994)
10. Su, H.J., McCarthy, J.M.: A Polynomial Homotopy Formulation of the Inverse Static Analysis of Planar Compliant Mechanisms. ASME Journal of Mechanical Design 128(4), 776–786 (2006)
11. Erdman, A.G., Sandor, G.N., Kota, S.: Mechanism Design - Analysis and Synthesis, vol. 1. Prentice Hall, New Jersey (2001)
12. Salamon, B.A.: Mechanical Advantage Aspects in Compliant Mechanism Design. MS Thesis, Purdue University (1989)
13. Howell, L.L., Midha, A.: A Method for the Design of Compliant Mechanisms with Small-Length Flexural Pivots. Journal of Mechanical Design Trans. ASME 116(1), 280–290 (1994)
14. Howell, L.L.: The Design and Analysis of Large-Deflection Members in Compliant Mechanisms. MS Thesis, Purdue University (1991)
15. Howell, L.L.: Compliant Mechanisms. John Wiley and Sons, New York (2001)
16. Pauly, J., Midha, A.: Improved Pseudo-Rigid-Body Model Parameter Values for End-Force-Loaded Compliant Beams. In: Proceedings of the 28th Biennial ASME Mechanisms and Robotics Conference, Salt Lake City Utah, pp. DETC 2004-57580-1–57580-5 (2004)

17. Midha, A., Christensen, M.N., Erickson, M.J.: On the Enumeration and Synthesis of Compliant Mechanisms using the Pseudo-Rigid-Body Four-Bar Mechanism. In: Proceedings of the 5th National Applied Mechanisms & Robotics Conference, Cincinnati Ohio, vol. 2, pp. 93-0–93-08 (1997)
18. Mettlach, G.A., Midha, A.: Using Burmester Theory in the Design of Compliant Mechanisms. In: Proceedings of the 24th Biennial Mechanisms Conference, CD-ROM Paper No. 96-DETC: MECH-1181 (1996)
19. Hall Jr., A.S.: Notes on Mechanism Analysis. Waveland Press Inc. Prospect Heights Illinois (1981)
20. Norton, T.W.: On the Nomenclature and Classification, and Mobility of Compliant Mechanisms. MS Thesis, Purdue University (1991)
21. Annamalai, Y.: Compliant Mechanism Synthesis for Energy and Torque Specifications. MS Thesis, University of Missouri - Rolla (2003)

9 Understanding Speed and Force Ratios for Compliant Mechanisms

Thomas G. Sugar and Matthew Holgate

Abstract. Active, compliant mechanisms with powered joints and compliant, spring-based links are beneficial at reducing input loads and power requirements by changing both the speed and force ratios. In these mechanisms, the speed and force ratios are a function of mechanism geometry and the load applied to the output link. Both ratios are analysed for a classic slider crank linkage.

1 Introduction

In the Human Machine Integration Laboratory, we have been studying compliant, spring-based mechanisms to develop wearable, powered, robotic orthoses, prostheses, and exoskeletons [1-6]. Others have been studying compliant actuators for safe human interaction [7-9]. Our goal is to use mechanism geometry with the addition of springs to reduce both the load and power requirements, and alter the input speed for an actuator, typically a high speed, low torque electric motor. Human joints typically require low speed and high torque which necessitates a gearbox ratio that decreases motor speed, and increases output torque with a fixed ratio.

Instead of a fixed ratio, smart design allows speed and force ratios to change based on mechanism geometry and required output load. Benefits will include high force ratios when high load is required and low force ratios when low load is required. Also, the speed ratio can be changed when a spring stores or releases energy into the system.

We develop a general method for understanding the force and speed ratios for a classic slider crank linkage used in a powered, robotic ankle. This method helps to illustrate how both the speed and force ratios can be changed using linkage design.

Thomas G. Sugar · Matthew Holgate
Arizona State University and SpringActive, Inc.
e-mail: Thomas.Sugar@asu.edu

V. Kumar et al. (Eds.): *Adv. in Mech., Rob. & Des. Educ. & Res.*, MMS 14, pp. 117–129.
DOI: 10.1007/978-3-319-00398-6_9   © Springer International Publishing Switzerland 2013

2   Developing Speed and Force Ratios

The slider crank linkage in Figure 1 is analysed using standard loop closure, but in this case the load, M_2, defines the spring force, F_3, in the connecting link, l_3. The input force, F, and position, x, define the actuator force and position. We are assuming a standard linear actuator such as a ball screw is used as the input. Because the moment, M_2, alters the link length, l_3, and the angle, θ_3, analytical solutions to the system of equations are difficult and a numerical solution is used. Six equations and six unknowns are used to solve for l_3, θ_3, x, Δ, F_3, and F. The deflection of the spring is defined by Δ. The output moment and position, M_2 and θ_2, initial length for the third link, l_o, stiffness, K, and length of the second link, l_2, are all given. The six equations are found using a loop closure and force analysis.

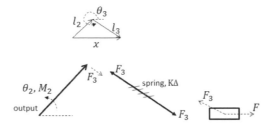

Fig. 1 A slider crank linkage is defined with a spring that allows the length of the connecting link to vary based on the output load, M_2

$$l_2 \cos(\theta_2) + l_3 \cos(\theta_3) = x$$

$$l_2 \sin(\theta_2) + l_3 \sin(\theta_3) = 0$$

$$l_3 = l_o + \Delta$$

$$F_3 = K\Delta$$

$$M_2 = Fl_2(\sin(\theta_2) - \cos(\theta_2)\tan(\theta_3))$$

$$F = F_3\cos(\theta_3)$$

$$l_2 = 0.04, l_o = 0.06$$

In a rigid slider crank linkage, the speed ratio and force ratios are a function of geometry only, and they are equal and opposite in sign. In this analysis, the speed ratio is $\frac{\dot{x}}{\dot{\theta}_2}$ with units of length and the force ratio is $\frac{M_2}{F}$ with units of length. These ratios change based on mechanism geometry and can be thought of as variable lever arms. If the pitch of the ball screw is given, then the lever arm lengths can be converted to dimensionless numbers and thought of as variable gear ratios.

$$\left(\frac{M_2}{F}\right) = -\left(\frac{\dot{x}}{\dot{\theta}_2}\right) = l_2(\sin(\theta_2) - \cos(\theta_2)\tan(\theta_3))$$

$$M_2 \dot{\theta}_2 + F\dot{x} = 0$$

In a compliant, spring-based linkage, the force ratio is a function of geometry and load only. The angle, θ_3, changes based on load. The speed ratio cannot be determined because the spring power, or $\dot{l}_3$ is not known. For a compliant linkage:

$$M_2 \dot{\theta}_2 + F\dot{x} - \frac{F_3 \dot{F}_3}{K} = 0$$

$$\left(\frac{\dot{x}}{\dot{\theta}_2}\right) + \left(\frac{M_2}{F}\right) - \frac{F_3}{F}\left(\frac{\dot{l}_3}{\dot{\theta}_2}\right) = 0$$

If the output load, M_2, does not vary with time, then $\dot{l}_3$ equals zero and the speed ratio can be found. It is equal in magnitude and opposite in sign to the force ratio. For the general case if $\frac{\dot{M}_2}{\dot{\theta}_2}$ is given, which is reasonable because the output load and positions are given with time, then the speed ratio can be found. For the general case, the force ratio is a function of the geometry of the linkage and the desired output load. The speed ratio is a function of the geometry, desired output load, and the rate of change of the output load. Three independent equations are found to solve for the ratios:

$$-l_2 \sin(\theta_2)\dot{\theta}_2 - l_3 \sin(\theta_3)\dot{\theta}_3 + \dot{l}_3 \cos(\theta_3) = \dot{x}$$

$$l_2 \cos(\theta_2)\dot{\theta}_2 + l_3 \cos(\theta_3)\dot{\theta}_3 + \dot{l}_3 \sin(\theta_3) = 0$$

$$\dot{M}_2 = K\, l_2 l_3 (\sin(\theta_2)\cos(\theta_3) - \cos(\theta_2)\sin(\theta_3))$$
$$+ K l_2 \Delta \dot{\theta}_3 \left(-\sin(\theta_2)\sin(\theta_3) + \cos(\theta_2)\sin(\theta_3)\tan(\theta_3)\right.$$
$$\left. - \frac{\cos(\theta_2)}{\cos(\theta_3)}\right) + K l_2 \Delta \dot{\theta}_2 (\cos(\theta_2)\cos(\theta_3) + \sin(\theta_2)\sin(\theta_3))$$

$$\begin{bmatrix} a & b & c \\ \cos(\theta_3) & -1 & -l_3\sin(\theta_3) \\ \sin(\theta_3) & 0 & l_3\cos(\theta_3) \end{bmatrix} \begin{pmatrix} \frac{\dot{l}_3}{\dot{\theta}_2} \\ \frac{\dot{x}}{\dot{\theta}_2} \\ \frac{\dot{\theta}_3}{\dot{\theta}_2} \end{pmatrix} = \begin{pmatrix} d \\ l_2\sin(\theta_2) \\ -l_2\cos(\theta_2) \end{pmatrix}$$

$$a = K\,(\sin(\theta_2)\cos(\theta_3) - \cos(\theta_2)\sin(\theta_3))$$

$$b = 0$$

$$c = K\Delta\left(-\sin(\theta_2)\sin(\theta_3) + \cos(\theta_2)\sin(\theta_3)\tan(\theta_3) - \frac{\cos(\theta_2)}{\cos(\theta_3)}\right)$$

$$d = \frac{\dot{M}_2}{l_2\dot{\theta}_2} - K\Delta(\cos(\theta_2)\cos(\theta_3) + \sin(\theta_2)\sin(\theta_3))$$

These ratios are quite useful because they define the speed ratio as well as power:

$$output\ power\ =\ M_2\dot{\theta}_2\ =\ \left(\frac{M_2}{F}\right)*(F\ \dot{\theta}_2)$$

$$input\ power\ =\ F\dot{x}\ =\ \left(\frac{\dot{x}}{\dot{\theta}_2}\right)*(F\ \dot{\theta}_2)$$

$$spring\ power\ =\ \frac{F_3\dot{F}_3}{K}\ =\ \left(\frac{\dot{l}_3}{\dot{\theta}_2}\right)*(F_3\ \dot{\theta}_2)$$

3   Optimizing a Linkage for a Tensile Spring Load

In one example where the spring is extended, the output angle and load are defined in Figures 2 and 3. The positive output moment causes the spring to be stretched and the geometry to change. The initial starting position of the output angle can be varied drastically changing the results. The starting position was chosen so that speed and force ratios have beneficial properties. The rate of change of the load is used to develop a non-singular matrix to calculate the speed ratio.

By choosing a compliant mechanism, the x movement of the input actuator is altered, see Figure 4. For a stiff linkage, the input movement changes direction at 0.25s when the desired moment is large and increasing. The compliant mechanism moves in the same direction at 0.25s. Experimentally, it is difficult for a motor to change direction under load.

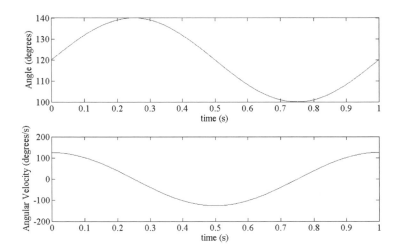

Fig. 2 The desired output angle and angular velocity are given for the problem

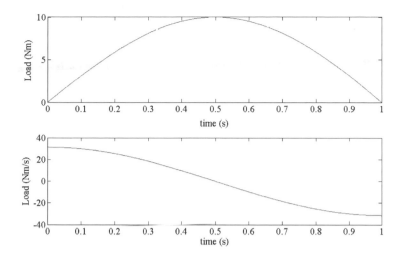

Fig. 3 The desired output load, a moment, and rate of change of the load are given for the problem

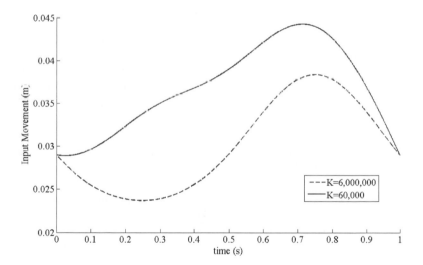

Fig. 4 The movement of the input actuator is altered for a compliant mechanism

The input force is reduced by 17.6% for the compliant mechanism (582.6 N vs 707.0N), see Figure 5.

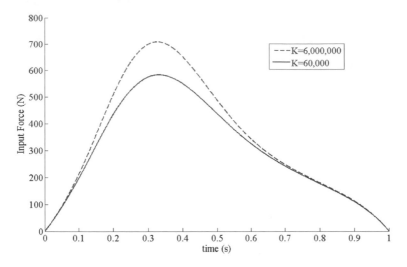

Fig. 5 The input force is reduced using a compliant spring

The power profiles for the stiff and rigid examples are shown in Figure 6. The output power for the compliant and stiff simulations is the same, but the peak input power is lowered for the compliant example. The spring stores energy in the beginning of the cycle and releases it in the middle of the cycle. In these problems, the input power plus the output power minus the spring power equals 0. The output power is considered a load and usually has an opposite sign compared to the input power (typically true for a rigid system unless overrunning occurs).

The load ratio during the cycle is shown in Figure 7. The load ratio varies based on the geometry of the linkage. For the rigid system, it varies between 0.01 and 0.035 m. The load ratio can be thought of as a gear ratio by multiplying it by 2π and dividing by the pitch, for example 1mm. In this case, the gear ratio would vary between 68.8 and 219.9. It is increased for the compliant mechanism at 0.5s which is beneficial. For the rigid system, the load ratio is symmetric over the cycle shown in the bottom graph of Figure 7. For the compliant case, the fluctuation is not symmetric during the cycle. The spring is being stretched during the first half of the cycle ($\dot{M}_2$ is positive) and is collapsing during the second half of the cycle, ($\dot{M}_2$ is negative).

The speed ratio is shown in Figure 8 and is defined as the input movement divided by the output angular speed. For the rigid system, the ratio hovers around (-0.02) with jumps at 0.5s and 0.75s when the output speed is zero. For a rigid system, as x moves to the right, the output angle rotates clockwise necessitating a negative sign. For the compliant case, the speed ratio varies between negative and positive numbers. For the compliant system, x can move to the right and the output angle can rotate counterclockwise when the spring is stretched allowing for a positive sign. For the compliant system the ratio varies between -0.06 and 0.06. As before, for the compliant system, the speed ratio is not symmetric over the cycle as shown in the bottom graph of Figure 8.

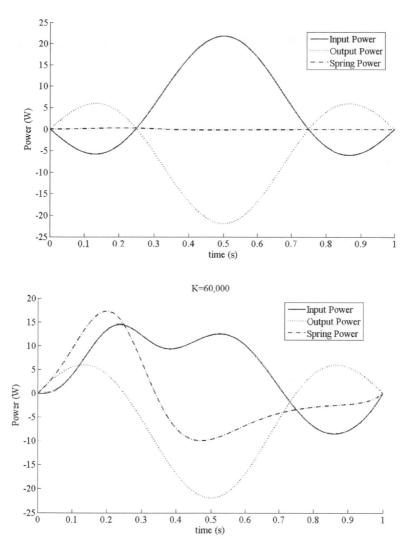

Fig. 6 Power profiles for the rigid case, K= 6,000,000, in the top graph, and power profiles for the compliant case, K=60,000, in the bottom graph

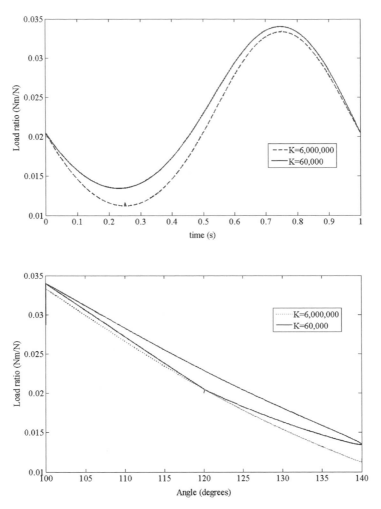

Fig. 7 The load ratio defines the desired output moment compared to the input force. It varies over the cycle due to the geometry and load. For the compliant case, the load ratio is increased at 0.5s.

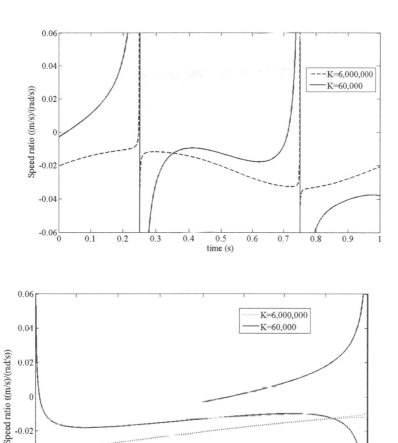

Fig. 8 The speed ratio is defined as the input speed divided by the output angular speed. For the compliant case, the ratio can be negative and positive.

4   Speed and Load Ratios over a Range of Values

The speed and load ratios are calculated over a range of output angles and output moments. A compliant K value of 150,000 N/m was chosen and different fixed values of $\frac{\dot{M}_2}{\dot{\theta}_2}$ are used. High values for $\frac{\dot{M}_2}{\dot{\theta}_2}$ caused the most change in the surface plots. In Figure 9, with a high rate of change in the load, the speed ratio can become positive as well as negative.

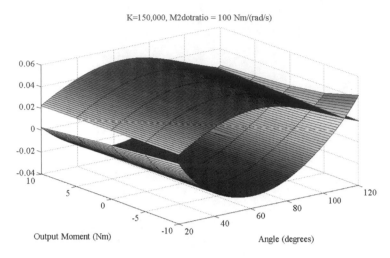

Fig. 9 The load ratio is the surface plot on the top and the speed ratio is the surface on the bottom. In this case, it can vary between -0.03 and 0.04.

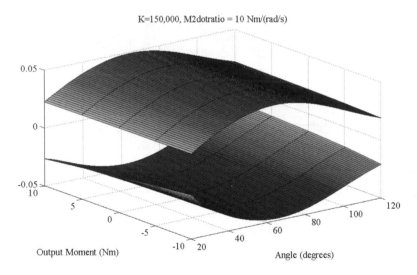

Fig. 10 The load ratio is the surface plot on the top and the speed ratio is the surface on the bottom. In this case, it can vary between -0.05 and -0.1.

In Figures 10 and 11, the rate of change of the load versus speed is reduced to 10 and 0. As the rate of change of the load is reduced the spring power is reduced and the speed ratio becomes a mirror of the load ratio.

K=150,000, M2dotratio = 0 Nm/(rad/s)

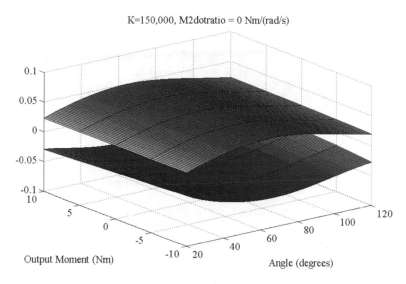

Fig. 11 The load ratio is the surface plot on the top and the speed ratio is the surface on the bottom. In this case, it can vary between -0.05 and -0.02.

In Figures 12 and 13, the speed ratio becomes more negative at large output angles. This slider crank linkage has beneficial properties when the spring is primarily used in tension but the properties are not beneficial when the spring is primarily used in compression.

K=150,000, M2dotratio = -10 Nm/(rad/s)

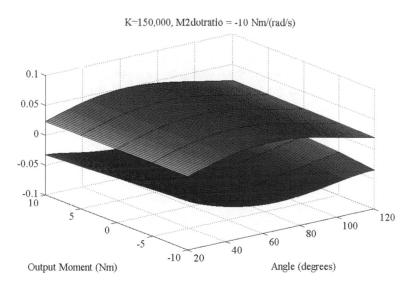

Fig. 12 The load ratio is the surface plot on the top and the speed ratio is the surface on the bottom. In this case, it can be as low as -0.053.

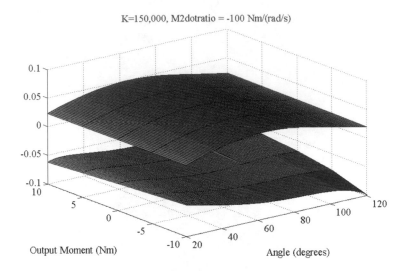

Fig. 13 The load ratio is the surface plot on the top and the speed ratio is the surface on the bottom. In this case, the speed ratio can be as low as -0.097.

5   Conclusions

Our goal is to develop speed and force ratios that change based on mechanism geometry and required output load. In the example in Section 3, benefits include high force ratios when high load is required and low force ratios when low load is required. Essentially, the "gear ratio" varies based on load. Also, the speed ratio can be changed to both negative and positive when a spring stores or releases energy into the system.

A general method for understanding the force and speed ratios for a classic slider crank linkage is shown. If the rate of change of the output load is defined, the speed ratio for a compliant linkage can be calculated. Surface plots are given to better understand the ratios for a range of angles and loads. The surface plots help to illustrate how both the speed and force ratios can be changed using linkage design. Our goal is to develop custom linkages for the ankle, knee, and elbows which require a limited range of output motion, but need high output forces in some scenarios while still requiring high speed with low output forces.

References

1. Hollander, K.W., Ilg, R., Sugar, T.G.: Design of the Robotic Tendon. In: Design of Medical Devices (2005)
2. Hitt, J., Sugar, T., Holgate, M., Bellman, R., Hollander, K.: Robotic transtibial prosthesis with biomechanical energy regeneration. Industrial Robot: An International Journal 36, 441–447 (2009)

3. Ham, R., Sugar, T., Vanderborght, B., Hollander, K., Lefeber, D.: Compliant actuator designs. IEEE Robotics & Automation Magazine 16, 81–94 (2009)
4. Hitt, J.K., Sugar, T.G., Holgate, M., Bellman, R.: An Active Foot-Ankle Prosthesis With Biomechanical Energy Regeneration. Journal of Medical Devices 4 (2010)
5. Ward, J., Sugar, T., Standeven, J., Engsberg, J.R.: Stroke survivor gait adaptation and performance after training on a Powered Ankle Foot Orthosis. In: 2010 IEEE International Conference on Robotics and Automation (ICRA), pp. 211–216 (2010)
6. Ward, J., Sugar, T.G., Hollander, K.W.: Optimizing the Translational Potential Energy of Springs for Prosthetic Systems. In: MSC 2011 (2011)
7. Bicchi, A., Tonietti, G., Schiavi, R.: Safe and fast actuators for machines interacting with humans. In: First IEEE Technical Exhibition Based Conference on Robotics and Automation, TExCRA 2004, pp. 17–18 (2004)
8. Tonietti, G., Schiavi, R., Bicchi, A.: Design and Control of a Variable Stiffness Actuator for Safe and Fast Physical Human/Robot Interaction. In: Proceedings of the 2005 IEEE International Conference on Robotics and Automation, ICRA 2005, pp. 526–531 (2005)
9. Schiavi, R., Bicchi, A., Flacco, F.: Integration of active and passive compliance control for safe human-robot coexistence. In: IEEE International Conference on Robotics and Automation, ICRA 2009, pp. 259–264 (2009)

10 Precision Flexure Mechanisms in High Speed Nanopatterning Systems

M.J. Meissl, B.J. Choi, and S.V. Sreenivasan

Abstract. Semiconductor wafer nanolithography machines integrate nano-scale precision mechanisms with advanced optical and mechatronic modules to achieve highly controlled nanopatterning capability for fabricating advanced integrated circuits, photonics, and optoelectronic devices. The machines achieve their precision by combining the use of highly repeatable nano-resolution actuators and motion systems, and on-tool precision calibration that is typically required for assembled systems since the sub-system machining precision is inherently insufficient.

In this article, precision flexure mechanisms used in advanced photolithography and imprint lithography systems are discussed and compared. The main difference between them is that nanoimprint lithography mechanisms need to possess high dynamic load carrying capability in addition to precision performance. The article first provides a brief review of quasi-static precision mechanisms in advanced optical instruments and tools. The article next discusses flexure mechanisms used for calibration and nano-precision real-time alignment for a UV nanoimprint process. These mechanisms require specialized designs as they need to support high dynamic loading encountered during the high-speed separation (demolding) process in imprint lithography. Finally, imprint template flexures – that have to satisfy very stringent precision and dynamic loading requirements – are described in detail. Specific design requirements of template flexures include (a) providing selectively compliant tilting about remote compliant centers, (b) possessing an order of magnitude higher stiffness in other axes, and (c) avoiding mechanical and thermal distortions from mounting of the template chuck to the supporting column.

Keywords: Precision Flexure Mechanics, Selective Compliance, Photolithography, Nanoimprint Lithography.

M.J. Meissl · B.J. Choi · S.V. Sreenivasan
Department of Mechanical Engineering, University of Texas,
Austin, TX, USA
e-mail: sv.sreeni@mail.utexas.edu

V. Kumar et al. (Eds.): *Adv. in Mech., Rob. & Des. Educ. & Res.*, MMS 14, pp. 131–144.
DOI: 10.1007/978-3-319-00398-6_10   © Springer International Publishing Switzerland 2013

1   Introduction

Flexure-based mechanisms are widely used to provide friction free motion with a typical motion range of a few microns to a few millimeters. Some of them are used for precision assembly purposes where fine calibration can be performed without wearing down mating surfaces. In general, precision calibrations need to be performed without altering surface flatness or other geometric specification. The benefit of particle free motion and good repeatability of flexures have been widely adapted for precision instruments and tools, including photolithography tools.

Two examples of quasi-static flexures used in precision systems are discussed first. A gimbal flexure is shown in Figure 1. The flexure between bodies 32 and 34 in Figure 1 enables relative tilting between two bodies while the center of the fiber (26) remains aligned with the focal point 48. In general, as shown here, the flexure unit is used for precision calibration as quasi static fixtures [1]. High compliance of the thin wall (50) provides two DOF gimbal motions while the remaining motions are constrained.

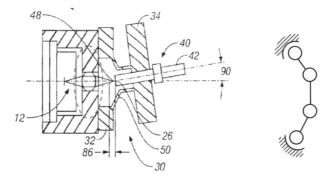

Fig. 1 Left: a static flexure unit to calibrate optical coupling (from [1]); Right: flexure portion within the dotted region can be represented by linkages and lumped revolute joints

Watson [2] presented a flexure apparatus for establishing motion about a center remote from the physical body of the device, known as a Remote Center Compliance (RCC) device. The device shown in Figure 2 is for part assembly tasks where initial tilting and location error with respect to a receiving part or body can be compensated by this flexure unit. In this design, vertical motion is structurally constrained.

One of the most widely used applications of flexures is in piezo based small motion range stages, where limited motion ranges of piezo stacks are significantly amplified via flexure mechanisms. Flexure units are also used as thermal-decoupling fixtures for precision optical units such as photolithography lens stacks or semiconductor mask stages.

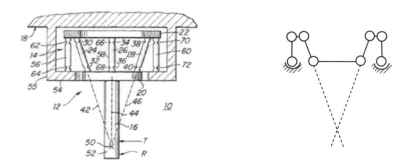

Fig. 2 Left: a flexure based RCC device (from [2]). Right: a simplified representation of the flexure unit as linkages and lumped joints.

For nanoimprint lithography tools, precision flexure units were developed to generate nano-scale precision parallelism between the template and substrate. Several articles have presented the design concept, analytical model, and experimental data of flexure based mechanisms for nanoimprint lithography [5, 6, 7] as well as micro-contact printing [8]. Reference [9] provides extensive design principles and sensitivity analyses of notch type flexures. Unlike quasi-static flexure mechanisms typically encountered in high precision instruments, template flexures for nanoimprint lithography undergo high dynamic loading. In addition to supporting high dynamic loading, specific design requirements of template flexures were (a) to provide compliant tilting about two remote compliant centers without generating particles, (b) to possess high stiffness in all other axes, (c) minimize mechanical distortion from mounting the template chuck to the supporting column, and (d) avoid thermal distortion to the template and template chuck. Three generations of template flexures were developed. The first generation was made of two separate four bar flexures that were serially assembled, while the second unit was a monolithic flexure system which still included a serial configuration of two four bar flexures. The third and last generation flexure is a monolithic parallel flexure. Next, we will present an overview of the nanoimprint lithography process, followed by specific designs of the template flexures along with various enhancements to achieve a near perfect parallel template-wafer configuration.

Figure 3 illustrates Jet and Flash Imprint Lithography (J-FIL) process steps [3, 4]. The glass template contains nano-scale resolution features. (a) Ideally, the gap and parallelism between the template and wafer are precisely controlled. (b) UV curable organic monomer material is delivered on the substrate by an array of piezo inkjet nozzles and filling is performed in the presence of in-liquid contact between the template and substrate. (c) UV exposure crosslinks the monomer leading to a solidified polymer layer. (d) The template is separated from the UV cured layer, and (e) the imprinted features are etched to transfer the pattern to the substrate.

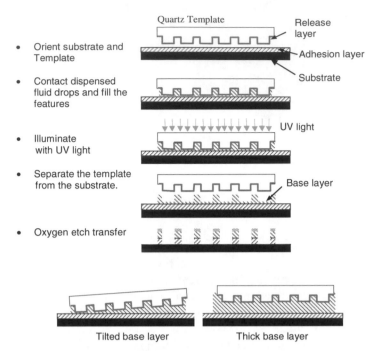

Fig. 3 Jet and Flash Imprint Lithography process steps; residual layer needs to be uniform and thin for etching process (from [5])

As illustrated in the second step in Figure 3, it is important to maintain near perfect parallelism between the template and wafer during the fluid filling step. Tilting error will cause an uneven imprinted layer, which causes a process problem at the etching step. A tilt error of even 2.5 nm across 25 mm (100 nrad) can cause undesirable process errors after etching. Further, an undesirably thick residual layer can form when the dispensed fluid volume is not optimized. For 25 nm lithography, and a 2:1 aspect ratio patterned polymer, the mean residual layer thickness needs to be < 15 nm.

Using best design and calibration practices in precision systems, a calibration between the template and wafer can achieve better than about 10 μrad (100 nm rise over 10 mm distance) using a high resolution gap sensor. However, it is necessary to compensate for the remaining tilting error to achieve tilting of 50 nrad or less during the fluid filling step.

Template flexure units were developed mainly to compensate the above mentioned residual tilting error *passively* in the presence of in-liquid contact between the template and wafer. Another important function of the template flexure is to either limit or eliminate the lateral motion when the tilting motion is performed. Tilting motion occurs not only in the fluid filling step but also in the separation step. In the presence of unpredictable localized tilting between the template and wafer, residual lateral motion errors can occur if the tilting correction motion is not fully decoupled from the lateral motion. During the separation step, the imprinted features are already cross-linked. Therefore, excessive lateral motion of the template during the separation step will cause catastrophic shear induced large area defects. In order to decouple the tilting from the later motion sequences, it is therefore necessary to precisely locate the tilting axes of the template at or very close to the template surface.

For the template flexure, not only does the center of tilting need to be at the template surface, but compliances also need to be selectively controlled. Since flexure units need to handle frequent high vertical loads during the separation step, it is important to minimize the vertical compliance, in the direction of separation. Compliance in both lateral directions and about the vertical axis (theta) can lead to limited pattern placement accuracy. Thus, compliances for the three translation directions and about the theta direction need to be substantially smaller than those about the horizontal tilting directions.

2   First Generation Template Flexure

Templates used for early development were made of UV transparent 25 mm square blocks diced from a 6025 semiconductor mask substrate. A square field boundary was defined by the diced edges of the template. Two serially assembled flexure units provided tilting alignment between the template and substrate about two orthogonal directions. A seating area for the template was integrated into the lower flexure unit. Figure 4 shows a template, template flexure assembly, and a calibration stage [10]. The template was mechanically secured using set screws and an intermediate adaptor (not shown in Figure 4) that distributed point-loading from the set screws into surface loading on the template. Each flexure unit is a single degree of freedom four-bar-linkage with four notch-type joints. Linkage angles were selected so that the rotation axis is aligned with the template surface. Joint compliances were balanced so that stored energy at each joint is equally distributed. The geometry of the semi-circular notches was designed so that when a 20 N load is applied at a distance of 9 mm from the center, one flexure rotates 0.5 mrad while the five remaining motion directions are constrained by high structural stiffness of the flexure. The second, orthogonally mounted flexure, due to geometric constraints, rotated 0.9 mrad for the same moment load, providing the second tilt correction. It was found based on finite element analyses that the compliance in the undesirable directions increases less than in the desired tilt direction when the notch flexure's minimum thickness is reduced. A minimum fabrication thickness of 0.5 mm was selected, considering the increased importance of available machining tolerances on small dimensions.

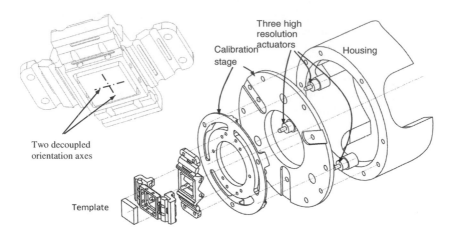

Fig. 4 Serially assembled first generation flexures holding a 25 mm square template (from [10])

3   Second Generation Template Flexure

In the 1st Gen template flexure where a 25 mm square template is seated into the lower flexure by compression loads, an undesirable nano-scale flatness error was induced to the template active area. Another noticeable problem was a poor field boundary since the boundary was defined by diced edges of the template. Therefore, a new template format was adopted, consisting of 65 mm square templates cut from 6025 semiconductor mask substrate. High resolution patterns were formed in the middle of the template, with the pattern boundary defined by a wet-etched mesa. Figure 5 shows this 65 mm square format template with an active area within the mesa.

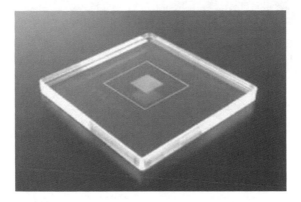

Fig. 5 65 mm format nano imprint template with an active area within the mesa (from [11])

The larger size of the template enabled its backside to be vacuum mounted onto a template chuck, eliminating the flatness distortion from compression loading. The template chuck was polished to be optically flat (to maintain the vacuum-chucked templates flat) and rigidly connected to the underside of the 2nd Gen template flexure. The template flexure was a monolithic part where the two serially stacked tilting flexures units were made of a single block of aluminum alloy or stainless steel using electric discharge machining (EDM). A benefit of the monolithic design is that the flexure joints can be machined very accurately relative to one another, so that the tilt axes can be placed at the template surface with higher precision and minimal offset. The mating surface of the flexure with the template chuck was also polished flat to maintain the template chuck flatness. The center of the flexure unit was machined out in order to provide a UV exposure path and also to house alignment microscopes for registering multiple levels of an integrated nanodevice. Figure 6 shows an assembly of the template flexure, template chuck and template.

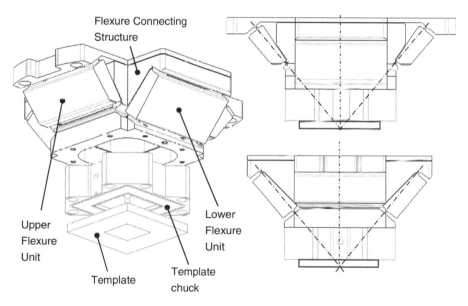

Fig. 6 Second generation template flexure and template chuck assembly with a 65 mm square template

A major objective of the 2nd generation template flexure was to further decrease the compliance in the undesirable motion directions as compared to the tilt directions. It was recognized that both deformation of the structure connecting the flexure joints (the links), as well as localized deformation of the joints themselves in undesirable directions (other than the desired joint rotation) was contributing to this compliance. By placing the four-bar linkages' upper joints further upwards, which increases the flexure's height, it is possible to obtain the same template rotation with less rotation of each joint. Therefore, the joints' rotation stiffness can be increased, which in turn stiffens them in undesirable directions. Specifically, the width of the joints was increased and the flexure notches' radius decreased, both of which lead to reduced overall lateral compliance at the template surface. It was also determined that the angle of the side links affects vertical compliance. The 1st generation template flexure's side links were angled at $25°$ and $16°$ for the upper and lower four-bar, respectively. This angle was increased to $50°$ in both four-bars of the 2nd Gen template flexure, which caused the side links to be placed more in compression and less in bending for vertical loads, which results in less vertical compliance of the flexure. More than 20x improvement was achieved in the vertical direction and 60% improvement in the lateral directions by optimizing the joint geometry, placement, and stiffening of the linkages as discussed above. The design was also optimized to provide the same compliance about both tilt axes. These geometry changes made this flexure much less compact, and as a result overall mass of the template flexure was increased significantly as compared to the 1st generation flexure. A significant portion of the mass increase can be attributed to the connecting structure between the two flexure units.

4   Third Generation Template Flexure

As an enhancement of the imprint tool, the slow ball screw driven imprint head was replaced with a three voice-coil driven imprint head. In order to improve the control bandwidth of the imprint head, the inertia of its moving components had to be reduced significantly, including the template flexure. However, compliance and tilting characteristics of the template flexure had to remain similar to the second generation, despite a significant decrease in mass.

The 3rd generation template flexure is also a monolithic unit with two orthogonal four-bar flexures, but rather than serial assembly of these four-bars, the linkages are now connected in parallel. It consists of four arms connecting to top and bottom plates. The top plate is connecting the flexure to the imprint head, while the bottom plate is connecting the flexure to the back side of the template chuck. Each arm has two sets of flexure joints where axes of one set of joints are orthogonal to those of the other set of joints. Figure 7 shows a 3D model of the template flexure assembly along with two projection views.

This new, parallel flexure design eliminated the need for a stiff, heavy connecting structure between the previously serially connected four-bar. A total mass reduction of 80% was achieved for the 3rd generation flexure as compared to the prior generation. A thinner template chuck design, fabricated from stiff silicon carbide ceramic, allowed the lower joints of the four-bar linkages to be placed closer to the template. This causes less joint rotation for a given template tilt rotation, akin to placing the upper joints higher, thereby creating an opportunity to further increase flexure joint stiffness. Instead of designing all joints to store equal energy, which requires different joint stiffnesses, the joints in each four-bar linkage of the 3rd generation template flexure were initially designed with equal geometry. This allowed each joint to be fabricated to the minimum manufacturable thickness while also maximizing its width. The minimum joint thickness was lowered from 0.5 mm in the 2nd generation template flexure to 0.4 mm for the 3rd generation, due to improved EDM manufacturing techniques. The design was ultimately revised to shorten the width of the lower joints using a minimum stress criterion and finite element analysis (FEA) in order to provide space for a microscope imaging the template from the top; this change was found to increase the tilt, vertical, and lateral compliances approximately proportionally. An interesting benefit of the parallel flexure design is that the upper joints are effectively separated into two halves spaced far apart at the corners of the template flexure. This leads to a significant compliance reduction in the presence of lateral loads on the template surface, since they create a moment about the upper flexure joints which is much better supported along the joint axes with the spaced apart joint halves. Lastly, some reduction of undesirable compliance was also achieved simply because the load path between the template and imprint head is shorter in the parallel flexure design compared to a serial flexure design. The combined result of these measures is a further improvement of the compliance ratio between the desired (tilt) and undesired motion directions. Tilt compliance was increased 50% while a further reduction of lateral compliance by 50% was achieved as compared to the 2nd generation template flexure, despite the significant removal of mass.

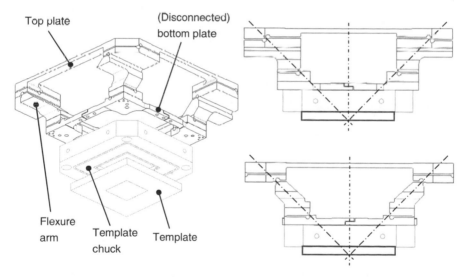

Fig. 7 Third generation template flexure assembly and its two projection views, showing RCCs on the template surface

In order to maintain the template surface flat, it is important to maintain the flatness of the template chuck both during assembly and tool operation. As previously mentioned, the chuck is typically made of silicon carbide material and the template flexure is made of aluminum alloy. When these two units are mated during assembly, it is important to minimize the distortion to the template chuck. Distortion can be caused not only by the fabrication errors in interfacing surfaces but also by thermal expansion differences. At the early development of this flexure system, we noticed that the flatness of the template chuck did not remain consistent. At that point, the bottom portion of the flexure that is interfacing with the chuck was a connected plate, which was necessary to maintain the flexure's structural integrity during machining. In order to eliminate thermal distortion of the flexure-chuck assembly, the design was improved by making additional cuts at the end of fabrication. The bottom plate was separated between each arm as shown in Figure 7 and simultaneously clamped to preserve the precise position of each flexure arm (which determines the spatial location of the tilt axes). At assembly, each disconnected bottom portion was secured to the back of the template chuck and the clamps removed. The template chuck itself then acted as the connecting portion; the precision of the flexure joint positions and overall stiffness characteristics of the flexure were retained. Figure 8 illustrates the thermal distortion reduction of the chucking region when the template chuck is attached to the disconnected flexure base. The width of each fringe in this finite element result is 10 nm wide in both cases shown. The flatness error was reduced by 10x when the flexure bottom was disconnected.

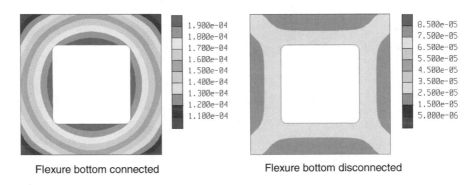

Flexure bottom connected   Flexure bottom disconnected

Fig. 8 FEA result of the template chuck flatness error before and after disconnecting the flexure bottom plate

The voice coil imprint head exhibited improved calibration between template and wafer in conjunction with non-contact sensing of template and wafer surface profiles. The resulting smaller tilt range required of the 3rd generation flexure also allowed the joints to be optimized to handle the expected tension/compression and shear loads as compared to the 2nd generation design. This, however, required a safety limit to prevent the joints from bending excessively in the presence of unexpected high tilting loading. Otherwise, the notch joint will fail due to excessive bending-induced stress. Figure 9 shows the integrated safety limits in this flexure.

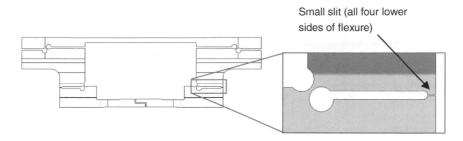

Fig. 9 Safety limit to prevent joints from bending excessively

5   Imprint Samples

As emphasized in Section 1, it is important to generate imprint fields with uniform residual layer, as measured by the residual layer thickness (RLT). A precision drop on demand dispensing scheme that provides highly controlled picoliter volume monomer drops is a key enabler of uniform RLT. However, as discussed earlier, it is also important to maintain the parallelism between the template and wafer as fluid drops fills the gap. Figure 10 shows three images of fluid spreading steps where the spreading front forms well controlled symmetric geometries. In this case, the template was gently deformed to form a micron-scale convex shape to obtain a central initial liquid contact between template and substrate, followed by an expanding symmetric contact that finally results in a fully conformed template-substrate sandwich encapsulating a uniform liquid RLT. Any tilt misalignment between template and wafer — even at the ~100 nrad level – is passively compensated by the liquid induced deformation of the template flexure system resulting in the symmetric fluid spreading characteristics shown in Figure 10.

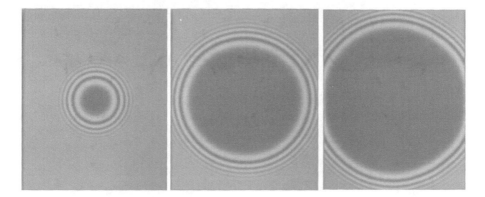

Fig. 10 Fluid spreading with symmetric fluid front

Figure 11 shows cross-section scanning electron micrographs (SEMs) of imprinted layers. RLT was measured at the edges and center of fields showing that the desired uniform RLT could be achieved.

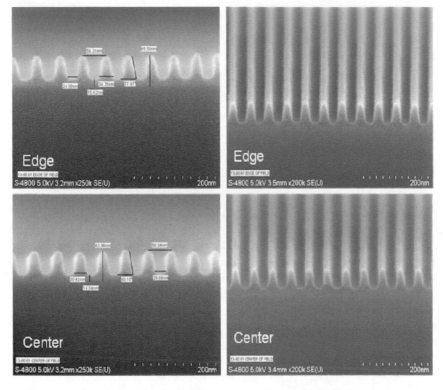

Fig. 11 Cross-section SEM of imprinted layer showing uniform RLT at edges and the center of field

The J-FIL process has been used to develop a lithography stepper technology that can process 300 mm silicon wafers [12, 13]. Recently, the J-FIL process capability has been expanded to successfully generate imprinted 450 mm silicon wafers. Figure 12 shows a fully patterned 450 mm wafer made by J-FIL technology demonstrating 26 nm half-pitch structures. This is the first demonstration of any lithography technology on 450 mm wafers.

6   Summary

Three generations of template flexures have been developed for nanoimprint lithography. The template flexure units resemble four-bar linkage joints with a remote compliance center precisely placed at the template active area surface. The flexures tilt as the template and wafer come into contact, aligning the template to the wafer. The tilting motion in this flexure design is decoupled from lateral motion, which is critical during the separation step to avoid pattern shear defects.

Unlike quasi-static applications, template flexures in imprint lithography are subjected to high dynamic loading from separation. To mitigate these loads as well as disturbance forces in other directions, template flexures must have high selective compliance. The compliance about the desired two tilt axes must be at least an order of magnitude higher than that in the remaining four degrees of freedom.

The first generation template flexure was a very compact, lightweight design consisting of two stacked parts that provided tilt about two orthogonal axes. The second generation template flexure retained the stacked, serial flexure design but both flexure units were integrated into one monolithic body. This template flexure also had much higher vertical stiffness for higher dynamic loading and reduced lateral compliance for improved template position stability in the presence of in-liquid disturbance forces. Finally, the third generation template flexure had two four-bar linkage flexure units in a novel parallel monolithic configuration. The joint thicknesses were optimized and safety stops added. This template flexure also has a disconnecting feature that significantly reduces thermal distortion of the template chuck, enabling templates to remain flat in the presence of temperature fluctuations. The parallel template flexure retained the high vertical stiffness of the prior generation, and a further reduction in lateral compliance was achieved. A significant weight reduction was obtained with the new geometry, which resulted in better dynamic performance.

Images of the spreading fluid front and cross-section SEMs were included here to illustrate the verification of the desirable passive (in-liquid) alignment behavior of the flexures. The fluid fronts exhibited symmetric spreading, and the SEMs revealed a uniform residual layer thickness both at the center and at the edges.

Fig. 12 Patterned 450 mm wafer generated with nanoimprint lithography

References

[1] US Patent 6925234, Flexure apparatus and method for achieving efficient optical coupling
[2] US Patent 4098001, Remote center compliance system
[3] Sreenivasan, S.V.: Nanoscale manufacturing enabled by imprint lithography. MRS Bulletin 33, 854–863 (2008)
[4] Colburn, M., et al.: Step and Flash Imprint Lithography: A Novel Approach to Imprint Lithography. In: SPIE's 24th Annual International Symposium on Microlithography, Santa Clara, CA (1999)
[5] Choi, B.J., et al.: Design of orientation stages for step and flash imprint lithography. Precision Engineering Journal of the International Societies for Precision Engineering and Nanotechnology 25, 192–199 (2001)
[6] Choi, K.B., Lee, J.J.: Passive compliant wafer stage for single-step nano-imprint lithography. Review of Scientific Instruments 76(7), Art. No. 075106 (July 2005)
[7] Shilpiekandula, V., Youcef-Toumi, K.: Characterization of Dynamic Behavior of Flexure-based Mechanisms for Precision Angular Alignment. In: 2008 American Control Conference, Seattle, Washington, USA, June 11-13 (2008)
[8] Kendale, A.M.: Automation of soft lithographic microcontact printing. SM thesis, Department of Mechanical Engineering, Massachusetts Institute of Technology, Cambridge MA (2002)
[9] Smith, S.T.: Flexures: elements of elastic mechanisms. Gordon and Breach, Amsterdam (2000)
[10] US Patent 6922906, Apparatus to orientate a body with respect to a surface
[11] Hiraka, T., et al.: Progress of UV-NIL template making. Proc. SPIE 7379, 73792S (2009)
[12] Higashiki, T., Nakasugi, T., Yoneda, I.: Nanoimprint lithography for semiconductor devices and future patterning innovation. Proc. SPIE 7970, 797003 (2011)
[13] Sreenivasan, S.V., et al.: Status of UV Imprint Lithography for Nanoscale Manufacturing. Comprehensive Nanoscience and Technology 4, 83–116 (2011)

11 Decomposition of Planar Burmester Problems Using Kinematic Mapping

Q.J. Ge, Ping Zhao, and Anurag Purwar

Abstract. In this paper, we revisit the classical Burmester problem of the exact synthesis of a planar four-bar mechanism with up to five task positions. Instead of assuming the joint type (revolute or prismatic) a priori, we seek to extract both the dimensions and joint types of a four-bar linkage from the given tasks. Kinematic mapping of plane kinematics has been used to formulate the Burmester problem as a manifold fitting problem in the image space. Instead of finding the design parameters of planar dyads directly, this paper seeks to determine a set of eight homogeneous coefficients for the constraint manifold in the null space associated with the five given tasks. Two additional constraints on these coefficients are then applied to finalize the synthesis process. The result is a novel algorithm that is simple and efficient and allows for task driven design of four-bar linkages with revolute and prismatic joints.

1 Introduction

The exact synthesis of a planar four-bar linkage such that its coupler link guides through up to five specified task positions is a classical problem known as the Burmester Problem. The five-position synthesis problem is typically solved by finding the intersections, via geometric or algebraic means, of two cubic curves known as the circle-centerpoint curves that are obtained from four-position problem. Kinematic mapping [1] has shown to be a very useful tool for linkage synthesis. While earlier work in this area [2, 3, 4, 5, 6] employed optimization based methods to obtain numerical solutions, Hayes et al. [7] sought to use polynomial methods to solve five quadratic design equations directly. Recently, unified methods for five

Q.J. Ge · Ping Zhao · Anurag Purwar
Computational Design Kinematics Lab,
Department of Mechanical Engineering, Stony Brook University,
Stony Brook, New York, 11794-2300
e-mail: Qiaode.Ge@stonybrook.edu

V. Kumar et al. (Eds.): *Adv. in Mech., Rob. & Des. Educ. & Res.*, MMS 14, pp. 145–157.
DOI: 10.1007/978-3-319-00398-6_11   © Springer International Publishing Switzerland 2013

positions synthesis have been developed that can handle four-bar linkages with not only revolute joints but also prismatic joints [8, 9, 11, 12]. Instead of finding the design parameters directly, this paper determines first the coefficients of the constraint manifolds in the image space of the mapping. This leads to a novel algorithm that decomposes the Burmester problem into a task driven data fitting problem, which can be solved using an SVD algorithm, and a dyad constraint satisfaction problem, which is equivalent to solving a quartic equation with one unknown. In addition, this formulation leads to truly task-driven, simultaneous joint-type and dimensional synthesis, without having to treat each case separately.

The organization of this paper is as follows. Section 2 reviews the concept of kinematic mapping in so far as necessary for the development of this paper. Section 3 and 4 present a unified representation of constraint manifolds of planar dyads including RR, PR, RP, PP. Section 5 outlines an novel algorithm for solving the five-position Burmester problem. Section 6 and 7 show how the same algorithm can be used for four- and three-position Burmester problems, respectively. Three examples are presented to illustrate the task-driven synthesis algorithm and how the prismatic joint can be obtained directly from the given tasks.

2   Kinematic Mapping in Plane Kinematics

Shown in Figure 1 is a planar displacement with translation parameters (d_1, d_2) and rotation angle ϕ. Let M denote a coordinate frame attached to the moving body and F be a fixed reference frame. Introduce the following kinematic mapping from Cartesian space parameters (d_1, d_2, ϕ) to Image Space coordinates $\mathbf{Z} = (Z_1, Z_2, Z_3, Z_4)$ (see [2, 4]),

$$Z_1 = \frac{1}{2}(d_1 \sin \frac{\phi}{2} - d_2 \cos \frac{\phi}{2}), \quad Z_2 = \frac{1}{2}(d_1 \cos \frac{\phi}{2} + d_2 \sin \frac{\phi}{2}), \tag{1}$$

$$Z_3 = \sin \frac{\phi}{2}, \quad Z_4 = \cos \frac{\phi}{2}.$$

In this way, a point in M in homogeneous coordinates $\mathbf{x} = (x_1, x_2, x_3)$ (with $x_3 \neq 0$) and its corresponding coordinates in F, $\mathbf{X} = (X_1, X_2, X_3)$ (with $X_3 \neq 0$), are related by the following homogeneous transform:

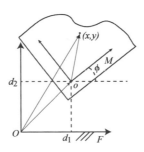

Fig. 1 A planar displacement of a moving frame M with respect to the fixed frame F

$$\mathbf{X} = [H]\mathbf{x}, \quad [H] = \begin{bmatrix} Z_4^2 - Z_3^2 & -2Z_3Z_4 & 2(Z_1Z_3 + Z_2Z_4) \\ 2Z_3Z_4 & Z_4^2 - Z_3^2 & 2(Z_2Z_3 - Z_1Z_4) \\ 0 & 0 & Z_3^2 + Z_4^2 \end{bmatrix}, \quad (2)$$

where $Z_3^2 + Z_4^2 = 1$. Similarly, for a line with homogeneous coordinates $\mathbf{l} = (l_1, l_2, l_3)$ in M and its corresponding coordinates $\mathbf{L} = (L_1, L_2, L_3)$ in F, we have

$$\mathbf{L} = [\overline{H}]\mathbf{l}, \quad [\overline{H}] = \begin{bmatrix} Z_4^2 - Z_3^2 & -2Z_3Z_4 & 0 \\ 2Z_3Z_4 & Z_4^2 - Z_3^2 & 0 \\ 2(Z_1Z_3 - Z_2Z_4) & 2(Z_2Z_3 + Z_1Z_4) & Z_3^2 + Z_4^2 \end{bmatrix}. \quad (3)$$

The four-dimensional coordinates $\mathbf{Z} = (Z_1, Z_2, Z_3, Z_4)$ are said to define a point in a projective three-space called the *Image Space* of planar displacement, denoted as Σ. In this way, a planar displacement is represented by a point in Σ; a single degree of freedom (DOF) motion is represented by a curve and a two DOF motion is represented by a surface in Σ [2].

3 Geometric Constraints of Planar Dyads

The goal of the classical Burmester problem is to find the geometric parameters of a four-bar linkage for a given set of task positions. This problem is commonly reduced to the exact synthesis of a planar dyad for up to five task positions. In this paper, we consider a dyad to include not just with revolute (R) joints but could also one or more prismatic (P) joints, i.e., RR, PR, RP and PP dyads (Figure 2). The end link of such a dyad is subject to one of the following four types of geometric constraints involving points, lines and circles:

1. For a RR dyad, one of its moving points stays on a circle;
2. For a PR dyad, one of its moving points stays on a line;
3. For a RP dyad, one of its moving lines stays tangent to a given circle, or equivalently when radius of the circle is zero, one of its moving lines passes through a fixed point;
4. For a PP dyad, one of its moving lines maintains a fixed angle with respect to a fixed line.

Let $\mathbf{a} = (a_1, a_2, a_0)$, where $a_0 \neq 0$, denote homogeneous coordinates of the center of a circle C in F. Then a point with homogeneous coordinates, $\mathbf{X} = (X_1, X_2, X_3)$, lies on C if

$$2a_1X_1 + 2a_2X_2 + a_3X_3 = a_0 \left(\frac{X_1^2 + X_2^2}{X_3} \right). \quad (4)$$

The radius r of the circle is given by

$$r^2 = (a_1/a_0)^2 + (a_2/a_0)^2 + a_3/a_0. \quad (5)$$

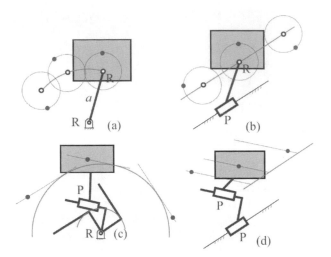

Fig. 2 Geometric constraints of planar dyads (a) *RR*, (b) *PR*, (c) *RP*, and (d) *PP*

When $a_0 = 0$, Eq.(4) becomes linear,

$$L_1X_1 + L_2X_2 + L_3X_3 = 0, \tag{6}$$

which represents a line with homogeneous coordinates $\mathbf{L} = (2a_1, 2a_2, a_3)$. Thus, Eq.(4) is a unified presentation for both a circle and a line, and therefore, could lead to a unified representation of the constraints of *RR* and *PR* dyads.

For a *RP* dyad, a line with homogeneous coordinates $\mathbf{L} = (L_1, L_2, L_3)$ passes through a fixed point $\mathbf{X} = (X_1, X_2, X_3)$. In other words, they also satisfy (6). Due to the duality between point and line in the projective plane, the point coordinates $\mathbf{X} = (X_1, X_2, X_3)$ may be considered as coordinates of a fixed line. Therefore, for the *PP* dyad, Eq. (6) can be reinterpreted as representing a moving line $\mathbf{L}$ that intersects with a fixed line $\mathbf{X}$ at a fixed angle.

Thus, we may conclude that all four dyadal constraints can be represented by Eq. (4) and that when $a_0 = 0$, the dyad has at least one prismatic joint.

4 Unified Representation of Constraint Manifolds

By substituting (2) into (4), Ge et al. [13] has shown that the constraint manifold of a *RR* dyad is the following quadric surface

$$p_1(Z_1^2 + Z_2^2) + p_2(Z_1Z_3 - Z_2Z_4) + p_3(Z_2Z_3 + Z_1Z_4) + p_4(Z_1Z_3 + Z_2Z_4)$$
$$+ p_5(Z_2Z_3 - Z_1Z_4) + p_6Z_3Z_4 + p_7(Z_3^2 - Z_4^2) + p_8(Z_3^2 + Z_4^2) = 0, \tag{7}$$

where the eight coefficients p_i are not independent but must satisfy

$$p_1 p_6 + p_2 p_5 - p_3 p_4 = 0, \quad 2p_1 p_7 - p_2 p_4 - p_3 p_5 = 0. \tag{8}$$

This is because p_i are related to the geometric parameters of the dyad by

$$
\begin{aligned}
&p_1 = -a_0, \quad p_2 = a_0 x \quad p_3 = a_0 y, \quad p_4 = a_1, \quad p_5 = a_2, \\
&p_6 = -a_1 y + a_2 x, \quad p_7 = -(a_1 x + a_2 y)/2, \quad p_8 = (a_3 - a_0(x^2 + y^2))/4,
\end{aligned}
\tag{9}
$$

where (a_0, a_1, a_2, a_3) are the homogeneous coordinates of the constraint circle and (x, y) are the coordinates of the circle point. For a PR dyad, we have $a_0 = 0$ and therefore, $p_1 = p_2 = p_3 = 0$. Eqns. (7) and (8) are said to define the constraint manifold of RR and PR dyads.

By substituting (3) into (4), it is found that for RP dyad, the constraint manifold has the same form as Eqns. (7) and (8), however we now have $p_1 = p_4 = p_5 = 0$. Similarly, for a PP dyad, we have $p_1 = p_2 = p_3 = p_4 = p_5 = 0$. Thus, all planar dyads can be represented in the same form by Eqns. (7) and (8), and we can determine the type of a planar dyad by looking at the zeros in the coefficients p_i.

Lastly, Eq.(9) may be inverted to obtain the coordinates of a circle (or a line), (a_0, a_1, a_2, a_3), as well as the circle point (x, y). Let, $u = p_4^2 + p_5^2$. For RR and PR dyads, we have $u \neq 0$ and

$$
\begin{aligned}
&a_0 : a_1 : a_2 : a_3 = -p_1 u : p_4 u : p_5 u : (4p_8 u - p_1(p_6^2 + 4p_7^2)), \\
&x : y : 1 = (p_6 p_5 - 2p_7 p_4) : -(p_6 p_4 + 2p_7 p_5) : u.
\end{aligned}
\tag{10}
$$

For an RP dyad, we have $u = 0$ and

$$
\begin{aligned}
&a_0 : a_1 : a_2 = (p_2^2 + p_3^2) : (-p_3 p_6 - 2p_2 p_7) : 2(p_2 p_6 - 2p_3 p_7), \\
&l_1 : l_2 : l_3 = p_2 : p_3 : 2p_8,
\end{aligned}
\tag{11}
$$

where $\mathbf{l} = (l_1, l_2, l_3)$ are the homogenous line coordinates of a line in M, which passes through a fixed point (a_1, a_2, a_0) in F.

In this paper, instead of seeking directly the dyad parameters (a_0, a_1, a_2, a_3) and (x, y), we first obtain the homogeneous coordinates p_i and then compute the dyad parameters using (10) or (11).

5 Five-Position Synthesis

Let $\mathbf{Z}_i = (Z_{i1}, Z_{i2}, Z_{i3}, Z_{i4})$ $(i = 1, 2, \ldots, 5)$ denote the image points associated with five specified task positions of a rigid body. Substituting them into (7) yields five linear equations in p_i. Assemble these equations in matrix form to obtain:

$$\begin{bmatrix} A_{11} & A_{12} & A_{13} & A_{14} & A_{15} & A_{16} & A_{17} & A_{18} \\ A_{21} & A_{22} & A_{23} & A_{24} & A_{25} & A_{26} & A_{27} & A_{28} \\ A_{31} & A_{32} & A_{33} & A_{34} & A_{35} & A_{36} & A_{37} & A_{38} \\ A_{41} & A_{42} & A_{43} & A_{44} & A_{45} & A_{46} & A_{47} & A_{48} \\ A_{51} & A_{52} & A_{53} & A_{54} & A_{55} & A_{56} & A_{57} & A_{58} \end{bmatrix} \mathbf{p} = 0 \qquad (12)$$

where $\mathbf{p}$ is a column vector with coordinates p_i ($i = 1, 2, \ldots, 8$) and

$$A_{i1} = Z_{i1}^2 + Z_{i2}^2, \;\; A_{i2} = Z_{i1}Z_{i3} - Z_{i2}Z_{i4}, \;\; A_{i3} = Z_{i2}Z_{i3} + Z_{i1}Z_{i4}, \;\; A_{i4} = Z_{i1}Z_{i3} + Z_{i2}Z_{i4},$$
$$A_{i5} = Z_{i2}Z_{i3} - Z_{i1}Z_{i4}, \;\; A_{i6} = Z_{i3}Z_{i4}, \;\; A_{i7} = Z_{i3}^2 - Z_{i4}^2, \;\; A_{i8} = Z_{i3}^2 + Z_{i4}^2.$$
$$\qquad (13)$$

define the elements of the 5×8 matrix $[A]$.

Instead of solving the five linear equations together with the two quadratic equations (8), we first compute the null space solution $\mathbf{p} = (p_1, p_2, \ldots, p_8)$ from the linear system (12) to obtain the *candidate* solutions for the five position Burmester problem. We then find such column vectors in the null space that satisfy (8). This effectively decomposes the Burmester problem into two much simpler subproblems. The null-space problem is linear and can be readily solved using the Singular Value Decomposition (SVD) method and the second subproblem can be reduced to a quartic equation with one unknown.

Since the rank of $[A]$ is five, the matrix $[A]^T[A]$ has three zero eigenvalues and the corresponding eigenvectors, $\mathbf{v}_\alpha$, $\mathbf{v}_\beta$ and $\mathbf{v}_\gamma$, define the basis for the null space. Let α, β, γ denote three real parameters. Then, any vector in the null space is given by

$$\mathbf{p} = \alpha \mathbf{v}_\alpha + \beta \mathbf{v}_\beta + \gamma \mathbf{v}_\gamma. \qquad (14)$$

For vector $\mathbf{p}$ to satisfy Eq. (8), we substitute (14) into (8) and obtain two homogeneous quadratic equations in (α, β, γ):

$$K_{10}\alpha^2 + K_{11}\beta^2 + K_{12}\alpha\beta + K_{13}\alpha\gamma + K_{14}\beta\gamma + K_{15}\gamma^2 = 0,$$
$$K_{20}\alpha^2 + K_{21}\beta^2 + K_{22}\alpha\beta + K_{23}\alpha\gamma + K_{24}\beta\gamma + K_{25}\gamma^2 = 0, \qquad (15)$$

where K_{ij} are defined by components of the three eigenvectors obtained from SVD algorithm. These two quadratic equations can be further reduced to a single quartic equation in one unknown in terms of the ratio of two of the three parameters (α, β, γ) and thus can be analytically solved. Since a quartic equation may have four real roots, two real roots or no real roots, there could be four solutions, two solutions, or no solutions for the coefficients $\mathbf{p}$ of the constraint manifold of planar dyads. As coefficients $\mathbf{p}$ are homogeneous, in this paper, we normalize them such that $\mathbf{p} \cdot \mathbf{p} = 1$.

Furthermore, by investigating whether the solution $\mathbf{p}$ falls into one of the following four patterns, we can determine the type of the resulting dyads:

1. if $p_1 = p_2 = p_3 = p_4 = p_5 = 0$, the resulting dyad is a PP dyad;
2. if $p_1 = p_2 = p_3 = 0$, the resulting dyad is a PR dyad;
3. if $p_1 = p_4 = p_5 = 0$, the resulting dyad is a RP dyad;
4. if none of the above, the resulting dyad is a RR dyad.

Table 1 Example 1: Five task positions

	1	2	3	4	5
d_1	3.6700	2.7965	2.5562	3.7451	4.5797
d_2	0.6457	1.5640	1.7066	1.1415	0.5694
ϕ	$-77.5362°$	$-56.6879°$	$-35.2713°$	$-38.6715°$	$-63.1693°$

Example 1: Now consider five task positions given in Table 1. The substitution of the data in the table into (1) yields five image points Z_i ($i = 1,2,3,4,5$), which are then substituted into (12) to obtain the matrix $[A]$. The application of SVD algorithm to $[A]$ yields the following eigenvalues:

$$0,\ 0,\ 0,\ 0.0038,\ 0.0407,\ 0.8890,\ 1.4832,\ 107.5677$$

as well as eight eigenvectors from the matrix $[A]^T[A]$. Listed in Table 2 are three of the eigenvectors associated with zero eigenvalues defining the null space of $[A]$. We note that *these eigenvectors in general do not define the constraint manifolds of planar dyads*. Instead, these three orthonormal eigenvectors define the null space that may yield the constraint manifolds. We use (8) to define the following deviation from the constraint manifolds of planar dyads:

$$e = \sqrt{[p_1 p_6 + p_2 p_5 - p_3 p_4]^2 + [2p_1 p_7 - p_2 p_4 - p_3 p_5]^2}. \tag{16}$$

Listed in the last column of Table 2 are these deviations and they happen to be all non-zero for this example, which means that none of the three eigenvectors represent the constraint manifolds associated with the five specified task positions.

To obtain the coefficient vectors **p** that define the constraint manifolds, we follow the procedure leading to (15) and solve the resulting two quadratic equations to obtain four real solutions:

$$(\alpha/\gamma)_1 = -0.3766,\ (\alpha/\gamma)_2 = -0.0084,\ (\alpha/\gamma)_3 = 1.6626,\ (\alpha/\gamma)_4 = 77.0848;$$
$$(\beta/\gamma)_1 = 0.3288,\ (\beta/\gamma)_2 = -2.1759,\ (\beta/\gamma)_3 = -0.2646,\ (\beta/\gamma)_4 = 224.3660. \tag{17}$$

Substituting them into (14), we obtain the homogeneous coordinates (listed in Table 3) of four constraint manifolds (shown in Figure 3) associated with four feasible dyads for the five given positions (listed in Table 1). The last coefficient vector $\mathbf{p}_4$ in Table 3 has the special feature that its first three coordinates are identically zero (up to floating point error), and thus represents a *PR* dyad. The other three are all *RR* dyads. They define three planar 4*R* linkages as well as three slider-crank mechanisms. Three constraint circles and one constraint line as well as their respective circle points are computed using (10) and are shown in Table 4.

Table 2 Example 1: Three eigenvectors defining the null space with deviation e

	1	2	3	4	5	6	7	8	e
$\mathbf{v}_\alpha$	0.0076	-0.3896	-0.1690	-0.3351	0.3048	-0.0558	0.7320	0.2747	0.1884
$\mathbf{v}_\beta$	-0.0035	0.1329	0.0558	0.2911	0.2460	0.8939	0.1864	-0.0102	0.0554
$\mathbf{v}_\gamma$	0.2023	0.2060	0.5127	-0.2219	-0.1643	0.0816	-0.0829	0.7510	0.1363

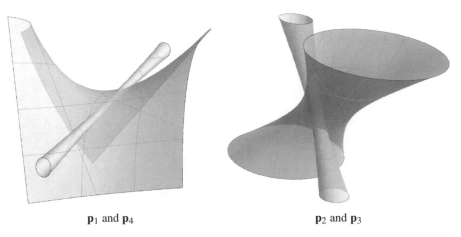

$\mathbf{p}_1$ and $\mathbf{p}_4$ $\mathbf{p}_2$ and $\mathbf{p}_3$

Fig. 3 Four constraint manifolds associated with Table 3

Table 3 Example 1: Four normalized coefficient vectors of the constraint manifolds of four planar dyads

	1	2	3	4	5	6	7	8	Dyad
$\mathbf{p}_1$	0.1773	0.3546	0.5319	-2×10^{-13}	-0.1773	0.3546	-0.2659	0.5762	RR
$\mathbf{p}_2$	0.0876	-0.0334	0.1640	-0.3560	-0.2934	-0.7779	-0.2066	0.3219	RR
$\mathbf{p}_3$	0.1103	-0.2435	0.1108	-0.4372	0.1417	-0.1265	0.5540	0.6182	RR
$\mathbf{p}_4$	-3×10^{-13}	-3×10^{-13}	-8×10^{-13}	0.1655	0.3310	0.8276	0.4138	0.0828	PR

6 Four-Position Synthesis

In this case, the matrix $[A]$ in (12) becomes a 4×8 matrix and thus the matrix $[A]^T [A]$ is of rank 4. The application of SVD algorithm yields four zero eigenvalues. Let $\mathbf{v}_\alpha$, $\mathbf{v}_\beta$, $\mathbf{v}_\gamma$ and $\mathbf{v}_\mu$ denote four eigenvectors associated with the zero eigenvalues. They form the basis of the four-dimensional null space of $[A]$. Any vector $\mathbf{p}$ in the null space as given by

$$\mathbf{p} = \alpha\mathbf{v}_\alpha + \beta\mathbf{v}_\beta + \gamma\mathbf{v}_\gamma + \mu\mathbf{v}_\mu \tag{18}$$

defines a candidate constraint manifold of a planar dyad that is compatible with the four specified task positions. We need to select the real parameters $(\alpha, \beta, \gamma, \mu)$ such that both constraints in Eqs.(8) are satisfied. This leads to

Table 4 Example 1: Homogeneous coordinates of the constraint circle (or line) and the circle point

	(a_0, a_1, a_2, a_3)	(x, y)
$\mathbf{p_1}$	$(1, 0, -1, 0)$	$(-2, -3)$
$\mathbf{p_2}$	$(1, -4.0639, -3.3470, 11.0521)$	$(0.3807, -1.8715)$
$\mathbf{p_3}$	$(1, -3.9639, 1.2843, 16.5279)$	$(2.2084, -1.0049)$
$\mathbf{p_4}$	$(0, 1, 2, 2)$	$(1, -3)$

Table 5 Example 2: Four task positions from [7]

	1	2	3	4
d_1	-3.339	-2.975	-3.405	-7.435
d_2	1.360	7.063	9.102	11.561
ϕ	$150.94°$	$-114.94°$	$100.22°$	$-74.07°$

$$K_{10}\alpha^2 + K_{11}\beta^2 + K_{12}\mu^2 + K_{13}\alpha\beta + K_{14}\alpha\mu + K_{15}\beta\mu + K_{16}\alpha\gamma$$
$$+K_{17}\beta\gamma + K_{18}\mu\gamma + K_{19}\gamma^2 = 0,$$
$$K_{20}\alpha^2 + K_{21}\beta^2 + K_{22}\mu^2 + K_{23}\alpha\beta + K_{24}\alpha\mu + K_{25}\beta\mu + K_{26}\alpha\gamma \qquad (19)$$
$$+K_{27}\beta\gamma + K_{28}\mu\gamma + K_{29}\gamma^2 = 0,$$

where K_{ij} are obtained from $\mathbf{Z}_i$ ($i = 1, 2, 3, 4$). Thus, the homogeneous parameters $(\alpha, \beta, \gamma, \mu)$ have ∞^1 many solutions .

An easy way to select one of the solutions without increasing the complexity of the problem is to impose a linear relationship among these parameters, i.e.,

$$k_1\alpha + k_2\beta + k_3\gamma + k_4\mu = 0. \qquad (20)$$

By varying the choice of k_i, one obtains different solutions for $(\alpha, \beta, \gamma, \mu)$. An indirect way of selecting k_i is to select the fifth task position and convert it into image point $\mathbf{Z}_5$. The substitution of $\mathbf{Z}_5$ and (18) into Eq.(7) will lead to a linear equation in the form of (20) and thus yields a set of k_i from the four eigenvectors and the fifth position $\mathbf{Z}_5$.

Example 2: Consider four task positions shown in Table 5, which is taken from [7]. The application of SVD algorithm yields the following eight eigenvalues:

$$0, 0, 0, 0, 0.0602, 3.2629, 20.9944, 3173.6955.$$

As expected, there are four zero eigenvalues. The four eigenvectors defining the basis of the four-dimensional null space are listed in Table 6.

We select the fifth position such that $d_1 = -9.171, d_2 = 11.219, \phi = 68.65°$, which is the fifth position from [7]. This leads to the following linear relation:

Table 6 Example 2: Four basis vectors for the null space and deviation e

	1	2	3	4	5	6	7	8	e
v_α	0.0321	0.1655	-0.0515	0.3788	0.1200	-0.0801	0.8964	-0.0326	0.0368
v_β	0.0151	0.0365	-0.0061	0.1131	-0.1060	0.9860	0.0467	-0.0043	0.0121
v_γ	0.1612	0.0102	0.6817	-0.0143	-0.6982	-0.0780	0.1242	0.0032	0.5162
v_μ	0.0244	0.1725	0.0004	0.3279	-0.0148	-0.0354	-0.1391	0.9172	0.0634

Table 7 Example 2: Coefficients $\mathbf{p}$ of the constraint manifolds of two feasible RR dyads

	1	2	3	4	5	6	7	8	Dyad
$\mathbf{p_1}$	0.0147	-0.0432	0.1183	-0.1177	-0.0004	-0.9457	0.1709	-0.2156	RR
$\mathbf{p_2}$	0.0572	0.2048	0.0249	0.4576	-0.0001	0.1995	0.8189	0.1856	RR

Table 8 Example 2: Constraint circles and circle points

	(a_0, a_1, a_2, a_3)	(x, y)
$\mathbf{p_1}$	$(1, -7.9879, -0.0279, -131.5185)$	(2.9323 , -8.0241)
$\mathbf{p_2}$	$(1, 7.9968, -0.0009, -0.0232)$	(-3.5794,-0.4356)

$$-0.0375\alpha + 0.0230\beta + 0.1948\gamma + 0.0281\mu = 0. \tag{21}$$

Solving (19) together with (21), we obtain only one pair of real solutions:

$$\begin{aligned} (\alpha/\gamma)_1 = 0.9986, \ (\beta/\gamma)_1 = -5.2500, \ (\mu/\gamma)_1 = -1.3070; \\ (\alpha/\gamma)_2 = 8.4764, \ (\beta/\gamma)_2 = 2.7108, \ \ (\mu/\gamma)_2 = 2.1755. \end{aligned} \tag{22}$$

The other pair are complex solutions. Table 7 shows homogeneous coefficients $\mathbf{p}$ of the constraint manifolds of two feasible dyads. The constraint circles and circle points are listed in Table 8.

7 Three-Position Synthesis

In this case, there are only three linear equations in the form of (7) and the null space of the resulting coefficient matrix $[A]$ is five dimensional. There are five zero eigenvalues from the matrix $[A]^T[A]$. The corresponding eigenvectors are denoted by $v_\alpha, v_\beta, v_\gamma, v_\mu, v_\eta$. A vector in the null space is given by

$$\mathbf{p} = \alpha v_\alpha + \beta v_\beta + \gamma v_\gamma + \mu v_\mu + \eta v_\eta \tag{23}$$

and only those $\mathbf{p}$ that satisfy (8) define the constraint manifolds of feasible dyads. In this case, there are ∞^2 solutions for $\mathbf{p}$. We may use two linear equations of the form (20) to obtain $\mathbf{p}$. This can be done by selecting two new task positions.

Table 9 Example 3: Three task positions from [14]

	1	2	3
d_1	-2.7037	−7.30565	-12.1993
d_2	0.6508	-0.1698	-1.8456
ϕ	$1-13.31°$	$−19.94°$	$−17.31°$

Table 10 Example 3: Five basis vectors of the null space and deviation e

	1	2	3	4	5	6	7	8	e
v_α	-0.0011	0.1140	-0.0834	0.0273	0.0550	0.0017	-0.6357	-0.7564	0.0090
v_β	-0.0114	-0.1673	0.6916	0.2545	-0.3664	0.0481	-0.4665	0.2733	0.3275
v_γ	-0.0113	0.3662	-0.1064	0.2131	-0.1691	0.8813	0.0595	0.0143	0.1091
v_μ	-0.0268	0.2043	-0.3752	-0.0114	-0.8550	-0.2924	0.0078	0.0024	0.3619
v_η	0.0098	0.6170	0.1333	0.6399	0.2059	-0.3642	0.1298	0.0064	0.4215

Table 11 Example 3: Coefficients **p** and the dyad types

	1	2	3	4	5	6	7	8	Dyad
p_1	0.0053	0.0487	0.0411	0.0893	0.0752	0.0021	0.7008	0.7008	RR
p_2	-6×10^{-11}	5×10^{-10}	-3×10^{-10}	-0.0340	0.1132	0.8130	0.4032	0.4032	PR
p_3	-0.0020	0.0158	0.0018	-0.0391	0.1215	0.9818	0.0988	0.0988	RR
p_4	-0.0032	0.0762	0.0289	0.0473	0.0275	0.2286	-0.6849	-0.6849	RR

By varying these two extra positions, we may obtain different solutions for **p** and thus the resulting dyads.

Example 3: Consider three task positions in Table 9 that are taken from [14]. The SVD algorithm yields the following eigenvalues:

$$0, 0, 0, 0, 0, 0.3563, 8.1559, 1734.2819.$$

as well as five basis vectors of the null space of $[A]$ (Table 10). In this case, there are ∞^2 solutions for **p** that satisfy (8). By choosing two extra positions: $(0,0,0°)$ and $(-16.2362 - 4.1004, -6.53°)$, we obtain the following two linear relations:

$$0.1207\alpha - 0.7397\beta + 0.0452\gamma + 0.0054\mu + 0.1234\eta = 0,$$
$$0.1114\alpha - 0.6423\beta + 0.2869\gamma + 0.4688\mu + 0.1966\eta = 0. \tag{24}$$

This leads to four real solutions for $(\alpha/\gamma, \beta/\gamma, \mu/\gamma, \eta/\gamma)$ and consequently four real solutions for **p**, which are listed in Table. 11. The four constraint circles and their circle points are listed in Table 12. It is clear that $\mathbf{p}_2$ represents a *PR* dyad, which is consistent with the result in [14].

Table 12 Example 3: Constraint circles (or line) and the circle points

	(a_0, a_1, a_2, a_3)	(x, y)
p_1	$(1, 16.8232, 14.1685, 383.8700)$	$(-9.1666, -7.7440)$
p_2	$(0, 1, -3.3305, -47.4661)$	$(8.5524, -4.5581)$
p_3	$(1, 19.1667, -59.5833, -255.4526)$	$(7.8000, 0.8825)$
p_4	$(1, -14.7401, -8.5819, 209.6250)$	$(23.7473, 8.9897)$

8 Conclusions

We presented a novel algorithm that uses specified task positions to obtain "candidate" manifolds and then find feasible constraint manifolds among them. The first part is solved using an SVD algorithm for null space analysis. The second part is reduced to the solution of a quartic equation. This algorithm has two advantages: it can synthesize both joint type and dimensions of a four-bar linkage simultaneously and it can handle the synthesis of three, four, and five positions in the same way.

Acknowledgements. The work has been supported by National Science Foundation under grant CMMI-0856594 to Stony Brook University. All findings and results presented in this paper are those of the authors and do not represent those of the funding agencies.

References

1. Bottema, O., Roth, B.: Theoretical Kinematics. North-Holland, Amsterdam (1979)
2. Ravani, B., Roth, B.: Motion Synthesis Using Kinematic Mappings. ASME J. Mech., Transm., Autom. Des. 105(3), 460–467 (1983)
3. Ravani, B., Roth, B.: Mappings of Spatial Kinematics. ASME J.Mech., Trans., Auto., Des. 106(3), 341–347 (1984)
4. McCarthy, J.M.: Introduction to Theoretical Kinematics. MIT, Cambridge (1990)
5. Bodduluri, R.M.C., McCarthy, J.M.: Finite Position Synthesis Using the Image Curve of a Spherical Four-bar Motion. ASME J. Mech. Des. 114(1), 55–60 (1992)
6. Larochelle, P.: Approximate Motion Synthesis of Open and Closed Chains via Parametric Constraint Manifold Fitting: Preliminary Results. In: 2003 ASME Design Automation Conference, Chicago, Illinois, USA, September 26, pp. 1049–1057, Paper no. DETC2003/DAC-48814 (2003)
7. Hayes, M.J.D., Murray, P.J.: Solving Burmester Problem Using Kinematic Mapping. In: Proc. 2002 ASME Design Engineering Technical Conferences, Montreal, Quebec, Canada, September 29-October 02, Paper number DETC/CIE2002/DAC-1234 (2002)
8. Hayes, M.J.D., Luu, T., Chang, X.-W.: Kinematic Mapping Application to Approximate Type and Dimension Synthesis of Planar Mechanisms. In: Lenarci, C.J., Galletti, C. (eds.) Advances in Robotic Kinematics, pp. 41–48. Kluwer Academic Publishers, Dordrecht (2004)
9. Brunnthaler, K., Pfurner, M., Husty, M.: Synthesis of Planar Four-Bar Mechanisms. Transactions of CSME 30(2), 297–313 (2006)
10. Husty, M.L., Pfurner, M., Schrocker, H.-P., Brunnthaler, K.: Algebraic methods in mechanism analysis and synthesis. Robotica 25, 661–675 (2007)

11. Chen, C., Bai, S., Angeles, J.: A Comprehensive Solution of the Classical Burmester Problem. Transactions of the CSME 32(2), 137–154 (2008)
12. Chen, C., Bai, S., Angeles, J.: The Synthesis of Dyads With One Prismatic Joint. ASME J. of Mechanical Design 130, 034501 (2008)
13. Ge, Q.J., Zhao, P., Purwar, A., Li, X.: A novel approach to algebraic fitting of a pencil of quadrics for planar 4R motion synthesis. ASME Journal of Computing and Information Science in Engineering 12(4), 7 (2012)
14. Angeles, J., Bai, S.: Some Special Cases of The Burmester Problem for Four and Five Poses. In: Proc. 2005 ASME Design Engineering Technical Conferences, Paper number DETC2005-84871 (2005)

12 Kinematics Analysis and Design Considerations of the Gear Bearing Drive

Elias Brassitos and Constantinos Mavroidis

1 Introduction

The development of high performance and efficient power trains is necessary to meet the radical design requirements of demanding next generation robotic systems, particularly in human centered applications where weight, efficiency and compact forms are decisive for the application functionality (e.g., powered portable bionics; humanoid manipulators; robotic rehabilitative devices for restoring human motor functions; and systems for robot-therapist collaboration during patient care). Such robotic applications require a new breed of actuators that have compact, reconfigurable hardware and inherent mechanical compatibility and adaptability to human-robot interaction applications. Existing power trains for this class of actuators have been dominated by the Harmonic Drives, offering compact mechanisms with high-speed reductions. However, problems in its non-linear dynamics and structural strength have limited their use to very specific applications in robotics.

The Gear Bearing Drive or GBD is a newly developed actuation concept that was co-invented between Northeastern University and NASA Goddard Space Center [1]. The system incorporates NASA's planetary gearbox technology and new brushless 'outrunner' motor technology to produce one of the highest torque density actuators. Due to its unique arrangement of planetary transmission and drive motor, the GBD is able to combine the motor, transmission and position sensing elements into a space that is volumetrically smaller than a human elbow joint. This results in ultra-compact actuators with incredible high torque output (more than 100 Nm), micro-precision accuracy, and strong and rugged structural

Elias Brassitos · Constantinos Mavroidis
Department of Mechanical and Industrial Engineering
Northeastern University, 360 Huntington Ave, Boston, MA 02115
e-mail: elias.brassitos1@gmail.com,
        mavro@coe.neu.edu

V. Kumar et al. (Eds.): *Adv. in Mech., Rob. & Des. Educ. & Res.*, MMS 14, pp. 159–175.
DOI: 10.1007/978-3-319-00398-6_12   © Springer International Publishing Switzerland 2013

integrity. These unique characteristics of the Gear Bearing Drive facilitate the development of compact powerful robots that are otherwise unattainable with traditional actuators. This book chapter describes the fundamental kinematic mechanism of the Gear Bearing Drive, and shed light on its novel actuation characteristics, and concludes with a design guideline for building a custom GBD.

2   Background

In the area of conventional compact high torque/force density actuators, hydraulic and pneumatic actuators are ranked with the highest force to weight and force to volume ratios. However, they require complementary sources of energy power, such as pressurized air source or pump, which adds weight, reduces efficiency and makes it difficult to function as powered joints for fully portable robotic systems, such as in wearable exoskeletons. Alternatively, the DC motor technology is considered the most mature and promising source of actuation that is likely to permit the development of fully portable, compact, and efficient high-torque density actuated robotic joints. The well-proven DC motor technology exhibits a number of desirable features such as linearity, high bandwidth, accuracy, efficiency, low friction and low-cost. However, DC motors suffer from limited torque density due to the low electromagnetic forces attainable at small scales. To overcome the problem of low torque density for DC motors, high reduction speed reducers are used in combination with electric motors to increase their torque output. There are two main mechanisms that offer high speed reduction in a compact configuration: A) Harmonic Drives and B) Planetary Gears.

2.1   Harmonic Drives

The use of Harmonic Drive transmissions in conjunction with high performance DC motors has often been regarded as the state of the art in actuated robotic joints such as in industrial manipulators. The principle of operation of Harmonic Drives is based on a unique type of transmission mechanism comprising three co-centric components, denoted by the Wave Generator, Flexpline, and Circular Spline. The Wave Generator consists of a bearing that is press fitted within an elliptically shaped steel disk and inserted within the Flexpline. The Flexpline is a compliant thin-walled steel cup that conforms to the shape of the wave generator, and has teeth on its external diameter that mates with the Circular Spline. The Circular Spline consists of a rigid steel ring with teeth on the internal diameter and represents the output. Harmonic Drives are designed such that the Flexpline has two teeth less than the Circular Spline, so that when the Wave Plug rotates one revolution, the Circular Spline is shifted by two teeth yielding very high torque advantages.

Harmonic Drives can exhibit large non-linear behavior under high dynamic loads due to their flexible gear component being in series inside the transmission. This elastic element creates a low stiffness medium inside the transmission that deforms under load and introduces instabilities under high gain feedback loops which further deteriorate the control system performance of the joint [2]. Additionally, Harmonic Drives are only transmission systems and require specialty motors to perform as actuators.

2.2   Planetary Gear Trains

Planetary gear trains first appeared around 2600 BC in ancient China in a device referred to as the south-pointing chariot [3]. The device was used as a terrain navigation tool to help ancient Chinese keep track of their direction through deserts and open plains. The next appearance of planetary gearboxes was recorded in the Antikythera machine, discovered off the coast of Greece around 1901 [4]. The device had been identified as a mechanical calculator for tracking eclipses and astrological objects, and is predicted to have existed around 80BC. Today, planetary gearboxes are widely used across several industries, from automobiles transmission and differentials, to power splitting devices in hybrid vehicles and continue to impact the industry and advance the state of the art in power transmission and delivery.

Overall, planetary gearboxes offer a number of advantages over ordinary serial gear trains. They provide a good combination of compactness, strength, and high power transmission efficiencies with a nominal outcome of 3% per stage. As a result, only small amounts of energy are dissipated into frictional losses inside the gearbox. The planetary gear arrangements are also more versatile than ordinary gear trains and allow for multiple degrees of freedom and multiple gear ratios in various operating modes. The elementary planetary gear train is defined as any gear train containing at least one gear that orbits by revolving about the axis of an arm, or carrier, and also around its own axis. The elementary train consists of three gears, the sun and planet gears, and a third outer ring that meshes and constrains the planets. A fourth component of the train is referred to as the planet carrier or arm. Planetary gears can be realized in any of the twelve elementary arrangements set forth by Zoltan Lévai [5]. The basic planetary gear train and its simplified kinematic representation are shown in Fig. 1.

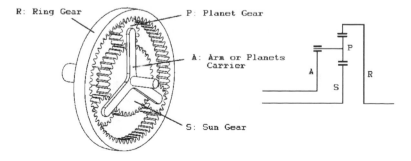

Fig. 1 The elementary epicyclic gear train and its kinematic representation

The key advantage of planetary gears is their ability to produce very high reduction ratios with small number of gear components. Less number of gear elements result in less friction inside the gearbox and further improve the system efficiency. Planetary gear trains are superior to ordinary gear trains in design versatility and compactness, higher reduction ratios in more compact forms, multi-degrees of freedom, less backlash, higher efficiencies and variable gear ratios.

3   The Gear Bearing Drive Concept

The Gear Bearing Drive (GBD) is a compact mechanism with two key abilities. It can operate as an actuator providing torque and as a joint providing joint support. This is possible because of the novel combination of external rotor (outrunner type) DC motor technology and Gear Bearing technology.

The GBD has two principle components, the Gear-Bearing and the DC "outrunner" drive motor. A Gear Bearing is a novel bearingless high-reduction ratio planetary gear system which places a rolling surface at the pitch diameter of each gear to maintain gearset alignment and support thrust, radial and bending loads [6]. The "outrunner" motor is a compact external rotor 3-phase DC motor that is commonly used in model airplanes. These motors fix the coils to the end bell (grounded stator) and place the magnets on the rotating can (rotor). This motor design has a higher torque output, greater heat dissipation, and a lower part count when compared to standard DC motor designs.

The novelty of the GBD mechanism lies in embedding the 'outrunner' motor in the gearbox by inscribing it within the input stage. The motor essentially behaves as an actuated gear within the gearbox saving significant space and volume on the assembly.

To get these high reduction ratios, the gear bearing is arranged as a two stage single tooth difference planetary system with the planets from the first stage rigidly attached to the planets of the second stage. The high torque output of brushless outrunner motors combined with the high reduction ratios that are possible using gear bearings produce compact actuators with incredible high torque output. A simplified model of the GBD is presented in Fig. 2.

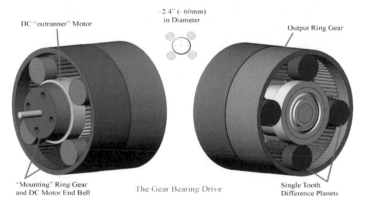

Fig. 2 The Gear Bearing Drive (GBD) Concept

Furthermore, a roller-bearing design is used to lock the assembly and eliminate the need for external bearings, shafts and journals to frame the assembly. This technique reduces the number of moving parts, increases reliability, and lowers on manufacturing cost. The GBD places the rolling surfaces at the gears pitch diameters and in parallel with the gears so that gear engagement and rolling motion are coherently synchronized. The rollers perform thrust bearing function to provide exceptional strength in axial and bending loadings, and add traction drive to the gear meshing to minimize the torque drag. The rollers are offset from the gear action so that bearing friction adds parasitic torque to the gear action. The assembly is held together such that if an axial tension force is applied, the planet rollers are blocked by the abutment of the planet rollers against the teeth of the input and output rings. A similar connection prevents the axial sliding of the sun gear by the constraints imposed by the teeth and rolling contact between the sun and input stage planets. The roller mate connection and assembly locking features can be seen in Fig. 3.

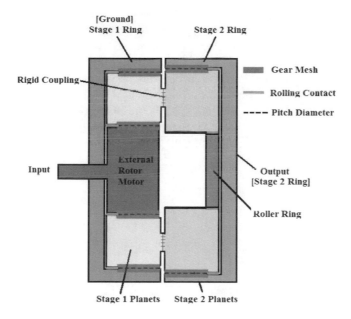

Fig. 3 A longitudinal cut through the GBD showing the roller bearing concept

4 The Gear Bearing Drive Inrunner / Outrunner Configuration

In order to gain a deeper understanding of the GBD's high-mechanical advantage mechanism, a free body diagram of the forces and torques governing the motion of its two-stage planetary gearbox is shown in Fig. 4. The diagrams are illustrated

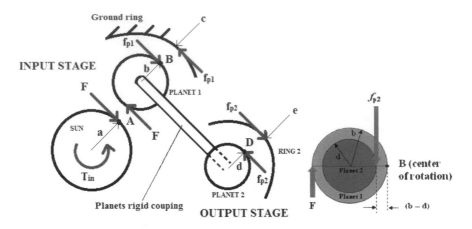

Fig. 4 Forces and moments acting on the GBD assembly

using the pitch diameters of the gears where the effective rolling contact occurs in the gears. It is assumed that there is neither slip nor frictional losses for the purpose of calculating the mechanical torque advantage.

The input to this mechanism is the torque of the first stage sun gear, and the output is the perceived torque at the second stage ring gear. The first stage ring is fixed and treated as the ground. The equilibrium of the forces acting on the planets yields the overall torque advantage of the mechanism as illustrated in Fig. 4. The torque advantage is calculated by summing the torques with respect to the instantaneous center of rotation (point B):

$$F \cdot (2b) = f_{p2} \cdot (b - d) \tag{1}$$

$$\frac{T_{out}}{T_{in}} = \frac{2 \cdot b \cdot e}{a \cdot (b - d)} \tag{2}$$

Where a, b, d, and e, denote the pitch diameters of the sun gear, input planets, output planets, and output ring respectively; T_{in} denotes the input torque from the motor and T_{out} is the amplified torque at the output ring.

In this configuration, it is the very slight difference in the pitch diameters of the planets that creates large torque advantages between the input sun and output ring. This is due to the fact that the input tangential force resulting from the motor torque, F, acts on a moment arm of $2b$ that is much larger than the moment arm $(b-d)$, equal to the difference in the planets pitch diameters, acted upon by the perceived ring force f_{p2}. Furthermore, the direction of motion of the output relative to the input is also dictated by the size difference of the planets. When the pitch diameter of the input stage planet is greater than the pitch diameter of the output stage planet, the denominator of Eqn. 1 takes a positive value and causes the output to rotate in the same direction as the input and vice versa. The forward and reverse configurations are illustrated in Fig. 5.

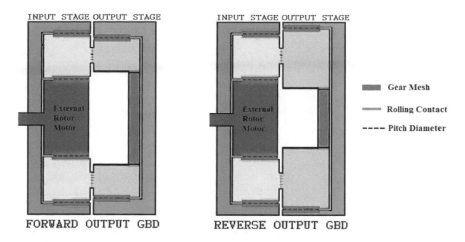

Fig. 5 Forward / Reverse configurations of the GBD

Hence, incorporating small variations in the planet diameters can produce very high torque ratios without altering the overall form factor of the mechanism. This configuration is referred to as the *inrunner / outrunner* mode due to the inner driving input sun gear and outer driven output ring gear, and it is adopted to develop the Gear Bearing Drive given its reconfigurable high reduction ratios and applicability for embedding the external rotor motor within the input stage sun gear.

5   Kinematic Modeling of the Inrunner / Outrunner GBD

The kinematic model of the *inrunner / outrunner* GBD configuration is further developed into relating the torque ratio with the permissible number of teeth on each gear element. The kinematic equations of motion are developed with respect to an imaginary arm speed, and then inverted with respect to ground. The arm constitutes an imaginary axis passing through the pinions axis of rotation and revolving around the sun gear. N_1 through N_5 denote the number of teeth on the first stage sun, first stage planets, first stage ring, second stage planets, and second stage ring respectively as shown in Fig. 6.

The overall angular speed of the input (first stage sun) with respect to the ground can be written as:

$$w_{Input} = w_{Input / Arm} + w_{Arm} \qquad (3)$$

The angular speed of the first stage pinion with respect to the arm is written as:

$$w_{2/ Arm} = -\frac{N_1}{N_2} w_{Input / Arm} \qquad (4)$$

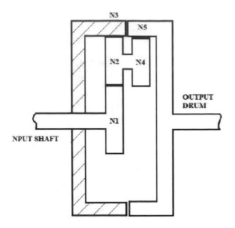

Fig. 6 Kinematic representation of the GBD assembly

The angular speed of the first stage ring (i.e. ground) with respect to ground is zero. Hence,

$$w_3 = w_{3/Arm} + w_{Arm} = 0 \tag{5}$$

The speed of ground with respect to the arm is:

$$w_{3/Arm} = \frac{N_2}{N_3} w_{2/Arm} \tag{6}$$

From Eqns (4) and (6), the arm speed can be extracted as:

$$w_{Arm} = -w_{3/Arm} = -\frac{N_2}{N_3} w_{2/Arm} = \frac{N_1}{N_3} w_{Input/Arm} \tag{7}$$

Substituting Eqn. (7) back into Eqn. (3), the input speed with respect to ground is:

$$w_{Input} = w_{Input/Arm} + \frac{N_1}{N_3} w_{Input/Arm} = \left(1 + \frac{N_1}{N_3}\right) w_{Input/Arm} \tag{8}$$

The pinions of both stages have equal angular velocity since they are rigidly coupled, hence:

$$w_{2/Arm} = w_{4/Arm} \tag{9}$$

The output (second stage ring) speed with respect to the arm is determined as:

$$w_{Output/Arm} = \frac{N_4}{N_5}w_{4/Arm} = \frac{N_4}{N_5}w_{2/Arm} = -\frac{N_1 N_4}{N_2 N_5}w_{Input/Arm} \tag{10}$$

The output speed with respect to ground is determined as:

$$w_{Output} = w_{Output/Arm} + w_{Arm} = \left(-\frac{N_1 N_4}{N_2 N_5} + \frac{N_1}{N_3}\right)w_{Input/Arm} \tag{11}$$

Dividing Eqn. (8) by Eqn. (11), the final angular velocity ratio, or alternatively torque ratio, is determined as:

$$\frac{w_{Input}}{w_{Output}} = \frac{1 + \dfrac{N_3}{N_1}}{1 - \dfrac{N_3 N_4}{N_2 N_5}} \tag{12}$$

Eqn. (12) dictates the permissible number of teeth imposed on each gear element for a given torque ratio. An additional geometrical condition on the first stage requires that the number of teeth on the ground ring be equal to the sum of sun gear teeth and twice of the input pinion teeth according to Eqn. (13).

$$N_3 = N_1 + 2N_2 \tag{13}$$

Furthermore, gears that mesh with each other must have the same diametral pitch and pressure angle properties in order to engage in pure rolling contact. The diametral pitch, denoted by P, is a measure of the number of teeth per inch and is related to the number of teeth, N, and pitch diameter, D, by Eqn. (14).

$$P = \frac{N}{D} \tag{14}$$

The pressure angle defines the shape of the tooth involutes of the spur gears such as the addendum, dedendum, whole depth, and base circle. Another condition is imposed on the pitch diameters since the planets of both stages must orbit at the same radial distance from the sun, as described by Eqn. (15).

$$\frac{N_1}{P_1} + \frac{N_2}{P_1} = \frac{N_5}{P_2} - \frac{N_4}{P_2} \tag{15}$$

Eqns. (12), (13) and (15) reduce the total number of unknowns to three that were iteratively solved to develop the planetary gearbox of the GBD. Also, we note that Eqn. (12) can be obtained by substituting Eqn. (15) into Eqn. (2).

Additional mathematical constraints are dictated on the permissible number of teeth on each gear since gear teeth must be integer numbers. The discussion will be limited to the GBD equation, as written in Eqn. (12); however the same process can be applied to any of the other configurations.

The terms of Eqn. (12) are re-arranged in the following form, where R is the desired gear ratio.

$$\frac{N_4}{N_5} = \frac{RN_1N_2 - 2N_2(N_1 + N_2)}{RN_1(N_1 + 2N_2)} = \frac{p}{q} \tag{16}$$

For integer values of R, the resulting p and q must also be integer numbers, hence N_4 and N_5 are calculated as:

$$N_4 = \frac{p}{GCF(p,q)} \tag{17}$$

$$N_5 = \frac{q}{GCF(p,q)} \tag{18}$$

Where $GCF(p,q)$ is the greatest common factor of p and q. Therefore, only the gear ratio and input stage sun and pinion gears number of teeth needs to be known (R, N_1, N_2) in order to calculate the remaining values for the number of the teeth on the GBD assembly.

6   The Gear Bearing Drive Other Kinematic Configurations

In addition to the *inrunner / outrunner* configuration of the GBD, it is also possible to design the Gear Bearing Drive in different configurations in order to vary the functionality of the drive based on the application requirements, such as in the case where a low torque ratio actuator is needed or when a torque reducer or speed increaser device is desired. These other GBD configurations are presented in this section.

6.1   Inrunner / Inrunner GBD

An *inrunner / inrunner* configuration consists of adding a sun gear to the output stage to form the actuator output, and setting the output stage ring gear idle, as shown in Fig. 7. This mode of operation can produce customizable low torque ratio gearheads, and also has the capability to perform speed-augmenting functions such as speed increaser devices.

Since the output gear is now shifted to the second stage sun gear, as opposed to the second stage ring gear, Eqns. (10-12) are modified to account for this change in the power path.

Eqn. (10) is re-written as:

$$w_{Output/Arm} = -\frac{N_4}{N_6} w_{4/Arm} = -\frac{N_4}{N_6} w_{2/Arm} = \frac{N_1 N_4}{N_2 N_6} w_{Input/Arm} \tag{19}$$

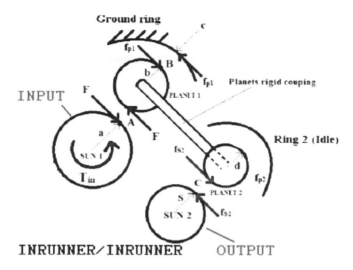

Fig. 7 Inrunner / *Inrunner* planetary configuration

Eqn. (11) is then calculated as:

$$w_{Output} = w_{Output/Arm} + w_{Arm} = \left(\frac{N_1 N_4}{N_2 N_6} + \frac{N_1}{N_3} \right) w_{Input/Arm} \tag{20}$$

The final angular speed ratio is obtained by dividing Eqn. (8) by Eqn. (20).

$$\frac{w_{Input}}{w_{Output}} = \frac{1 + \dfrac{N_3}{N_1}}{1 + \dfrac{N_3 N_4}{N_2 N_6}} \tag{21}$$

Furthermore, since the input and output planets must orbit at the same radial distance, the pitch diameters must satisfy the following relationship as expressed by Eqn. (22).

$$\frac{D_1}{2} + \frac{D_2}{2} = \frac{D_4}{2} + \frac{D_6}{2} \tag{22}$$

By substituting Eqn. (14) into (22), Eqn. (22) can be re-written using the gears number of the teeth and diametral pitch selected:

$$\frac{N_1}{P_1} + \frac{N_2}{P_1} = \frac{N_6}{P_2} + \frac{N_4}{P_2} \tag{23}$$

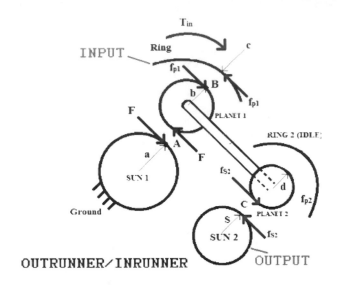

Fig. 8 Outrunner / *Inrunner* planetary configuration

6.2   *Outrunner / Inrunner GBD*

An *outrunner / inrunner* configuration, shown in Fig 8, is very similar in functionality as the *inrunner / outrunner* presented in the previous sections, with the difference that the input stage ring gear is driven as opposed to the input stage sun gear while the output of this mechanism is the output stage sun gear.

The overall angular speed of the input (first stage ring) with respect to the ground can be written as:

$$w_{Input} = w_3 = w_{Input\,Arm} + w_{Arm} \tag{24}$$

The angular speed of the first stage pinion with respect to the arm is written as:

$$w_{2/\,Arm} = \frac{N_3}{N_2} w_{Input\,/\,Arm} \tag{25}$$

The angular speed of the first stage sun (i.e. ground) with respect to ground is zero. Hence,

$$w_1 = w_{1/\,Arm} + w_{Arm} = 0 \Rightarrow w_{arm} = -w_{1/\,Arm} = \frac{N_2}{N_1} w_{2/\,Arm} = \frac{N_3}{N_1} w_{Input\,/\,Arm} \tag{26}$$

Substituting Eqn. (26) in Eqn. (24), the input angular speed is written as:

$$w_{Input} = \left(1 + \frac{N_3}{N_1}\right) w_{Input\,/\,Arm} \tag{27}$$

The angular speed of the output (second stage sun) can be written with respect to the arm as:

$$W_{output/arm} = W_{6/Arm} = -\frac{N_4}{N_6}W_{4/Arm} = -\frac{N_4}{N_6}W_{2/Arm} = -\frac{N_4 N_3}{N_6 N_2}W_{Input/Arm} \quad (28)$$

The output speed with respect to the ground is computed by adding the arm speed to Eqn. (28).

$$W_{output} = W_{output/Arm} + W_{arm} = \left(-\frac{N_4 N_3}{N_6 N_2} + \frac{N_3}{N_1}\right)W_{input/Arm} \quad (29)$$

The angular speed ratio is calculated by dividing Eqn. (27) by Eqn. (29):

$$\frac{W_{Input}}{W_{Output}} = \frac{1+\dfrac{N_1}{N_3}}{1-\dfrac{N_1 N_4}{N_2 N_6}} \quad (30)$$

The kinematic conditions described in Eqns. (22) and (23) for the *Inrunner / Inrunner* mode also hold for the *Outrunner / Inrunner* mode due to the similar gear arrangement of their second stage.

6.3   *Outrunner / Outrunner GBD*

The *outrunner / outrunner* uses the input stage ring gear as the actuated gear while it assigns the output stage ring gear as the output, as shown in Fig. 9.

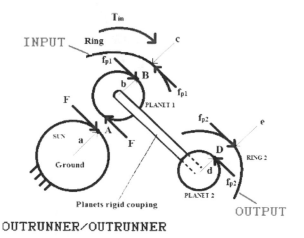

OUTRUNNER/OUTRUNNER

Fig. 9 Outrunner / Outrunner planetary configuration

Since the output gear is now shifted to the second stage ring gear, as opposed to the second stage sun gear in the *Outrunner / Inrunner* mode (see previous section), Eqns. (28-30) are modified to account for this change in the power path.

Eqn. (28) can be re-written as:

$$w_{output\,/\,arm} = w_{5\,/\,Arm} = \frac{N_4}{N_5} w_{4\,/\,Arm} = \frac{N_4}{N_5} w_{2\,/\,Arm} = \frac{N_4 N_3}{N_5 N_2} w_{Input\,/\,Arm} \tag{31}$$

The output speed with respect to ground is written as:

$$w_{output} = w_{output\,/\,Arm} + w_{arm} = \left(\frac{N_4 N_3}{N_5 N_2} + \frac{N_3}{N_1} \right) w_{input\,/\,Arm} \tag{32}$$

The final angular velocity ratio is thus calculated by dividing Eqn. (27) by Eqn. (32).

$$\frac{w_{Input}}{w_{Output}} = \frac{1 + \dfrac{N_1}{N_3}}{1 + \dfrac{N_1 N_4}{N_2 N_5}} \tag{33}$$

Similarly, the kinematic condition described in Eqn. (15) for the *Inrunner / Outrunner* also holds for the *Outrunner / Outrunner* as they both have a similar planetary gear arrangement.

Table 1 Various configurations of the GBD mechanism

Operating Mode	Gear Ratio (Based on Number of Teeth)	Kinematic Constraints	Applications
Inrunner / Outrunner	$\dfrac{1 + \dfrac{N_3}{N_1}}{1 - \dfrac{N_3 N_4}{N_2 N_5}}$	$\dfrac{N_1}{P_1} + \dfrac{N_2}{P_1} = \dfrac{N_5}{P_2} - \dfrac{N_4}{P_2}$ $N_3 = N_1 + 2N_2$	Very high speed reducers. High torque ratio transmission.
Inrunner / Inrunner	$\dfrac{1 + \dfrac{N_3}{N_1}}{1 + \dfrac{N_3 N_4}{N_2 N_6}}$	$\dfrac{N_1}{P_1} + \dfrac{N_2}{P_1} = \dfrac{N_6}{P_2} + \dfrac{N_4}{P_2}$ $N_3 = N_1 + 2N_2$	Low speed reducers. Speed increasers
Outrunner / Outrunner	$-\dfrac{1 + \dfrac{N_1}{N_3}}{1 + \dfrac{N_1 N_4}{N_2 N_5}}$	$\dfrac{N_1}{P_1} + \dfrac{N_2}{P_1} = \dfrac{N_5}{P_2} - \dfrac{N_4}{P_2}$ $N_3 = N_1 + 2N_2$	Low speed reducers. Speed increasers
Outrunner / Inrunner	$\dfrac{1 + \dfrac{N_1}{N_3}}{1 - \dfrac{N_1 N_4}{N_2 N_6}}$	$\dfrac{N_1}{P_1} + \dfrac{N_2}{P_1} = \dfrac{N_6}{P_2} + \dfrac{N_4}{P_2}$ $N_3 = N_1 + 2N_2$	Very high speed reducers. High torque ratio transmission.

A summary of the applications and modes of operation for each GBD configuration is presented in Table 1, which is also used as a preliminary design tool for choosing a suitable embodiment for a given application i.e. high or low-speed reducer, or speed increaser.

Unlike conventional methods of ordinary gearhead design, this technology is superior in terms of versatility and freedom of design when developing compact actuators requiring exceptional size, power and torque requirements.

7   Design Considerations for the *Inrunner / Outrunner* GBD

Based on Eqns (12) to (18), there are a total of 13 parameters that fully define the Gear Bearing Drive, as summarized in Table 2. Of these 13 parameters, only 4 are fully independent. Consequently, the 4 critical parameters, called *'floating variables,'* are assigned to the most critical design constraints of the Gear Bearing Drive. The four main parameters are: *Desired gear ratio (R)*, *Input Stage Sun Number of Teeth N_1*, *Input Stage Planet Number of Teeth N_2* and *Input Stage Sun Diameter D_1*. The remaining 9 parameters are consequently calculated as a function of the 4 critical parameters as shown in Table 2.

Table 2 Summary of the kinematic relationships of the *Inrunner / Outrunner* GBD

Parameter	Symbol	Value
Gear Ratio	R	R (Floating variable)
Input Sun Teeth	N_1	N_1 (Floating variable)
Input Planet Teeth	N_2	N_2 (Floating variable)
Input Ring Teeth	N_3	$N_1 + 2N_2$
Output Planet Teeth (minimum optimal)	N_4	$\dfrac{RN_1N_2 - 2N_2(N_1 + N_2)}{GCF[RN_1N_2 - 2N_2(N_1 + N_2), RN_1(N_1 + 2N_2)]}$
Output Ring Teeth (minimum optimal)	N_5	$\dfrac{RN_1(N_1 + 2N_2)}{GCF[RN_1N_2 - 2N_2(N_1 + N_2), RN_1(N_1 + 2N_2)]}$
Input Sun Diameter	D_1	D_1 (Floating variable)
Input Stage Diameteral Pitch	P_1	$\dfrac{N_1}{D_1}$
Output Stage Diameteral Pitch	P_2	$P_1\left(\dfrac{N_5 - N_4}{N_1 + N_2}\right)$
Input Planet Diameter	D_2	$\dfrac{N_2}{P_1}$
Input Gear Diameter	D_3	$\dfrac{N_3}{P_1}$
Output Planet Diameter	D_4	$\dfrac{N_4}{P_1}$
Output Gear Diameter	D_5	$\dfrac{N_5}{P_1}$

In order to develop a custom GBD, we first start by selecting the *desired gear ratio (R)*. This selection is based on torque and power output requirements defined by a specific task / application. Once the range of output torque is known, a gear ratio is selected that amplifies the input DC motor torque up to the application required torque. In general, the GBD transmission efficiency can vary between 85% to 95% depending on the gear type and lubrication used. Therefore, it is recommended to use a 10% to 20% larger torque ratio margin to ensure that the GBD fully matches the application torque output requirements, by compensating for the frictional losses inside the gearbox.

Once the gear ratio R is known, the next step is to determine the best combination of number of teeth for the input stage sun N_1 and input stage pinion N_2. Because the numbers of teeth on the remaining gears is a function of R, N_1 and N_2 (as shown in Table 1), it is critical to choose a correct combination of N_1 and N_2 that does not result in extreme numbers for the teeth on the remaining gears. For this, a parametric analysis can be run in Excel or MATLAB by solving all the possible values for N_3 and N_4 using Eqns. (17) and (18) where the values for N_1 and N_2 vary from 19 (i.e. the minimum required number of teeth to avoid gear undercut during manufacturing) up to 100 (upper limit is arbitrary and can be increased until an acceptable solution is attained.) The set of (N_1, N_2) for which Eqns. (17) and (18) produce the largest Greatest Common Factor will yield the smallest permissible integers N_4 and N_5, which are adopted to develop the planetary gearbox of the Gear Bearing Drive.

After populating the gear number of teeth, N_1 through N_5, the next step is to size the diameters of the gear component. To specify the scale of the GBD, only one dimension can be specified which can either be one of the gear diameters *or* one of diametral pitches for any of its stages. Once a parameter is defined, the remaining parameters are locked relative to that parameter. Our approach to develop the GBD uses the outrunner motor diameter as the lower limit for the pitch diameter of the input stage sun gear, as shown in Table 2. This is because the input stage sun gear must be larger than that motor rotor in order to fit the motor. Once the input sun pitch diameter is defined, the remaining gears' diameters and diametral pitches can be calculated using the relationships of Table 2.

Once the pitch diameters and numbers of teeth are known, the gears can be further developed in SolidWorks to generate their teeth profile and roller surfaces, and subsequently taken into manufacturing and components assembly.

8   Conclusions

A new compact actuator has been proposed and developed to improve the dexterity, modularity, strength, and torque output of modern robotic systems. Due to its unique arrangement of planetary transmission and 'outrunner' motor design, the actuator is able to combine the motor, transmission and position sensing elements into a space that is volumetrically smaller than a human elbow joint. This combination produces ultra-compact actuators with incredible high torque output, micro-precision accuracy, and strong and rugged structural integrity. These unique characteristics facilitate the development of high payload-to-weight robots that are otherwise unattainable with traditional actuators. Furthermore, the actuator can be realized in multiple configurations such as a low or high torque reducer, single-input-multi-output power transmission, thereby adding tremendous versatility and design freedom for designers. Future work will investigate the effects of the rolling contact inside the transmission and analysis of advanced lubrication, heat dissipation, and failure analysis.

References

1. Weinberg, B., Vranish, J., Mavroidis, C.: Gear Bearing Drive, Patent 8,016,893 (September 13, 2011)
2. Sweet, L.M., Good, M.C.: Redefinition of the robot motion-control problem. IEEE Control Synthesis Magazine, 18–25 (August 1985)
3. Santander, M.: The Chinese south-seeking chariot: A simple mechanical device for visualizing curvature and parallel transport. American Journal of Physics 60(9), 782–787 (1992)
4. The Antikythera Mechanism Research Project,
   http://www.antikythera-mechanism.gr/
5. Levai, Z.: Bibliography of Planetary Mechanisms, Budapest, Budapesti Műszaki Egyetem Gépjárművek tanszéke (1969)
6. Vranish, J.: Gear Bearings, Patent 6,626,792 (September 30, 2003)

13 A Short Story on Long Pinions

Madhusudan Raghavan

Abstract. The design of automotive transmissions with multiple planetary gear sets leads to various interesting mathematical problems. Once a powerflow with favorable characteristics is established, there are many ways in which it can be mechanized. Factors such as manufacturability, packaging, availability of common component sets, etc., determine the selections leading to the final physical layout of a transmission. Several manufacturers use "long pinions" as building block elements in the design. This is an interesting architectural arrangement that combines certain packaging and functional attributes. In the present article, the algebraic use of the long pinion is demonstrated in the creation of vastly different transmission architectures from the same powerflow.

1 Introduction

A vehicle transmission delivers mechanical power from the engine to the remainder of the drive system, such as fixed final drive gearing, axles and wheels. The mechanical transmission allows some freedom in engine operation, usually through alternate selection of multiple drive ratios, a neutral selection that allows the engine to operate accessories with the vehicle stationary, and clutches and a torque converter for smooth transitions between driving ratios and to start the vehicle from rest with the engine turning. Transmission gear selection typically allows power from the engine to be delivered to the rest of the drive system with a ratio of torque multiplication/reduction and with a reverse ratio. An electrically variable transmission (EVT) transmission is a mechanical transmission augmented by one or more electric motor/generators. Typically, an EVT uses differential gearing to send a fraction of its transmitted power through an electric path to the final drive. The remainder of its power flows through another, parallel path that is all mechanical and direct, of fixed ratio, or alternatively selectable. One form of

Madhusudan Raghavan
Group Manager, Hybrid Systems & Global Energy Systems,
GM R&D, Warren, Michigan, USA
e-mail: madhu.raghavan@gm.com

V. Kumar et al. (Eds.): *Adv. in Mech., Rob. & Des. Educ. & Res.*, MMS 14, pp. 177–188.
DOI: 10.1007/978-3-319-00398-6_13   © Springer International Publishing Switzerland 2013

differential gearing is the well-known planetary gear set with the advantages of compactness and different torque and speed ratios among the various members of the gear set.

2   Background and Prior Work

Much of the theory of multi-speed transmission kinematic operation has been described in the language of lever diagrams [1]. Since the automative industry is generally moving in the direction of larger numbers of fixed speed ratios we briefly review recent work on multi-speed transmissions. Baran et al. [2] present the Hydra-Matic six-speed RWD automatic transmission family. The variants of this transmission family are created using built-in modularity, that allows tremendous parts-sharing and part-scaling. These designs improve fuel economy and acceleration performance relative to their four-speed predecessors. Borgerson et al., [3] present the design of a six-speed transmission having an input shaft connectable with an engine and planetary gear unit. A single carrier supports pinions from adjacent planes of gears. Lewis and Bollwahn [4] present the General Motors Hydra-Matic/Ford six-speed FWD automatic transmission family. Designed in modularity requires only changes to the second and third axis and case housings to achieve various torque requirements as stipulated by the specific vehicle application. Wittkopp [5] proposes a three planetary design with three brakes, three clutches, and three fixed interconnections between the gearsets.

In the electrified powertrain literature, Robinette and Powell [6], describe the use of a 12V start/stop system to turn the engine off and on during periods of vehicle idle. In particular integration issues such as start ability, noise and vibration, and vehicle launch are discussed in addition to the use of a correlated lump parameter modeling methodology. Hawkins et al. [7] describe General Motors' recently launched eAssist powertrain, which delivers approximately three times the peak electric boost and regenerative braking capability of GM's first generation 36V Belted Alternator Starter. Key elements include a water-cooled induction motor/generator, an accessory drive with a coupled dual tensioner system, air cooled power electronics integrated with a 115V lithium-ion battery pack, a direct-injection 2.4 liter 4-cylinder gasoline engine, and a modified 6-speed automatic transmission.

An example of a highly successful electrically variable transmission concept developed at GM is the Two-Mode Hybrid system produced for transit buses and SUVs. Schmidt [8] describes an embodiment with three planetary gear sets coaxially aligned. Gear members of the first and second planetary gear set are respectively connected to the two motor/generators. Their carriers are operatively connected to the output member. The Two-Mode system innovations provide performance and fuel economy improvements at highway speeds and better trailer towing ability. Grewe et al. [9], describe the GM Two-Mode Hybrid transmission for full-size, full-utility SUVs. This system integrates two electromechanical power-split operating modes with four fixed gear ratios and provides fuel savings from electric assist, regenerative braking and low-speed electric vehicle operation.

Miller et al. [10] describe the Voltec 4ET50 multi-mode electric transaxle, which introduces a unique two-motor EV driving mode that allows both the driving motor and the generator to simultaneously provide tractive effort while reducing electric motor speeds and the total associated electric motor losses. This new operating mode, however, does not introduce the torque discontinuities associated with a two-speed EV drive. For extended range operation, the Voltec transaxle provides both the completely decoupled action of a pure series hybrid, as well as a more efficient powerflow with decoupled action for driving at light loads and high vehicle speed.

3   Lever Diagrams and Planetary Geartrains

The primary building block of a planetary gear train is the planetary gear set shown in Fig. 1. This is comprised of a sun gear, a ring gear, and a set of planet gears (also known as pinion gears). Planetary gear sets are popular building blocks in transmissions because they allow compactness, the multiple pinions permit gear tooth loading to be shared. This allows finer pitch gears which are generally quieter than coarser pitch gears. The pinions are positioned equally about the carrier – so the sum of the gear loads is nominally zero. This eliminates large bearing loads except on the pinion shafts. This enables the supporting external structure to be lighter and permits the use of smaller journal-type bearings.

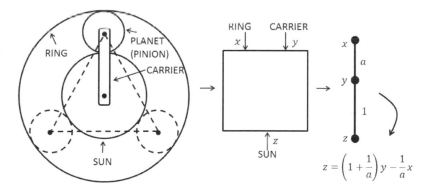

$$z = \left(1 + \frac{1}{a}\right) y - \frac{1}{a} x$$

Fig. 1 Planetary Gear Trains and Graph Representation

The planet gears are carried on a carrier member. The entire system of Fig. 1 may be represented by an edge-vertex graph as shown by the progression of drawings in Fig. 1, with the nodes labeled as ring R, carrier C, and sun S. Such a graph is also known as a "lever," because the relative rotational speeds of ring, carrier, and sun may be computed by treating rotational speeds as forces acting on

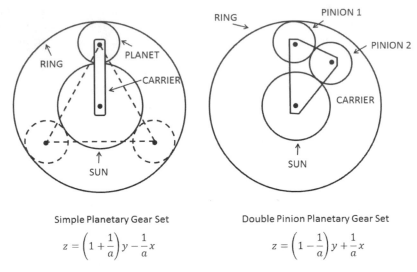

Simple Planetary Gear Set

$$z = \left(1 + \frac{1}{a}\right)y - \frac{1}{a}x$$

Double Pinion Planetary Gear Set

$$z = \left(1 - \frac{1}{a}\right)y + \frac{1}{a}x$$

Fig. 2 Simple Planetary and Double Pinion Planetary Gear Sets

the lever, and taking moments about appropriate nodes on the lever. An alternative to this geometric representation of levers is the following algebraic representation, which is applicable to all types of planetary gear trains (single pinion and double pinion). Let x, y and z be the speeds of the ring, the carrier, and the sun in Fig. 1. Then these speeds are related by the following equation:

$$z = (1 + \frac{1}{a})y - \frac{1}{a}x \qquad (1)$$

This relationship holds true by virtue of the mechanical interconnections and gear interactions in the planetary gear set (see Shigley and Uicker[11]). Eq. (1) contains a parameter a, which is equal to the ratio $\frac{n_S}{n_R}$, n_S and n_R being the numbers of teeth on the sun gear and the ring gear, respectively. In turn, $\frac{n_S}{n_R} = \frac{S}{R}$, the ratio of the diameters of the sun and ring gears. Eq. (1) is linear in the variables x, y and z. If we specify the values of any 2 of them, we may compute the remaining one, provided the value of the parameter a is specified. In other words, Eq. (1) represents a system, which requires 2 constraints to make it a well-defined algebraic problem. Fig. 2 shows simple and double-pinion planetary gear sets with the associated equations (note the sign changes).

4   Long Pinion Arrangements

Given 2 adjacent planetary gear sets, we say that they use a long-pinion if they share a common carrier and an integral pinion member. We can write expressions for the long-pinion arrangement as follows.

Case 1: For a combination of a simple planetary gear set (henceforth abbreviated as PG) connected to another simple PG via a long-pinion carrier, the constraint equations are:

$$z_1 = -k_1 x_1 + (1 + k_1)y_1, \qquad \text{(first planetary)} \qquad (2)$$

$$z_2 = -k_2 x_2 + (1 + k_2)y_2, \qquad \text{(second planetary)} \qquad (3)$$

$$y_2 = y_1, \qquad \text{(common carrier)} \qquad (4)$$

where $k_i = \dfrac{1}{a_i}, i = 1,2$

Additionally, the pinion speed of the first PG, $\Omega_1 = \dfrac{(x_1 - y_1)(2k_1)}{(k_1 - 1)}$, is set equal to the pinion speed of the second PG, $\Omega_2 = \dfrac{(x_2 - y_2)(2k_2)}{(k_2 - 1)}$, because the pinion is shared between the 2 PGs. The interested reader is referred to the work of Shigley and Uicker [11] for details on how to derive the kinematic equations relating the speeds of the ring, carrier, sun, and pinion gears.

$$\frac{(x_1 - y_1)(2k_1)}{(k_1 - 1)} = \frac{(x_2 - y_2)(2k_2)}{(k_2 - 1)}. \qquad \text{(common pinion)} \qquad (5)$$

Using Eqs. (4) and (5) to eliminate the variables x_2 and y_2 from Eqs. (2) and (3) in terms of x_1 and y_1, we get the following equations

$$z_1 = -k_1 x_1 + (1 + k_1)y_1, \qquad \text{(first PG)} \qquad (6)$$

$$z_2 = -\left(\frac{k_2 \xi_2}{\xi_1}\right) x_1 + \left(1 + \frac{k_2 \xi_2}{\xi_1}\right) y_1, \quad \text{(second PG)} \qquad (7)$$

where $\xi_i = (1 - a_i), i = 1,2..$

In the cases wherein one or both PGs are of the double-pinion type (see Fig. 2), the governing equations may be similarly derived as follows.

Case 2(a): PG1 is simple and PG2 is of the double-pinion type

$$z_1 = -k_1 x_1 + (1 + k_1)y_1, \qquad \text{(first planetary)} \qquad (8)$$

$$z_2 = k_2 x_2 + (1 - k_2)y_2, \qquad \text{(second planetary)} \qquad (9)$$

$$y_2 = y_1, \qquad \text{(common carrier)} \qquad (10)$$

The pinion speed of the first PG is to be set equal to the pinion speed of the second PG. However, the second PG has a double pinion arrangement and so we have choice of either Pinion 1 or Pinion 2 (see Fig. 2). If we use Pinion 1, its speed in terms of other parameters of PG2 is $\frac{R_2}{P_2}(x_2 - y_2)$ where P_2 and R_2 are respectively, the pinion diameter and the ring gear diameter on PG2. From Fig 2 we can see that for Pinion 1, $P = \xi \frac{R-S}{2}$, where $\xi < 1$. Using this result, the speed of Pinion 1 may be expressed as $\frac{(x_2-y_2)}{\frac{\xi}{2}(1-a_2)}$.

Setting this equal to the speed of one of the pinions on PG1, we get

$$\frac{(x_1-y_1)(2k_1)}{(k_1-1)} = \frac{(x_2-y_2)}{\frac{\xi}{2}(1-a_2)} \tag{11}$$

Upon rearrangement of terms in Eq. (11), we get

$$x_2 = \left(\xi \frac{\xi_2}{\xi_1}\right)x_1 + \left(1 - \xi \frac{\xi_2}{\xi_1}\right)y_1 \tag{12}$$

Substituting this expression for x_2 in Eq. (9), and using Eq. (10) to eliminate y_2, and finally, re-arranging terms we get

$$z_2 = \left(k_2\xi \frac{\xi_2}{\xi_1}\right)x_1 + \left(1 - k_2\xi \frac{\xi_2}{\xi_1}\right)y_1 \tag{13}$$

Eqs. (8) and (13) serve as constraints relating the rotational speeds of the 4 independent nodes $\{x_1, y_1, z_1, z_2\}$ of the 2-planetary system comprised of PG1 and PG2 together with the common carrier and common pinion fixed interconnection between the simple planetary gear set pinion and Pinion 1 on the double pinion planetary gear set.

Case 2(b)
PG1 is simple and PG2 is of the double-pinion type. The PGs share a common carrier. Further, a pinion on PG1 is integral with Pinion 2 on PG2 (see Fig. 2)

Eqs. (8), (9), and (10) apply to this case also. The speed of Pinion 2 on PG2 in terms of the other parameters of PG2 may be expressed as $\frac{S_2}{P_2}(z_2 - y_2)$. Further, from Fig. 2 and the fact that $P = \xi \frac{R-S}{2}$, we see that

$$\frac{P_2}{S_2} = \frac{\xi}{2}\left(\frac{R_2}{S_2} - 1\right).$$

As a result, we get $\frac{(z_2-y_2)}{\frac{\xi}{2}(k_2-1)}$ as the expression for the speed of Pinion 2 on PG2. Consequently, Eq. (11) becomes (in this case)

$$\frac{(x_1-y_1)(2k_1)}{(k_1-1)} = \frac{(z_2-y_2)}{\frac{\xi}{2}(k_2-1)}.$$

Upon rearrangement of terms this equation becomes

$$z_2 = \left(k_2\xi \frac{\xi_2}{\xi_1}\right)x_1 + \left(1 - k_2\xi \frac{\xi_2}{\xi_1}\right)y_1. \tag{14}$$

Eqs. (8) and (14) serve as the constraints relating the rotational speeds of the 4 independent nodes $\{x_1, y_1, z_1, z_2\}$ of the 2-planetary system comprised of PG1 and PG2 together with the common carrier and common pinion fixed interconnection between the pinion of the simple planetary gear set and Pinion 2 of the double pinion planetary gear set.

Equations for additional cases listed in the table below, e.g., double pinion planetary with double pinion planetary, Pinion 1 connected to Pinion 2, etc., may be derived in a similar manner (and are left as an exercise for the reader)

3(a)	PG1 is double pinion; PG2 is simple	Pinion 1-Pinion
3(b)	PG1 is double pinion; PG2 is simple	Pinion 2-Pinion
4(a)	PG1 is double pinion; PG2 is double pinion	Pinion 1-Pinion 1
4(b)	PG1 is double pinion; PG2 is double pinion	Pinion 1-Pinion 2
4(c)	PG1 is double pinion; PG2 is double pinion	Pinion 2-Pinion 1
4(d)	PG1 is double pinion; PG2 is double pinion	Pinion 2-Pinion 2

5   Long Pinion Transformations

Example 1

Let's take a look at the 3 PG design (Fig. 3) with 4 brakes and 2 clutches yielding 7 forward speeds and 1 reverse speed (see Raghavan et al.[12]) . If we focus on just the first 2 planetary gearsets with their fixed interconnections, the sub-system is as shown in Fig. 4(a). The associated equations are:

$$z_1 = -k_1 x_1 + (1 + k_1)y_1, \quad \text{(first planetary)} \tag{15}$$

$$z_2 = -k_2 x_2 + (1 + k_2)y_2, \quad \text{(second planetary)} \tag{16}$$

$$x_1 = z_2, \quad \text{(fixed interconnection 1)} \tag{17}$$

$$y_1 = y_2. \quad \text{(fixed interconnection 2)} \tag{18}$$

Substituting for x_1 from Eq. (17) and for y_1 from Eq. (18) into Eq. (15) we get

$$z_1 = -k_1 z_2 + (1 + k_1)y_2$$

$$z_1 = -k_1(-k_2 x_2 + (1 + k_2)y_2) + (1 + k_1)y_2$$

$$z_1 = (k_1 k_2)x_2 + (1 - k_1 k_2)y_2 \tag{19}$$

Therefore, Eqs. (19) and (16) taken together represent the physical system shown in Fig. 4(b), featuring a double pinion planetary gear set coupled to a simple planetary gear set via fixed interconnections, Ring to Ring, and Carrier to Carrier. Going a step further we may compare Eq. (19) with Eq. (13) which represents the governing equation of a double pinion planetary gear set which shares Pinion 1 and carrier with an adjacent simple planetary gear set. Equating corresponding coefficients (and noting that PG2 remains simple while PG1 gets transformed via the long pinion conversion) we get

$$k_{1_{original}} k_{2_{original}} = k_{1_{LP}} \xi_{LP} \frac{\xi_{1_{LP}}}{\xi_{2_{LP}}} \tag{20}$$

where the subscripts *original* and *LP* apply respectively to the original gear sets of Fig 4(a) and the Long Pinion gear sets that we wish to transform them into. Using the numerical values for the $\frac{Ring}{Sun}$ ratios from Fig. 4(a), we set $k_{1_{original}} = 1.87$, $k_{2_{original}} = 2.13$; also, setting $\xi_{LP} = 0.9$ and $k_{2_{LP}} = k_{2_{original}}$, we may compute $k_{1_{LP}}$ from Eq. (20). Its value is 3.34. By this process the arrangement in Fig. 4(b) with Ring-Ring and Carrier-Carrier fixed interconnections, transforms into the long pinion arrangement of Fig. 4(c), with $\left(\frac{Ring}{Sun}\right)_1 = 3.34$, $\left(\frac{Ring}{Sun}\right)_2 = 2.13$.

Kinematically, the 3 arrangements of Fig. 4 are equivalent. Therefore we may substitute the arrangement of Fig. 4(c) in place of the first 2 PGs in Fig. 4(a) to get Fig. 5 (see Raghavan[13]). Note that the speed ratios and clutch sequences are identical for the designs in Figs. 3 and 5.

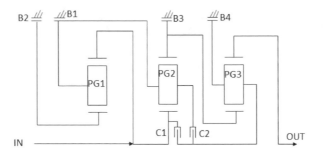

	Ratios	B1	B2	B3	B4	C1	C2
Reverse	-3.82		X		X		
Neutral	0.00				X		
1	3.98				X		X
2	2.04			X			X
3	1.35		X				X
4	1.00					X	X
5	0.78		X			X	
6	0.65			X		X	
7	0.56	X				X	

(X = engaged clutch)

Ratio Spread	7.11
Ratio Steps	
Rev/1	-0.96
1/2	1.95
2/3	1.51
3/4	1.35
4/5	1.27
5/6	1.21
6/7	1.16

$$\frac{R1}{S1} = 1.87, \frac{R2}{S2} = 2.13, \frac{R3}{S3} = 1.87$$

Fig. 3 Three Planetary Seven Speed Transmission

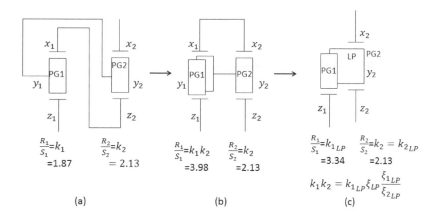

$$\frac{R_1}{S_1}=k_1 \qquad \frac{R_2}{S_2}=k_2$$
$$=1.87 \qquad =2.13$$

(a)

$$\frac{R_1}{S_1}=k_1k_2 \qquad \frac{R_2}{S_2}=k_2$$
$$=3.98 \qquad =2.13$$

(b)

$$\frac{R_1}{S_1}=k_{1LP} \qquad \frac{R_2}{S_2}=k_2=k_{2LP}$$
$$=3.34 \qquad =2.13$$

$$k_1k_2=k_{1LP}\xi_{LP}\frac{\xi_{1LP}}{\xi_{2LP}}$$

(c)

Fig. 4 Transformation to Long Pinion

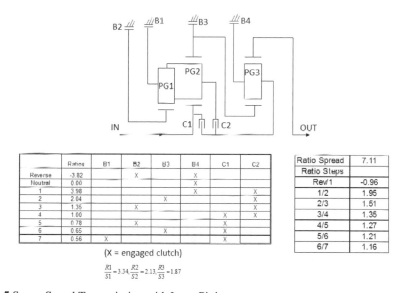

	Ratios	B1	B2	B3	B4	C1	C2
Reverse	-3.82		X		X		
Neutral	0.00				X		
1	3.98				X	X	
2	2.04			X		X	
3	1.35		X			X	
4	1.00					X	X
5	0.78		X			X	
6	0.65			X		X	
7	0.56	X				X	

(X = engaged clutch)

Ratio Spread	7.11
Ratio Steps	
Rev/1	-0.96
1/2	1.95
2/3	1.51
3/4	1.35
4/5	1.27
5/6	1.21
6/7	1.16

$$\frac{R1}{S1}=3.34, \frac{R2}{S2}=2.13, \frac{R3}{S3}=1.87$$

Fig. 5 Seven Speed Transmission with Long Pinion

Example 2

Let's take a look at another 3 PG design with 2 brakes and 3 clutches yielding 6 forward speeds and 1 reverse speed (see Bucknor et al.[14]). It is shown in Fig 6 along with the clutching table. If we focus on just the first 2 planetary gearsets with their fixed interconnections, the sub-system is as shown in Fig. 7(a). The associated equations are:

$$z_1 = -k_1x_1 + (1+k_1)y_1, \quad \text{(first planetary)} \tag{21}$$

$$z_2 = -k_2x_2 + (1+k_2)y_2, \quad \text{(second planetary)} \tag{22}$$

$$x_1 = y_2, \qquad \text{(fixed interconnection 1)} \qquad (23)$$

$$z_1 = z_2. \qquad \text{(fixed interconnection 2)} \qquad (24)$$

Substituting for y_2 from Eq. (23) and for z_2 from Eq. (24) into Eq. (22) we get

$$x_2 = \left(1 + \frac{(k_1+1)}{k_2}\right)x_1 - \left(\frac{k_1+1}{k_2}\right)y_1 \qquad (25)$$

Eqs. (21) and (25) taken together represent the physical system shown in Fig. 7(b), featuring a simple planetary gear set coupled to a double pinion planetary gear set via fixed interconnections, Ring to Ring, and Carrier to Carrier. Going a step further we may compare Eq. (25) with Eq. (13) which represents the governing equation of a double pinion planetary gear set which shares Pinion 1 and carrier with an adjacent simple planetary gear set. Equating corresponding coefficients we get

$$1 + \frac{\left(k_{1original}+1\right)}{k_{2original}} = k_{2LP}\xi_{LP}\frac{\xi_{2LP}}{\xi_{1LP}} \qquad (26)$$

where the subscripts *original* and *LP* apply respectively to the original gear sets of Fig 7(a) and the Long Pinion gear sets that we wish to transform them into.

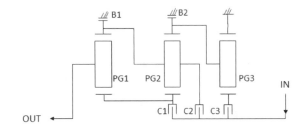

	Ratios	B1	B2	C1	C2	C3
Reverse	-3.81	X				X
Neutral	0	X				
1	3.9	X		X		
2	2.2		X	X		
3	1.52			X		X
4	1			X	X	
5	0.7				X	X
6	0.59		X		X	

(X = engaged clutch)

Ratio Spread	6.64
Ratio Steps	
Rev/1	-0.98
1/2	1.78
2/3	1.45
3/4	1.52
4/5	1.44
5/6	1.18

$$\frac{N_{R_1}}{N_{S_1}} = 2.90, \frac{N_{R_2}}{N_{S_2}} = 2.73, \frac{N_{R_3}}{N_{S_3}} = 1.67$$

Fig. 6 Three Planetary Six Speed Transmission

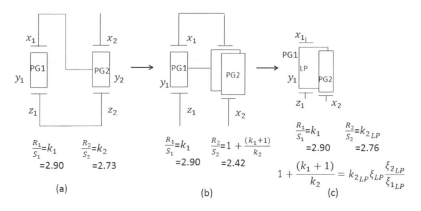

Fig. 7 Transformation to Long Pinion

Using the numerical values for the $\frac{Ring}{Sun}$ ratios from Fig. 6, we set $k_{1_{original}} =$ $1.57, k_{2_{original}} = 2.37$; also, setting $\xi_{LP} = 0.9$ and $k_{1_{LP}} = k_{1_{original}}$, we may compute $k_{2_{LP}}$ from Eq. (26). Its value is 2.76. By this process the arrangement in Fig. 7(b) with Ring-Ring and Carrier-Carrier fixed interconnections, transforms into the long pinion arrangement of Fig. 7(c), with $\left(\frac{Ring}{Sun}\right)_1 = 2.90, \left(\frac{Ring}{Sun}\right)_2 =$ 2.76. Kinematically, the 3 arrangements of Fig. 7 are equivalent. Therefore we may substitute the arrangement of Fig. 7(c) in place of the first 2 PGs in Fig. 6 to get Fig. 8 (see Raghavan[15]). Note that the speed ratios and clutch sequences are identical for Figs. 6 and 8.

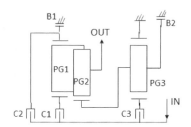

	Ratios	B1	B2	C1	C2	C3
Reverse	-3.81	X				X
Neutral	0	X				
1	3.9	X		X		
2	2.2		X	X		
3	1.52			X		X
4	1			X	X	
5	0.7				X	X
6	0.59		X		X	

(X = engaged clutch)

Ratio Spread	6.64
Ratio Steps	
Rev/1	-0.98
1/2	1.78
2/3	1.45
3/4	1.52
4/5	1.44
5/6	1.18

$$\frac{N_{R_1}}{N_{S_1}} = 2.90, \frac{N_{R_2}}{N_{S_2}} = 2.76, \frac{N_{R_3}}{N_{S_3}} = 1.67$$

Fig. 8 Six Speed Transmission with Long Pinion

6   Conclusions

In the preceding sections, we have attempted to demonstrate the algebraic use of long pinion arrangements in the creation of alternative architectures for automotive transmissions. Once a powerflow with favorable characteristics is established, there are many ways in which it can be mechanized. We have shown the use of long pinion arrangements to replace (Ring-Sun, Carrier-Carrier) connections as well as (Ring-Carrier, Sun-Sun) connections in adjacent planetary gear sets, while retaining functional equivalence. Various other permutations and combinations of fixed interconnections may also be worked out.

References

1. Benford, H., Leising, M.: The Lever Analogy: A New Tool in Transmission Analysis. Society of Automotive Engineers, Paper No. 810102 (1981)
2. Baran, J., Hendrickson, J., Solt, M.: General Motors New Hydra-Matic RWD Six-Speed Automatic Transmission Family, SAE 2006-01-0846
3. Borgerson, J., Maguire, J., Kienzle, K.: Transmission with Long Ring Planetary Gearset, U.S. Patent 7,029,417 (April 18, 2006)
4. Lewis, C., Bollwahn, B.: General Motors Hydra-Matic & Ford New FWD Six-Speed Automatic Transmission Family, SAE 2007-01-1095
5. Wittkopp, S.: Seven-Speed Transmission, U.S. Patent 6,910,986 (June 28, 2005)
6. Robinette, D., Powell, M.: Optimizing 12V Start-Stop for Conventional Powertrains, SAE 2011-01-0699
7. Hawkins, S., Billotto, F., Cottrell, D., Houtman, A., Poulos, S., Rademacher, R., Van Maanen, K., Wilson, D.: Development of General Motors' eAssist Powertrain. SAE Int. J. Alt. Power 5(1) (June 2012)
8. Schmidt, M.: Two-Mode, Compound-Split Electromechanical Vehicular Transmission, U.S. Patent 5,931,757 (August 3, 1999)
9. Grewe, T., Conlon, B., Holmes, A.: Defining the General Motors 2-Mode Hybrid Transmission, SAE 2007-01-0273
10. Miller, M., Holmes, A., Conlon, B., Savagian, P.: The GM Voltec 4ET50 Multi-Mode Electric Transaxle, SAE 2011-01-0887
11. Shigley, J.E., Uicker, J.J.: Theory of Machines and Mechanisms. McGraw-Hill (1980)
12. Raghavan, M., Kao, C.-K., Usoro, P.: Family of Multi-Speed Planetary Transmission Mechanisms having Fixed Interconnections and Six Torque-Transmitting Mechanisms, U.S. Patent 6,679,802 (January 20, 2004)
13. Raghavan, M.: Multi-Speed Transmissions with a Long Pinion and One Fixed Interconnection, US Patent 7,699,745 (April 20, 2010)
14. Bucknor, N., Raghavan, M., Usoro, P.: Family of Multi-Speed Transmission Mechanisms having Three Planetary Gearsets and Three Input Torque-Transmitting Mechanisms, US Patent 6,705,969 (March 16, 2004)
15. Raghavan, M.: Multi-Speed Transmissions with a Long Pinion and Four Fixed Interconnections, U.S. Patent 7,429,229 (September 30, 2008)

14 Time-Optimal Path Planning for the General Waiter Motion Problem

Francisco Geu Flores and Andrés Kecskeméthy

Dedicated to Professor Ken Waldron on occasion of his 70th birthday.

Abstract. This paper presents a direct solution approach for the so-called general waiter motion problem, which consists in moving a tablet as fast as possible from one pose to the other such that non of the objects resting on the tablet slides at any time. The question is akin to several industrial problems in which tangential forces are restricted due to functional reasons, such as suction grippers, motion of sensitive goods, etc. In contrast to existing approaches which parametrize the problem in configuration (joint) space, we decompose the overall task into two cascaded main components: shaping the optimal geometry of the spatial path, and finding the time optimal one-dimensional motion of the system along this path. The spatial path is parametrized using via poses in SE(3), making it possible to reduce the search space to significant physical subspaces, and to interact intuitively with the user. The overall optimization is subdivided into a series of subproblems with cost functions and search spaces of increasing fineness, such that each subproblem can be solved with the output of its predecessor. A solution of the waiter motion problem with four objects illustrates the applicability of the algorithm.

1 Problem Statement

Discussed in this paper is a solution procedure to the so-called general waiter motion problem, which consists in computing the time-optimal motion of a six-degrees-of-freedom manipulator carrying a tablet with an arbitrary number of objects resting on it from an initial pose $\mathcal{K}_{E0}$ to a final pose $\mathcal{K}_{Ef}$ as fast as possible, such that non of the objects slides at any time (see Fig. 1). This problem is akin to (offline) motion planning problems in which tangential forces at contacts are restricted due

Francisco Geu Flores · Andrés Kecskeméthy
Lehrstuhl für Mechanik und Robotik, Universität Duisburg-Essen,
Lotharstr. 1, 47057 Duisburg, Germany
e-mail: {francisco.geu,andres.kecskemethy}@uni-due.de

V. Kumar et al. (Eds.): *Adv. in Mech., Rob. & Des. Educ. & Res.*, MMS 14, pp. 189–203.
DOI: 10.1007/978-3-319-00398-6_14   © Springer International Publishing Switzerland 2013

to functional constraints (e.g. suction grippers, motion of sensitive goods). Most state-of-the-art motion planning algorithms use spline functions in *configuration coordinates* (i.e., robot joint coordinates) to transform the optimal control problem into a nonlinear optimization problem [1]. Hereby, the spline domain is defined either as a linear mapping of the cycle time [1] — which allows for a direct search of the optimal time intervals between the inner spline knots — or as a general spline parameter decoupled from time [2] — which requires the additional computation of the optimal time history of the spline parameter for a given set of spline coefficients and knots. However, such algorithms display difficulties to converge for coupled shape/time optimality problems such as the one presented in this paper, mainly because the overall optimization problem's extreme nonlinearity causes the optimizer to remain trapped in local minima or even in infeasible regions.

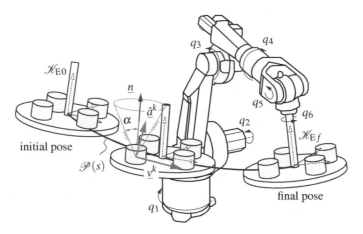

Fig. 1 General waiter-motion problem

In this paper, we propose a novel procedure for coupled shape/time optimality problems by (1) using target (i.e. task space) coordinates instead of configuration coordinates and (2) parametrizing the spline shape functions in terms of via-poses instead of using the spline coefficients and knots. This allows for optimizing over a more natural function basis and for restricting the search space during optimization to significant subspaces. Hence, the condition of the optimization problem is improved and less and computationally cheaper iterations are required for convergence. To this end, the robot's motion is decomposed in two cascaded main components: the spatial path $\mathscr{P}(s) \in \mathrm{SE}(3)$ traversed by the robot end-effector (the tablet) $\mathscr{K}_E$, and the one-dimensional motion $s(t)$ of the tablet along this path, where s denotes the path coordinate along the curve described by $\mathscr{K}_E$. The spatial path $\mathscr{P}(s)$ is parametrized using fifth and third-order fitted B-splines for translation and rotation, respectively, and the motion $s(t)$ along the path is computed by solving the time-optimal problem using the well-known and classical methods of robotics developed by [3], [4] and [5]. In order to warrant convergence, the overall optimal control problem is formulated as a sequence of nonlinear optimization problems of increasing fineness, such that each subproblem takes the solution of its predecessor as initial guess. Hereby, each subproblem is solved by standard gradient-based optimization routines. This shows that the decomposition of the problem into spatial shape and time-optimal motion is instrumental in finding the coupled shape/time optimal solution. The proposed method can be used to handle a broad spectrum of applications, ranging from optimal roller-coaster design up to minimal-cycle-time excavator operations under power constraints ([6], [7]).

The rest of the paper is organized as follows. Section 2 describes the model of a spatial path as a kinetostatic transmission element – termed "curved joint" – transmitting motion and forces. In Section 3, the time-optimal problem along a specified trajectory is solved using an extension of the well-known methods from robotics ([3], [4], [5]) to general multibody systems, which is easily possible due to the employed object-oriented approach. Section 4 finally describes the application of the developed methods to a 4-object waiter motion problem and the solutions obtained.

2 Generic Properties of Spatial Paths in Multibody Systems

Let a general spatial path be given by the pose of an output frame $\mathcal{K}_E = \mathcal{P}(s) \subset$ SE(3), with the translation part of $\mathcal{P}(s)$ described by a vector $\Delta \underline{r}(s)$ and the rotation part parametrized by a rotation matrix $\Delta \mathbf{R}(s)$, both measured with respect to a basis frame $\mathcal{K}_1$. Let the coordinate s be the path length of the spatial curve followed by $\mathcal{K}_E$, as shown in Fig. 2.

The pose of $\mathcal{K}_E$ can be computed as a function of the pose of the basis frame $\mathcal{K}_1$ and the path coordinate s as

$$\begin{aligned} {}^{0}\mathbf{R}_E &= {}^{0}\mathbf{R}_1 \Delta \mathbf{R}, && \text{with} \quad \Delta \mathbf{R} = \Delta \mathbf{R}(s) \\ \underline{r}_E &= \Delta \mathbf{R}^{\mathrm{T}} \left(\underline{r}_1 + \Delta \underline{r} \right), && \text{with} \quad \Delta \underline{r} = \Delta \underline{r}(s). \end{aligned} \tag{1}$$

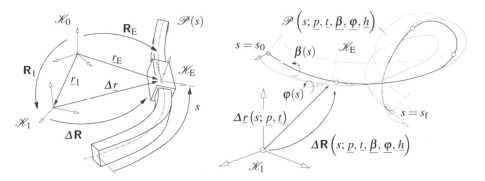

Fig. 2 Curve joint as a kinetostatic transmission element ([8])

The angular and linear absolute velocities of $\mathscr{K}_E$ can be computed as

$$
\underline{t}_E = \begin{bmatrix} \underline{\omega}_E \\ \underline{v}_E \end{bmatrix} = \underbrace{\begin{bmatrix} \Delta\mathbf{R}^T & 0 \\ -\Delta\mathbf{R}^T\widetilde{\Delta\underline{r}} & \Delta\mathbf{R}^T \end{bmatrix}}_{\mathbf{J}_g} \begin{bmatrix} \underline{\omega}_1 \\ \underline{v}_1 \end{bmatrix} + \mathbf{J}_{\mathscr{P}}\dot{s}, \tag{2}
$$

where $\mathbf{J}_g$ is the rigid-body Jacobian, and $\mathbf{J}_{\mathscr{P}}$ is the Jacobian mapping the path velocity $\dot{s}$ along the spatial path to the twist $\underline{t}_E$ at the output frame $\mathscr{K}_E$.

The angular and linear absolute accelerations of $\mathscr{K}_E$ can be computed as

$$
\dot{\underline{t}}_E = \begin{bmatrix} \dot{\underline{\omega}}_E \\ \underline{a}_E \end{bmatrix} = \mathbf{J}_g \begin{bmatrix} \dot{\underline{\omega}}_1 \\ \underline{a}_1 \end{bmatrix} + \begin{bmatrix} 0 \\ 2\widetilde{\underline{\omega}}_1^2 \Delta\mathbf{R}^T \Delta\underline{r} \end{bmatrix} + \mathbf{J}_{\mathscr{P}}\ddot{s} + \mathbf{J}_{\mathscr{P}}'\dot{s}^2 + \begin{bmatrix} 2\widetilde{\underline{\omega}}_1 & 0 \\ 0 & 2\,{}^2\widetilde{\underline{\omega}}_1 \end{bmatrix} \mathbf{J}_{\mathscr{P}}\dot{s}, \tag{3}
$$

where $(.)'$ denotes the derivative with respect to the path coordinate s.

Furthermore, the duality of velocity and force transmission yields

$$
\begin{bmatrix} \underline{\tau}_1 \\ \underline{f}_1 \\ Q_s \end{bmatrix} = \begin{bmatrix} \mathbf{J}_g^T \\ \mathbf{J}_{\mathscr{P}}^T \end{bmatrix} \begin{bmatrix} \underline{\tau}_E \\ \underline{f}_E \end{bmatrix}, \tag{4}
$$

where $\underline{\tau}_1$, $\underline{f}_1$ are the torque and force at frame $\mathscr{K}_1$ and $\underline{f}_1$ is the force projected along $\mathscr{P}(s)$. The functions $\Delta\underline{r}(s)$, $\Delta\mathbf{R}(s)$, and $\mathbf{J}_{\mathscr{P}}$ depend on the description of the spatial path $\mathscr{P}(s)$.

In this work, the parametrization $\Delta\underline{r}(s)$ of the spatial curve is computed by interpolating relative via-positions of the origin of the output frame $\mathscr{K}_E$ with respect to the input frame $\mathscr{K}_1$ using smoothing splines with end-point derivative constraints (curve-fitting routine concur [9]). The orientation of frame $\mathscr{K}_E$ with respect to $\mathscr{K}_1$, i.e. the rotation matrix $\Delta\mathbf{R}(s)$, is prescribed by means rotations about the tangential and normal directions of a DARBOUX frame ([8]), as shown in Fig. 2. In the figure, the via-points are collected in vector $\underline{p}$, the boundary conditions of $\Delta\underline{r}(s)$ are collected in vector $\underline{t}$, the horizon via-vectors are collected in vector $\underline{h}$, and the via-angles in tangential and normal directions are collected in vectors $\underline{\beta}$ and $\underline{\varphi}$, respectively.

3 Time-Optimal Motion Along a Given Spatial Path

3.1 Formulation of the Time-Optimal Problem

Let $\underline{\varphi}_q$ represent the direct kinematics of a multibody system consisting of r massive rigid bodies, p dependent joint coordinates β_j, n independent joint coordinates q_i, collected in the vector $\underline{q} \in \mathbf{R}^n$, as well as an end effector frame $\mathscr{K}_E$, as shown in Fig. 3. Let the spatial motion of the multibody system be given in target coordinates $\mathscr{K}_E = \mathscr{P}(s) \in SE(3)$ as a function of a the path coordinate s. The configuration of the system is uniquely defined by s, and its state is uniquely defined by the vector $[s, \dot{s}]^T$. If all configurations $\mathscr{P}(s)$ are reachable, both system and spatial path can be

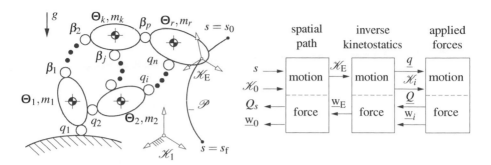

Fig. 3 Multibody system performing a task along the spatial path $\mathscr{P}$

regarded as a single kinetostatic transmission element mapping the path coordinate s to the joint motion q and end-effector motion $\mathscr{K}_E$ (see Fig. 3).

The corresponding joint motion is hereby described by the equations

$$q = \underline{\varphi}_q^{-1}(\mathscr{P}(s))$$

$$\dot{q} = \mathbf{J}_\varphi^{-1}\mathbf{J}_{\mathscr{P}}\dot{s}$$

$$\ddot{q} = \mathbf{J}_\varphi^{-1}\mathbf{J}_{\mathscr{P}}\ddot{s} + \left[\mathbf{J}_\varphi'^{-1}\mathbf{J}_{\mathscr{P}} + \mathbf{J}_\varphi^{-1}\mathbf{J}_{\mathscr{P}}'\right]\dot{s}^2 , \tag{5}$$

where $\mathbf{J}_\varphi = \partial\underline{\varphi}_q/\partial\underline{q}$ represents the transmission Jacobian of the multibody system. The prescribed spatial path $\mathscr{P}(s)$ as well as its corresponding Jacobian $\mathbf{J}_{\mathscr{P}}$ describe the kinematical transmission of a curve joint as presented in Sect. 2.

The velocities and accelerations at the independent joints and at the end-effector have the general form

$$\dot{q} = {}_q\mathbf{J}_s\,\dot{s}, \qquad \ddot{q} = {}_q\mathbf{J}_s\,\ddot{s} + {}_q\mathbf{J}_s'\,\dot{s}^2 ,$$

$$\underline{t}_E = {}_E\mathbf{J}_s\,\dot{s}, \qquad \underline{\dot{t}}_E = {}_E\mathbf{J}_s\,\ddot{s} + {}_E\mathbf{J}_s'\,\dot{s}^2 , \tag{6}$$

Let now the dynamics of the multibody system be described by the differential equations in minimal form

$$\mathbf{M}(\underline{q})\underline{\ddot{q}} + \underline{b}(\underline{q},\underline{\dot{q}}) - \underline{Q}_e(\underline{q},\underline{\dot{q}}) - \underline{Q}_G(\underline{q}) = \underline{Q}, \tag{7}$$

where $\mathbf{M}$ is the $n \times n$ mass matrix of the multibody system, $\underline{b}(\underline{q},\underline{\dot{q}})$ is the n-dimensional vector containing the centripetal and Coriolis terms, $\underline{Q}_G(\underline{q})$ is an n-dimensional vector containing the projection of the gravitational forces on the generalized coordinates, $\underline{Q}_e(\underline{q},\underline{\dot{q}})$ is an n-dimensional vector containing the projection of general external forces, and $\underline{Q}$ is an n-dimensional vector collecting the generalized actuator forces.

Let the velocities $\underline{\dot{q}}$, accelerations $\underline{\ddot{q}}$, and generalized actuator forces $\underline{Q}$ at the joints be constrained by equations of the form

$$\dot{q}^{\min}(q) \leq \dot{q} \leq \dot{q}^{\max}(q)$$
$$\ddot{q}^{\min}(q,\dot{q}) \leq \ddot{q} \leq \ddot{q}^{\max}(q,\dot{q}) \qquad (8)$$
$$\underline{Q}^{\min}(q,\dot{q}) \leq \underline{Q} \leq \underline{Q}^{\max}(q,\dot{q}),$$

and the velocities $\mathfrak{t}_E$ and accelerations $\mathfrak{i}_E$ at the end-effector be constrained by equations of the form

$$\mathfrak{t}_E^{\min}(\mathscr{K}_E) \leq \mathfrak{t}_E \leq \mathfrak{t}_E^{\max}(\mathscr{K}_E)$$
$$\mathfrak{i}_E^{\min}(\mathscr{K}_E,\mathfrak{t}_E) \leq \mathfrak{i}_E \leq \mathfrak{i}_E^{\max}(\mathscr{K}_E,\mathfrak{t}_E). \qquad (9)$$

With the relations Eq. 6, the equations of motion described in Eq. 7 can be written in terms of the motion coordinate s as

$$\underline{m}(s)\ddot{s} + \underline{c}(s,\dot{s}) + \underline{d}(s) = \underline{Q}, \qquad (10)$$

with $\underline{m}(s) = \mathbf{M}(s)_q \mathbf{J}_s$, $\underline{c}(s,\dot{s}) = [\mathbf{M}(s)_q \mathbf{J}_s' + \underline{b}(s)] \dot{s}^2 - \underline{Q}_e(s,\dot{s})$, and $\underline{d}(s) = -\underline{Q}_G(s)$, where the components m_i and c_i of vectors $\underline{m}$ and $\underline{c}$ represent the effective inertia and velocity forces at every independent joint, respectively, and the term $\underline{b}(q,\dot{q})$ in Eq. 7 can be written as $\underline{b}(s) \dot{s}^2$, with $\underline{b}(s)$ depending only on the configuration s.

Furthermore, Eq. 6 allows for all constraints of the form Eq. 8 and Eq. 9 to be collected in the vector inequality

$$\hat{\underline{b}}_1(s,\dot{s}) \leq \hat{\underline{m}}(s)\ddot{s} \leq \hat{\underline{b}}_2(s,\dot{s}), \qquad (11)$$

where $\hat{\underline{b}}_1, \hat{\underline{b}}_2$ and $\hat{\underline{m}}$ are vectors in $\mathbf{R}^l$ and l is the number of constraints.

The left and right sides of this inequality define the set of admissible states $[s,\dot{s}]^T$ and can be written compactly as the scalar inequality

$$G(s,\dot{s}) \leq 0, \qquad (12)$$

with $G(s,\dot{s}) = \max\{\hat{b}_{1j}(s,\dot{s}) - \hat{b}_{2j}(s,\dot{s})\}$, for all $j = 1,2,\cdots,\ell$.

For all constraints j for which $\hat{m}_j(s)$ does not vanish, Eq. 11 further limits the acceleration $\ddot{s}$ along the spatial path, since it must hold

$$\frac{\hat{b}_{1j}(s,\dot{s})}{|\hat{m}_j(s)|} \leq \text{sgn}[\hat{m}_j(s)]\ddot{s} \leq \frac{\hat{b}_{2j}(s,\dot{s})}{|\hat{m}_j(s)|}, \qquad (13)$$

or compactly $l_j(s,\dot{s}) \leq \ddot{s} \leq u_j(s,\dot{s})$, where $l_j(s,\dot{s})$ and $u_j(s,\dot{s})$ are the lower and upper bounds of the j-th constraint, functions of the state $[s,\dot{s}]^T$. These equations can be rewritten as the one dimensional inequality

$$L(s,\dot{s}) \leq \ddot{s} \leq U(s,\dot{s}), \qquad (14)$$

where $L(s,\dot{s}) = \max\{l_j(s,\dot{s})\}$ and $U(s,\dot{s}) = \min\{u_j(s,\dot{s})\}$ for all $j = 1,2,\cdots,\ell$ for which $\hat{m}_j(s) \neq 0$.

Sought is the optimal motion law $s(t)$ which minimizes the total cycle time t_f that the system needs to move from a state $[s_0,\dot{s}_0]^T$ to state $[s_f,\dot{s}_f]^T$ without violating the constraints defined by Eq. 14 and Eq. 12.

3.2 Computation of the Dynamic Constraints

All terms in Eq. 11 can be computed at every state $[s_{act}, \dot{s}_{act}]^T$ by using the object-oriented approach described in [10]. The residual forces resulting from the computation of the motion and force transmissions, namely

$$\overline{Q} = \varphi_S^{D^{-1}}(s, \dot{s}, \ddot{s}) = -\underline{m}(s)\,\ddot{s} - \underline{c}(s, \dot{s}) - \underline{d}(s), \tag{15}$$

can be used to generate $\underline{m}$, $\underline{c}$ and $\underline{d}$ of equation Eq. 10 at every configuration s by the following simplified procedure:

a) Computation of $\underline{d}$: Set, at the input of $\varphi_S^{D^{-1}}$, the configuration to $s = s_{act}$, the generalized velocity to $\dot{s} = 0$ and the generalized acceleration to $\ddot{s} = 0$. Then, the terms $\underline{m}\ddot{s}$ and $\underline{c}$ of Eq. 15 vanish and the residual force $\overline{Q}$ obtained at the input is exactly $-\underline{d}$.

b) Computation of $\underline{c}$: Eliminate the term $\underline{d}$ in the calculation of $\varphi_S^{D^{-1}}$ by 'switching off' the gravitational forces $\underline{Q}_G$, and set, at the input of $\varphi_S^{D^{-1}}$, the configuration to $s = s_{act}$, the generalized velocity to $\dot{s} = \dot{s}_{act}$ and the generalized acceleration to $\ddot{s} = 0$. Then, the term $\underline{m}\ddot{s}$ of Eq. 15 vanishes and the residual force $\overline{Q}$ obtained at the input is exactly $-\underline{c}$.

c) Computation of $\underline{m}$: Similarly, eliminate the terms $\underline{c}$ and $\underline{d}$ in the calculation of $\varphi_S^{D^{-1}}$ and set the input acceleration to $\ddot{s} = 1$. Then, the resulting force $\overline{Q}$ is exactly $-\underline{m}$.

The generation of the equations of motion by this approach requires 3 traversals of the inverse dynamics $\varphi_S^{D^{-1}}$ for one set of equations.

The Jacobians $_E J_s$ and $_q J_s$ of equations Eq. 6 can be computed similarly using the so-called kinematical differentials (see [10]).

3.3 Solution of the Time-Optimal Problem

It can be shown that the time-optimal motion along a given spatial path $\mathscr{P}(s)$ under the above mentioned constraints is composed exclusively of trajectories with maximal or minimal acceleration. Time-optimality implies that the time-optimal motion law $s(t)$ is a monotonically increasing time function, i.e. that the velocity $\dot{s}$ along the trajectory is further constrained by

$$\dot{s} \geq 0 \tag{16}$$

at every configuration s. Eq. 16 and Eq. 12 as well as Eq. 14 form a set of velocity and acceleration limits at every configuration s, which determines an admissible region in the plane $\dot{s} - \ddot{s}$ for each configuration s (see Fig. 4).

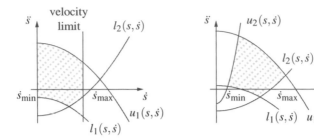

Fig. 4 Admissible acceleration region for given configuration s

If the functions $l_j(s,\dot{s})$, $u_j(s,\dot{s})$ are linear in $\dot{s}^2$, one simple interval $[\dot{s}_{\min}, \dot{s}_{\max}]$ of admissible velocities can be defined for every configuration s (Otherwise, the admissible regions may consist of not connected sub-regions leading to a set of not connected admissible velocity intervals.). In this case, the equations Eq. 11 can be written as

$$\underline{\hat{c}}_1(s)\dot{s}^2 + \underline{\hat{d}}_1(s) \leq \underline{\hat{m}}(s)\ddot{s} \leq \underline{\hat{c}}_2(s)\dot{s}^2 + \underline{\hat{d}}_2(s), \tag{17}$$

the resulting function $G(s,\dot{s})$ (see Eq. 12) is also linear in $\dot{s}^2$, the admissible regions in the plane $\dot{s} - \ddot{s}$ are simply connected, and the set of admissible states has no holes in its interior. This allows for the definition of the functions $\dot{s}_{\min}(s)$ and $\dot{s}_{\max}(s)$ describing the maximally and minimally allowed velocities $\dot{s}$ as a function of the motion coordinate s, called hereafter "lower limiting curve" and "upper limiting curve", respectively. The area between both curves contains the set of admissible states, as shown in Fig. 5.

The states lying on the upper limiting curve $\dot{s}_{\max}(s)$ are classified using Eq. 14 in:

a) sinks, if $U(s,\dot{s})$ and $L(s,\dot{s})$ are defined and

$$U(s,\dot{s}) = L(s,\dot{s}) > \dot{s}_{\max}\, d\dot{s}_{\max}/ds; \tag{18}$$

b) sources, if $U(s,\dot{s})$ and $L(s,\dot{s})$ are defined and

$$U(s,\dot{s}) = L(s,\dot{s}) < \dot{s}_{\max}\, d\dot{s}_{\max}/ds; \quad \text{or} \tag{19}$$

c) tangent points, elsewhere.

Moreover, the tangent points at which the velocity constraints described in Eq. 12 are active are called singular points, or singular arcs if they are connected.

The solution to the time-optimal motion is a sequence of branches of maximal acceleration and maximal deceleration which lies in the feasible region and touches tangentially the upper limiting curve. At states $[s,\dot{s}]^{\mathrm{T}}$ lying inside the feasible region, the solution consists of segments with maximal acceleration $U(s,\dot{s})$ and segments with maximal deceleration $L(s,\dot{s})$. At singular points, the maximal acceleration is further bounded by the upper limiting curve tangent $d\dot{s}_{\max}/ds$.

With these definitions, the following algorithm based on the one proposed in [5] has been constructed:

Step 0: Check if the initial state s_0 and the final state s_f are feasible for the given initial velocity $\dot{s}_0$ and final velocity $\dot{s}_f$ respectively. If not, the problem has no feasible solution.

Step 1: Set a counter k to 1. Integrate the equation $\ddot{s} = \max\{l_j(s,\dot{s})\}$ backwards in time from the final state $s = s_f$, $\dot{s} = \dot{s}_f$ until leaving the feasible region. Name the computed deceleration curve $\dot{s}^d(s)$.

Step 2: Integrate the equation $\ddot{s} = \min\{u_j(s,\dot{s})\}$ forwards in time from the initial state $s = s_0$, $\dot{s} = \dot{s}_0$ until leaving the feasible region. Name the computed acceleration curve $\dot{s}_k^a(s)$. If the acceleration curve $\dot{s}_k^a(s)$ crosses the lower limiting curve $\dot{s}_{\min}(s)$, the problem is not feasible and the algorithm should be terminated. Else, continue.

Step 3: If $\dot{s}_k^a(s)$ crosses the deceleration curve $\dot{s}^d(s)$ terminate the algorithm: the intersection of both curves is the only switching point S_k. Otherwise, continue.

Step 4: Search forwards on the upper limiting curve $\dot{s}_{\max}(s)$ for the next tangent point S_{k+1}. The point S_{k+1} is a switching point candidate.

Step 5: Integrate the equation $\ddot{s} = \max\{l_j(s,\dot{s}),\ \dot{s}_{\max}\,d\dot{s}_{\max}/ds\}$ backwards in time from the state S_{k+1} until crossing one of the acceleration curves $\dot{s}_\ell^a(s)$, with $1 \le \ell < k$. The intersection of both curves is the switching point S_ℓ. Set $k = \ell$. Disregard the candidates S_r, with $r \le \ell$.

Step 6: Integrate the equation $\ddot{s} = \min\{u_j(s,\dot{s}),\ \dot{s}_{\max}\,d\dot{s}_{\max}/ds\}$ forward in time from the state S_k until leaving the feasible region. Add one to the counter k. Name the computed acceleration curve $\dot{s}_k^a(s)$. If the acceleration curve $\dot{s}_k^a(s)$ crosses the lower limiting curve $\dot{s}_{\min}(s)$, the problem is not feasible and the algorithm should be terminated. Else, go to step (3).

Fig. 5 shows how a typical solution looks like.

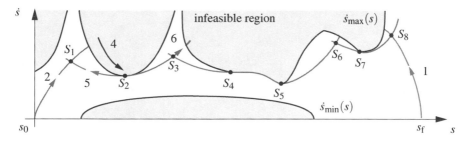

Fig. 5 Time-optimal solution algorithm (no islands)

4 General Waiter Motion Problem

4.1 Waiter Motion Problem along a Given Path

The waiter motion problem considers two set of constraints. One set is given as constant limits in the joint velocities and accelerations $\dot{q}_i^{\min}$, $\dot{q}_i^{\max}$, $\ddot{q}_i^{\min}$ and $\ddot{q}_i^{\max}$, which are typically provided by the manufacturer and are pre-programmed in the robot controller as soft-limits. A second set is given by the sticking ('no sliding') condition for every object k on the tablet

$$\frac{\hat{\underline{a}}^k \cdot \underline{n}}{\|\hat{\underline{a}}^k\|_2} \ge \cos\alpha, \quad \text{with} \quad \hat{\underline{a}}^k = \underline{a}^k + \underline{g}, \tag{20}$$

where $\underline{a}^k$ is the acceleration of object k, $\underline{g}$ is the gravity vector, $\underline{n}$ is the normal vector of the tablet plane, and $\mu_0 = \tan\alpha$ is the dry friction coefficient between the tablet and the objects. These k additional dynamic constraints can be rewritten as

$$\sqrt{[\hat{a}_x^k]^2 + [\hat{a}_y^k]^2} \le \mu_0 \hat{a}_z^k, \quad \text{with} \quad \hat{\underline{a}}^k = {}_k\mathbf{J}_s \ddot{s} + {}_k\mathbf{J}'_s \dot{s}^2 - \mathbf{R}_k^\mathrm{T}\underline{g}, \tag{21}$$

where ${}_k\mathbf{J}_s$ is the Jacobian mapping the linear velocities $\dot{s}$ along the spatial path to the velocities of object k, and $\mathbf{R}_k$ is the transformation matrix from the inertial frame to the local coordinate frame of object k.

Clearly, equations 21 are nonlinear in the unknowns $\ddot{s}$, which makes their treatment with the previous methods infeasible. However, it is possible to approximate these constraints by replacing the friction cone by a friction polyhedron given by the equations

$$\begin{aligned}
(1) \qquad & |\hat{a}_x^k| \le \mu_0 \hat{a}_z^k, \\
(2) \qquad & \left|\frac{\hat{a}_x^k}{\tan(\varphi_i)} + \hat{a}_y^k\right| \le \frac{\mu_0 \hat{a}_z^k \cos(\varphi_1)}{\sin(\varphi_i)},
\end{aligned} \tag{22}$$

defined by the discretization angles

$$\varphi_i = \frac{i\pi}{2^{p-1}}, \quad i = 1, 2, \ldots, 2^{p-1} - 1, \tag{23}$$

for each $\hat{a}_z^k$ (see Fig. 6).

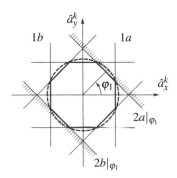

Fig. 6 Cone of friction for $p = 3$

Eq. 22 together with the joint velocity and acceleration limits form a system of constraints which are linear in $\dot{s}^2$, so that simply connected admissible acceleration regions are guaranteed. The approximation can be arbitrarily refined by choosing a sufficiently large integer p. Higher numbers p yield better cycle times, though increasing considerably the size of the problem and, hence, the computational effort required to solve it.

The solution for the case of four objects symmetrically distributed on the tablet, a friction polyhedron approximation with $p = 4$, and a given path is shown in Fig. 7.

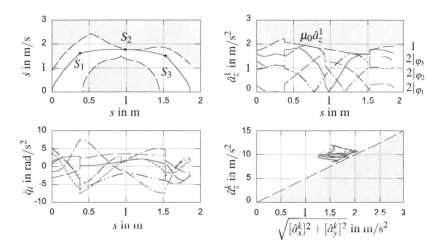

Fig. 7 Optimized waiter-motion along a given trajectory with four objects and $p = 3$ - phase plot, joint accelerations, linearized dry-friction constraints for object 1, and accelerations of all objects

4.2   Optimization of the Spatial Path Geometry

The method presented in Sect. 4.1 assigns an optimal cycle time to every feasible spatial path $\mathscr{P}(s; \underline{\zeta})$. allowing for a search for the optimal set $\underline{\zeta}^*$ of parameters yielding the smallest cycle time. In order to reduce the search to significant physical subspaces, we propose to define the set of optimization parameters

$$\underline{\xi} = \left[\underline{\varphi}^{\mathrm{T}}, \underline{\beta}^{\mathrm{T}}, \Delta\underline{\theta}_{z0}^{\mathrm{T}}, \Delta\underline{\theta}_{zf}^{\mathrm{T}}, \Delta\underline{\rho}_y\right]^{\mathrm{T}}, \tag{24}$$

where $\underline{\xi}$ describes the geometry of the spatial path $\mathscr{P}(s; \underline{\zeta})$ with respect to a reference geometry $\mathscr{P}(s; \underline{\zeta}_0)$. Vector $\Delta\underline{\rho}_y$ collects m center-line via-point displacements in transversal direction $\underline{e}_{yj}$, so that the modified via-points are defined as

$$\underline{p}_j = \underline{p}_j(\underline{\zeta}_0) + \Delta\rho_{yj} \cdot \underline{e}_{yj}(\underline{\zeta}_0) \quad \forall j = 1, 2, \ldots, m. \tag{25}$$

The angles $\Delta\theta_{z0}$ and $\Delta\theta_{zf}$ describe $m_t = 2$ displacements in the first-order boundary conditions such that

$$t_0 = \text{Rot}[z, \Delta\theta_{z0}] \cdot t_0(\underline{\zeta}_0) \tag{26}$$

$$t_f = \text{Rot}[z, \Delta\theta_{zf}] \cdot t_f(\underline{\zeta}_0), \tag{27}$$

where t_0 and t_f are the tangents of $\Delta\underline{r}(s)$ at its beginning and its end, respectively. Finally, $\underline{\beta}$ and $\underline{\varphi}$ collect m_β and m_φ via-angles describing rotations about the tangential and normal directions of the DARBOUX frame, respectively.

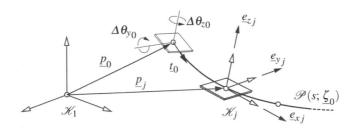

Fig. 8 Spatial path design parameters

The optimization problem is stated as

$$\text{minimize}$$
$$F(\underline{\xi}) = t_f, \tag{28}$$

where t_f is the optimal cycle time along the spatial path $\mathscr{P}(s; \underline{\zeta})$ as computed in Sect. 2. If $\mathscr{P}(s; \underline{\zeta})$ is not feasible, $F(\underline{\xi})$ is set to a very large number. The fineness of the optimization subproblems is improved by increasing the number of via-points m_β, m_φ and m. Each subproblem uses the results of the previous one as an initial guess and reference geometry and is repeated until no significant improvement of the cost function is achieved. The algorithm stops when neither a further attempt to solve a subproblem nor a refinement of the optimization parameters yields a cost improvement greater than the optimization tolerance.

Table 1 and Fig. 9 show the results for the case of a KUKA robot KR-15-2 carrying a tablet with 4 objects between two given tablet poses for a cone of friction approximation with $p = 4$. One can clearly see that the optimal solution is "pressed" almost during 50% of the path length to the constraint boundary (second plot of Fig. 9). This is an interesting behavior showing that, at least in this case, the optimal path shape also forms the constraint boundary such that it minimizes the area between optimal-time trajectory and upper limit curve (compare to Fig. 7, where the boundary is much more "ragged"). Moreover, one can appreciate the low curvatures of the solution path in target space. This shows that spatial path optimization in task space indeed allows for restricting the optimization search to lower-dimensional

Table 1 Multistep computation of the general waiter-motion problem with $p = 4$, with B-splines of degree $k = 3$ and the NAG routine e04unc ($dt = 1.0e^{-4}$, $relTol = absTol = 1.0e^{-10}$, $ds = 1.0e^{-4}$, $optimTol = 2.0e^{-2}$, $fPrec = 1.0e^{-2}$, $cDiffInt = fDiffInt = 1.0e^{-3}$, $stepLimit = 2.0e^{-2}$), on a processor Intel(R) Core(TM) i7-950 @ 3.07GHz

step	m_β	m_φ	m_t	t_f (s)	t_f (s)	CPU (s)
initial	2	2	-	-	2.0985	-
1 (2 ×)	2	2	-	-	1.6411	2872.41
2 (1 ×)	2	2	1	2	1.6409	483.70
3 (1 ×)	5	5	2	2	1.6404	1083.57

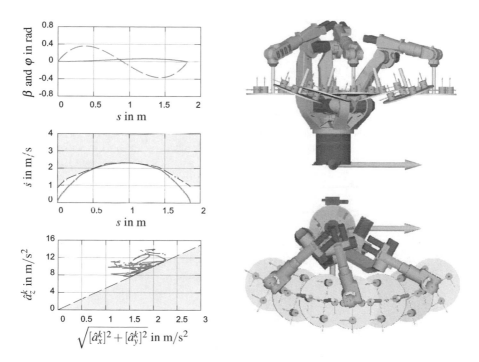

Fig. 9 Optimized waiter-motion along free trajectory with four objects and $p = 3$ - tablet orientation, phase plot and acceleration of all objects

significant subspaces. Table 1 shows the progress of the optimization search for a sequence of subproblems of increasing fineness. The first row shows the cycle time for an initial, slightly curved tablet displacement with purely horizontal attitude. The second row shows the results for the first subproblem: the optimization of the orientation of the tablet, which yields a cycle time improvement of 22%. The third and fourth lines show the results of the final subproblems, in which, in addition to orientation optimization, curve bending is allowed by setting free the two boundary tangents as well as one and then two via points, respectively. One can see that the last optimization yields only marginal improvements while consuming substantial CPU time. Altogether, the example shows that advantages may exist for certain applications when using task space coordinates for describing and shaping the optimal path instead of using joint coordinates.

5   Conclusions

In conclusion, the paper shows that the so-called general waiter motion problem — which is akin to several industrial problems in which tangential forces are restricted due to functional reasons — can be solved by decomposing the overall optimization problem into two cascaded optimization components: shaping the optimal geometry of the spatial path, and finding the time optimal one-dimensional motion of the system along this path. By parametrizing the spatial path using via poses in SE(3), a more natural and lower-dimensional search space could be obtained, improving the convergence behavior of the optimizer. Subdividing the problem in a sequence of cost functions and search spaces of increasing fineness yielded convergence, which was not possible by direct optimization. The completion of the method involved describing spatial motion by quintic and cubic B-Spline curves for translation and rotation, respectively, as well as the extension of the well-known time-optimal algorithm from robotics in three directions: (1) formulating it in an object-oriented multibody framework, (2) allowing for no-slip conditions to be considered, and (3) introducing the concept of a lower limiting curve in order to handle spatial paths for which some configurations are not feasible at rest.

Future research will focus on extending the method to more general types of constraints. Of special interest for robotic applications is the limitation of motor jerks as well as the consideration of non-conservative effects such as sliding friction and discontinuities caused by impacts. Moreover, genetic algorithms will be tested, in particular for generating better initial-value guesses.

References

1. Gasparetto, A., Zanotto, V.: Optimal trajectory planning for industrial robots. Advances in Engineering Software 41(4), 548–556 (2010)
2. Rana, A., Zalzala, A.: An evolutionary planner for near time-optimal collision-free motion of multi-arm robotic manipulators. In: UKACC International Conference on Control, vol. 1, pp. 29–35 (1996)
3. Bobrow, J.E., Dubowsky, S., Gibson, J.S.: Time-optimal control of robotic manipulators along specified paths. The International Journal of Robotics Research 4(3) (1985)
4. Pfeiffer, F., Johanni, R.: A concept for manipulator trajectory planning. IEEE Journal of Robotics and Automation 3, 115–123 (1987)
5. Shiller, Z., Lu, H.-H.: Robust computation of path constrained time optimal motions. In: Proceedings of the IEEE International Conference on Robotics and Automation, Cincinnati, OH, USA, vol. 1, pp. 144–149 (1990)

6. Geu Flores, F., Kecskemethy, A., Poettker, A.: Time-optimal motion planning along specified paths for multibody systems including power constraints and dry friction. In: Proceedings of the Multibody Dynamics 2011, Brussels, Belgium, July 4-7 (2011)
7. Geu Flores, F.: An Object-Oriented Framework for Spatial Motion Planning of Multibody Systems. Ph.D. thesis, Duisburg (2012)
8. Tändl, M., Kecskeméthy, A., Schneider, M.: A design environment for industrial roller coasters. In: CD Proceedings of the ECCOMAS Thematic Conference on Advances in Computational Multibody Dynamics, Milano, Italy, June 25-28 (2007)
9. Dierckx, P.: Curve and Surface Fitting with Splines. Clarendon Press, Oxford (1993)
10. Kecskeméthy, A., Hiller, M.: An object-oriented approach for an effective formulation of multibody dynamics. CMAME 115, 287–314 (1994)

15 Kinematic Analysis of Quadrotors with Manufacturing Errors

Yash Mulgaonkar, Caitlin Powers, and Vijay Kumar

Abstract. We discuss the problem of calibrating quadrotors fabricated using inexpensive printing techniques used by the do-it-yourself community. Although it is easy to create prototypes rapidly, the rotor axes and positions in these prototypes may not be according to specifications. In such a case, operating the motors at the nominal speeds will not result in stable hovering. The fundamental equations that govern hovering are similar to those encountered in objects suspended with cables in that they couple the position and orientation variables with the forces required for equilibrium. We develop the kinematics and statics and derive the conditions for stable equilibrium with a numerical example to illustrate the basic ideas and point to approaches in which adaptation through software can rectify shortcomings in inexpensive manufacturing processors.

1 Introduction

Aerial robotics is a growing field with tremendous civil and military applications. Potential applications for micro unmanned aerial vehicles include search and rescue, environmental monitoring, aerial transportation and manipulation, and surveillance [7]. Quadrotors designs, rotorcrafts whose propulsive force is provided by four rotors, make for flexible and adaptable platforms for aerial robotics. Further, quadrotors are easy to build and are quite robust. The blades have fixed pitch and the propellors rotate in one direction allowing for simple motors and controllers. There are no hinges or flaps. Thus it is possible to rapidly design and prototype large teams of quadrotors using inexpensive manufacturing processes, thus making them the platform of choice for the do it yourself community. However, a common problem encountered with inexpensive prototypes is the presence of manufacturing errors. In particular, if the rotor axes are not parallel and the motors are not symmetrically arranged, the control of the quadrotor can become quite difficult.

Yash Mulgaonkar · Caitlin Powers · Vijay Kumar
University of Pennsylvania, Philadelphia, USA
e-mail: {yashm,cpow,kumar}@seas.upenn.edu

V. Kumar et al. (Eds.): *Adv. in Mech., Rob. & Des. Educ. & Res.*, MMS 14, pp. 205–214.
DOI: 10.1007/978-3-319-00398-6_15   © Springer International Publishing Switzerland 2013

In this paper we address the modeling of imperfectly-manufactured quadrotors. We introduce the kinematics and equations of static equilibrium of quadrotors. We next formulate the direct kinematics problem, the problem of determining the equilibrium pose of the quadrotor. We show that this is related to the problem of determining equilibrium position for suspended objects[3] and to the kinematics of in-parallel manipulators[8]. Finally, we provide several numerical results illustrating the methodology.

2 Printed Robots

The maturation of 3-D printing technology has lowered the barrier to entry for fabrication. It is possible for a person with modest training to use Fused Deposition 3D Modeling (FDM) to rapidly design, customize, and print specialized variants of quadrotors.

The robot shown in Fig. 1(left) was printed with the above mentioned technique using an open source 3D Printer[2]. This unique design was chosen in order to maximize the central surface area in the hub to facilitate the mounting of electrical components and maximize the torsional and bending strength, while keeping the mass as low as possible. 3-D Printing allows the fabrication of sparse structures with monocoque frames, without significant loss of structural integrity.

The electronics for this quadrotor includes an ARM Cortex-M3 Processor, a 3 axis accelerometer, 3 axis gyroscope, a 900MHz communication module and motor controllers. With all its motors and battery, the quadrotor weighs 31 grams and has a 5 minute flight time. While significant improvements need to be made in extending the flight time, the 3-D printed airframe weighs in at a mere 7 grams, as compared to the 10 gram mass of it's more conventional solid counterpart. The 3 gram reduction is a significant 30% decrease in the mass of the frame and a 10% reduction of the overall mass.

An alternative low mass design is shown in Fig. 1(right). This robot, designed by KMEL Robotics[1], uses a printed circuit board as its structural element and essentially eliminates the need for an independent frame.

There are other alternatives to printing technologies. One can also imagine sheets of smart materials with embedded actuators and sensors that can be folded to create complex shapes using origami[4].

One major drawback of such techniques however, is that due to the tolerances and precision of inexpensive fabrication technologies, there can be variations from prototype to prototype. The dimensions of the frame can vary and the four motor axes may not be perfectly aligned. Because the quadrotor relies on all its axes being parallel and on symmetry, misalignments can cause the robot to function poorly. This raises a very interesting controls and calibration problem which is the main focus of this paper.

Fig. 1 Two examples of rapidly prototyped quadrotors: a 3-D printed robot (left); and a printed circuit board robot [1] (right)

3 Kinematics and Statics

3.1 Coordinate Systems and Reference Frames

The coordinate systems and free body diagram for the quadrotor are shown in Fig. 2. The inertial frame, $\mathscr{A}$, is defined by the triad $\mathbf{a}_1$, $\mathbf{a}_2$, and $\mathbf{a}_3$ with $\mathbf{a}_3$ pointing upward. The body frame, $\mathscr{B}$, is attached to the center of mass of the quadrotor with $\mathbf{b}_1$ coinciding with the preferred forward direction and $\mathbf{b}_3$ perpendicular to the plane of the rotors pointing vertically up during perfect hover (see Fig. 2). These vectors are parallel to the principal axes. The center of mass is C. Rotor 1 is a distance L away along $\mathbf{b}$, 2 is L away along $\mathbf{b}_2$, while 3 and 4 are similarly L away along the negative $\mathbf{b}_1$ and $\mathbf{b}_2$ respectively.

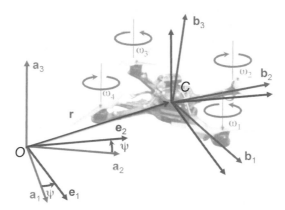

Fig. 2 The body-fixed and the inertial reference frames. A pair of motors spins counter clockwise while the other pair spins clockwise. The pitches on the corresponding propellers are reversed so that the thrust is always pointing in the $\mathbf{b}_3$ direction for all propellors. However, while the reaction moments on the frame of the robot are also in the vertical direction, the signs are such that they oppose the direction of the angular velocity of the propeller.

We will use $Z - X - Y$ Euler angles to model the rotation of the quadrotor in the world frame. To get from $\mathscr{A}$ to $\mathscr{B}$, we first rotate about $\mathbf{a}_3$ through the the the yaw angle, ψ, to get the triad $\mathbf{e}_i$. A rotation about the $\mathbf{e}_1$ through the roll angle, ϕ gets us to the triad $\mathbf{f}_i$ (not shown in the figure). A third pitch rotation about $\mathbf{f}_2$ through θ results in the body-fixed triad $\mathbf{b}_i$. The rotation matrix for transforming components of vectors in $\mathscr{B}$ to components of vectors in $\mathscr{A}$ is given by:

$$^{\mathscr{A}}[R]_{\mathscr{B}} = \begin{bmatrix} c\psi c\theta - s\phi s\psi s\theta & -c\phi s\psi & c\psi s\theta + c\theta s\phi s\psi \\ c\theta s\psi + c\psi s\phi s\theta & c\phi c\psi & s\psi s\theta - c\psi c\theta s\phi \\ -c\phi s\theta & s\phi & c\phi c\theta \end{bmatrix}. \tag{1}$$

3.2 Equilibrium Conditions

Let $\mathbf{r}$ denote the position vector of C in $\mathscr{A}$. The forces on the system are gravity, in the $-\mathbf{a}_3$ direction, and the forces from each of the rotors, F_i, in the $\mathbf{b}_3$ direction. For static equilibrium, we know that the forces must balance:

$$R^{\mathrm{T}} \begin{bmatrix} 0 \\ 0 \\ -mg \end{bmatrix} + \begin{bmatrix} 0 \\ 0 \\ F_1 + F_2 + F_3 + F_4 \end{bmatrix} = \mathbf{0} \tag{2}$$

In addition to forces, each rotor produces a moment perpendicular to the plane of rotation of the blade, M_i. Rotors 1 and 3 rotate in the $-\mathbf{b}_3$ direction while 2 and 4 rotate in the $+\mathbf{b}_3$ direction. Since the moment produced on the quadrotor is opposite to the direction of rotation of the blades, M_1 and M_3 act in the $\mathbf{b}_3$ direction while M_2 and M_4 act in the $-\mathbf{b}_3$ direction. L is the distance from the axis of rotation of the rotors to the center of mass of the quadrotor. For static equilibrium, the net moment on the aircraft frame must be zero, which gives us:

$$\begin{bmatrix} L(F_2 - F_4) \\ L(F_3 - F_1) \\ M_1 - M_2 + M_3 - M_4 \end{bmatrix} = \mathbf{0} \tag{3}$$

Each rotor has an angular speed ω_i and produces a vertical force F_i according to

$$F_i = k_F \omega_i^2. \tag{4}$$

The rotors also produce a moment according to

$$M_i = k_M \omega_i^2. \tag{5}$$

We can rewrite (2,3) as:

$$
\begin{bmatrix}
0 & 0 & 0 & 0 \\
0 & 0 & 0 & 0 \\
1 & 1 & 1 & 1 \\
0 & L & 0 & -L \\
-L & 0 & L & 0 \\
h & -h & h & -h
\end{bmatrix}
\begin{bmatrix}
F_1 \\ F_2 \\ F_3 \\ F_4
\end{bmatrix}
=
\begin{bmatrix}
R^{\mathrm{T}}\begin{bmatrix} 0 \\ 0 \\ -mg \end{bmatrix} \\
0 \\
0 \\
0
\end{bmatrix}
\tag{6}
$$

where $h = \frac{k_M}{k_F}$ is the relationship between lift and drag given by Equations (4-5). We can see that each column represents a unit wrench with pitch h or $-h$, and the non negative wrench intensities ($F_i \geq 0$) are proportional to the square of the propellor angular speed.

If the axes of the quadrotor are parallel and if the motors are symmetric (at a distance L from the center of mass C), the solution to this set of equations is given by:

$$
F_i = \frac{mg}{4} \tag{7}
$$

$$
R = I \tag{8}
$$

If this condition of symmetry is not satisfied, we must model the direction of each axis and its location. Let u_i be the unit vector along axis i and ρ_i be the position vector of suitably chosen point on the axis with C as an origin. Writing all the vectors in the body-fixed frame we get:

$$
\begin{bmatrix}
\begin{bmatrix} u_1 \\ u_1^0 + hu_1 \end{bmatrix} &
\begin{bmatrix} u_2 \\ u_2^0 - hu_2 \end{bmatrix} &
\begin{bmatrix} u_3 \\ u_3^0 + hu_3 \end{bmatrix} &
\begin{bmatrix} u_4 \\ u_4^0 - hu_3 \end{bmatrix}
\end{bmatrix}
\begin{bmatrix}
F_1 \\ F_2 \\ F_3 \\ F_4
\end{bmatrix}
=
\begin{bmatrix}
mgc\phi s\theta \\
-mgs\phi \\
-mgc\phi c\theta \\
0 \\
0 \\
0
\end{bmatrix}
\tag{9}
$$

where $u_i^0 = \rho_i \times u_i$.

The conditions for static equilibrium are no longer obvious. However, they can be obtained by solving this set of six nonlinear equations in six unknowns: F_1, F_2, F_3, F_4, θ, and ϕ.

4   Kinematics and Statics of Suspended Payloads

The problem discussed in Section 3 is related to the problem of finding equilibrium configurations of payloads suspended by n cables in three dimensions, which is described in [5] and [9]. This problem is also analyzed in [3, 6] in the context of co-operative transport of payloads as shown in Fig. 3. The $n = 6$ case is addressed in the literature on cable-actuated payloads, where the payload pose is fully specified[8].

When $n = 5$, if the line vectors are linearly independent and the cables are taut, the line vectors and the gravity wrench axis must belong to the same linear complex [5]. To visualize the twist that is reciprocal to these line vectors[11], Hunt describes the following experiment:

If a wooden model constructed from a chunk of timber, a few hooks, some string and an open frame were constructed you can place a rod along this dotted line and this rod, "would remain sharp in a time-exposure photograph in an otherwise blurred picture"[5].

The rod represents the reciprocal screw axis about which the payload is free to instantaneously twist. When $n = 4$, under similar assumptions on linear independence and positive tension, the line vectors and the gravity wrench must belong to the same linear congruence. The unconstrained freedoms correspond (instantaneously) to a set of twists whose axis lie on a cylindroid. The $n = 3$ case admits solutions where all three cables and the gravity wrench axis lie on the same regulus - the generators of a hyperboloid which is a ruled surface [9].

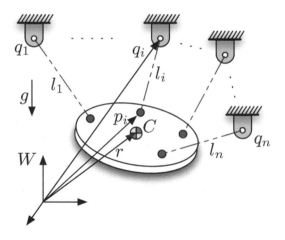

Fig. 3 A payload suspended by n cables and pivots. The pivots can represent aerial robots that are controlled to hover at designated points in the inertial frame.

The problem discussed in [3, 6] is the special case of three robots in three dimensions carrying an object (although the analysis readily extends to n robots). Consider point-model robots and a configuration space is given by $\mathscr{Q} = \mathbb{R}^3 \times \mathbb{R}^3 \times \mathbb{R}^3$. Each robot is modeled by $q_i \in \mathbb{R}^3$ with coordinates $q_i = [x_i, y_i, z_i]^T$ in an inertial frame, W (Fig. 2). Define the robot configuration as $q = [q_1, q_2, q_3]^T$. The ith robot cable with length l_i is connected to the payload at the point P_i with coordinates $p_i = [x_i^p, y_i^p, z_i^p]^T$ in W. Let $p = [p_1, p_2, p_3]^T$ denote all attachment points. P_2, and P_3 to be non-collinear and span the center of mass. The payload has mass m with the center of mass at C with position vector $r = [x_C, y_C, z_C]^T$. We denote the fixed Euclidean distance between attachment points P_i and P_j as $r_{i,j}$.

The equations of static equilibrium can be written as follows. The cables exert zero-pitch wrenches on the payload which take the following form after normalization:

$$\begin{bmatrix} u_i \\ u_i^0 \end{bmatrix} = \frac{1}{l_i} \begin{bmatrix} q_i - p_i \\ p_i \times (q_i - p_i) \end{bmatrix}.$$

The equations of equilibrium, like before, take the form

$$\begin{bmatrix} u_1 & u_2 & u_3 \\ u_1^0 & u_2^0 & u_3^0 \end{bmatrix} \begin{bmatrix} F_1 \\ F_2 \\ F_3 \end{bmatrix} = \begin{bmatrix} mgc\phi s\theta \\ -mgs\phi \\ -mgc\phi c\theta \\ 0 \\ 0 \\ 0 \end{bmatrix} \tag{10}$$

where $F_i \geq 0$ is the tension in the ith cable.

In order for (10) to be satisfied, the four line vectors or zero pitch wrenches, $\mathbf{w}_1$, $\mathbf{w}_2$, $\mathbf{w}_3$, and $\mathbf{g}$ must belong to the same *regulus*. The lines of a regulus are points on a 2-plane in $\mathbb{PR}^5$ [10], which implies that the body is under constrained and has three degrees of freedom. Instantaneously, these degrees of freedom correspond to twists in the *reciprocal screw system* that are reciprocal to $\mathbf{w}_1$, $\mathbf{w}_2$, and $\mathbf{w}_3$. They include zero pitch twists (pure rotations) that lie along the axes of the *complementary regulus* (the set of lines each intersecting all of the lines in the original regulus).

5   Direct Kinematics of Asymmetric Quadrotors

In this section we investigate solutions to Equation (9) to derive conditions for static equilibrium that characterize the hover condition. We solve for the equilibrium configuration in two steps. First, since the sum of the moments must be zero, we can write the bottom three equations in Equation (9):

$$\begin{bmatrix} [u_1^0 + hu_1] & [u_2^0 - hu_2] & [u_3^0 + hu_3] & [u_4^0 - hu_3] \end{bmatrix} \begin{bmatrix} F_1 \\ F_2 \\ F_3 \\ F_4 \end{bmatrix} = \begin{bmatrix} 0 \\ 0 \\ 0 \end{bmatrix} \tag{11}$$

The solution to this set of equations is found by finding the vector $\mathbf{F} = \begin{bmatrix} F_1 & F_2 & F_3 & F_4 \end{bmatrix}$ that lies in the nullspace of the matrix. Since the operation takes $\mathscr{R}^4$ to $\mathscr{R}^3$, there will be at least one vector in the nullspace. Since our coefficients are lift forces produced by propellors that can only spin in one direction, this vector must only have positive components. Thus we must find a null space vector with non negative components with the magnitude of mg. This is easily done using computation tools such as MATLAB.

Second, since the sum of the forces in (9) must be zero, we write:

$$
\begin{bmatrix} u_1 & u_2 & u_3 & u_4 \end{bmatrix}
\begin{bmatrix} F_1 \\ F_2 \\ F_3 \\ F_4 \end{bmatrix}
=
\begin{bmatrix} mgc\phi s\theta \\ -mgs\phi \\ -mgc\phi c\theta \end{bmatrix}
= R^T
\begin{bmatrix} 0 \\ 0 \\ -mg \end{bmatrix}
\tag{12}
$$

Since we know all of the variables on the left hand side of this equation by solving (11), we can now solve this set of nonlinear equations for the two Euler angles, θ and ϕ.

Example

We consider the example of the 31 gram 3-D printed quadrotor in Figure 1 (left). The pitch h is taken to be 0.1 meters. In practice this value is experimentally determined and $h = 0.1$ is a representative value for small quadrotors. Experimentation with a fixed rotor at steady-state yields an estimate of $k_F = 1.77 \times 10^{-6}$Newtons-s^2. If the motors are positioned so that their centers are 6.35 cm from the center of the hub as specified in the design, the position vectors are given by

$$
\rho_1 = \begin{bmatrix} 0 \\ 6.35 \\ 0 \end{bmatrix}, \rho_2 = \begin{bmatrix} 6.35 \\ 0 \\ 0 \end{bmatrix}, \rho_3 = \begin{bmatrix} 0 \\ -6.35 \\ 0 \end{bmatrix}, \rho_4 = \begin{bmatrix} -6.35 \\ 0 \\ 0 \end{bmatrix}
\tag{13}
$$

However, we consider a prototype with fabrication errors where the position vectors are:

$$
\rho_1 = \begin{bmatrix} 0.127 \\ 6.223 \\ 0 \end{bmatrix}, \rho_2 = \begin{bmatrix} 6.477 \\ -0.127 \\ 0 \end{bmatrix}, \rho_3 = \begin{bmatrix} -0.127 \\ -6.223 \\ 0 \end{bmatrix}, \rho_4 = \begin{bmatrix} -6.477 \\ 0.127 \\ 0 \end{bmatrix}
\tag{14}
$$

The axes of rotation for the rotors are specified to be perpendicular to the vehicle, or $u_i = \begin{bmatrix} 0 & 0 & 1 \end{bmatrix}^T$. However, we will consider the case where all motors are placed perfectly but the axes are slightly misaligned by a vector δu_i so that

$$
u_i = \frac{v_i}{\|v_i\|}, \quad v_i = \begin{bmatrix} 0 \\ 0 \\ 1 \end{bmatrix} + \delta u_i = \begin{bmatrix} 0 \\ 0 \\ 1 \end{bmatrix} + \begin{bmatrix} \delta u_{xi} \\ \delta u_{yi} \\ \delta u_{zi} \end{bmatrix} \quad 1 \le i \le 4
\tag{15}
$$

As an example, we choose δu_3 and δu_4 be zero while $\delta u_1 = \begin{bmatrix} 0.05 & 0.02 & 0.0 \end{bmatrix}$ and $\delta u_1 = \begin{bmatrix} 0.05 & 0.0 & 0.0 \end{bmatrix}$. Our condition for equilibrium (11) leads to

$$
\begin{bmatrix} 0.003 & 0.057 & 0.001 & -0.062 \\ -0.066 & -0.001 & 0.064 & 0.001 \\ -0.098 & -0.102 & 0.100 & 0.100 \end{bmatrix}
\begin{bmatrix} F_1 \\ F_2 \\ F_3 \\ F_4 \end{bmatrix}
=
\begin{bmatrix} 0 \\ 0 \\ 0 \end{bmatrix}
\tag{16}
$$

The null space is given by $F_1 = .470\alpha$, $F_2 = .524\alpha$, $F_3 = .483\alpha$, $F_4 = .519\alpha$, where α is any positive scalar. This means that the net force on the vehicle is

$$F_1 u_1 + F_2 u_2 + F_3 u_3 + F_4 u_4 = mg \begin{bmatrix} c\phi s\theta \\ -s\phi \\ -c\phi c\theta \end{bmatrix} \qquad (17)$$

We use the small angle assumption ($c \approx 1, s\theta \approx \theta$) and the first two components of the net force to solve for ϕ and θ. The third component, which is equal to $-c\phi c\theta$ can be used to check the validity of our assumption. Solving this equation for the magnitude leads to $\alpha = 0.152$ and motor speeds (in RPM):

$$\omega_1 = 12060, \ \omega_2 = 12738, \ \omega_3 = 12237, \ \omega_4 = 12681.$$

Solving the nonlinear equations (12) in MATLAB gives us $\theta = 0.077°$ and $\phi = -0.269°$. Clearly the rotor speeds required for equilibrium are different for the four rotors and the equilibrium pose has a non zero roll and pitch angle.

6 Conclusion

In this paper, we have formulated the conditions for equilibrium for a quadrotor with manufacturing errors. We have shown that even if the rotors are not perfectly positioned or aligned, the robot will still be able to hover but not in a perfectly horizontal position. Of course this analysis raises the obvious question: is it possible for the robot to learn the actual dimensions so that the correct motor speeds can be commanded. This is the kinematic calibration problem. Given the hovering configuration (ϕ, θ) and the motor speeds needed to hover (ω_1, ω_2, ω_3 ω_4), what are the position vectors ρ_i and directions u_i? If the ρ_i are assumed to be known, we can use the system identification is linear and standard techniques can be used to find the u_i. For the case where we wish to calibrate both ρ_i and u_i, we can take advantage of the fact that the net wrench on the quadrotor is a bilinear function of ρ_i and u_i. This means we can iteratively approach a solution for both vectors by alternately optimizing for ρ_i and then u_i. This line of research will make it possible to produce inexpensive autonomous quadrotors that can adapt to manufacturing errors and changes in parameters over time.

Acknowledgements. We gratefully acknowledge the support of NSF grant CCF-1138847 and ONR grant N00014-09-1-1051.

References

1. KMel Robotics, http://www.kmelrobotics.com
2. MakerBot, http://www.makerbot.com

3. Fink, J., Michael, N., Kim, S., Kumar, V.: Planning and control for cooperative manipulation and transportation with aerial robots. The International Journal of Robotics Research 30(3) (March 2011)
4. Hawkes, E., An, B., Benbernou, N., Tanaka, H., Kim, S., Demaine, E.D., Rus, D., Wood, R.J.: Programmable matter by folding. Proc. Nat. Acad. Sci. 107(28), 12441–12445 (2010)
5. Hunt, K.H.: Kinematic Geometry of Mechanisms. Oxford University Press (1978)
6. Jiang, Q., Kumar, V.: Determination and stability analysis of equilibrium configurations of objects suspended from multiple aerial robots. Journal of Mechanisms and Robotics 4(2), 021005 (2012)
7. Kumar, V., Michael, N.: Opportunities and challenges with autonomous micro aerial vehicles. In: Int. Symposium of Robotics Research, Flagstaff, AZ (August 2011)
8. Nanua, P., Waldron, K.J., Murthy, V.: Direct kinematic solution of a Stewart platform. IEEE Transactions on Robotics and Automation 6(4), 438–444 (1990)
9. Phillips, J.: Freedom in Machinery, vol. 1. Cambridge University Press (1990)
10. Selig, J.M.: Geometric Fundamentals of Robotics. Springer (2005)
11. Waldron, K.J., Hunt, K.H.: Series-parallel dualities in actively coordinated mechanisms. The Int. Journal of Robotics Research 10(5), 473–480 (1991)

16 Omnicopter: A Novel Overactuated Micro Aerial Vehicle

Yangbo Long and David J. Cappelleri

Abstract. This paper deals with a novel micro aerial vehicle (MAV) design, the Omnicopter MAV. The Omnicopter has two central counter-rotating coaxial propellers for thrust and yaw control and three perimeter-mounted variable angle ducted fans for lateral forces and roll/pitch control. It can work under two configurations, a fixed 90° ducted fan angle configuration and a variable angle ducted fan configuration. The latter one makes the Omnicopter overactuated. After a brief introduction of the Omnicopter platform and a comparison between different MAV configurations, we discuss the control design and allocation techniques for the variable angle ducted fan configuration. Simulation and experimental results verify the design of the Omnicopter, and compare the performance between the two configurations.

1 Introduction

Traditional vertical take-off and landing (VTOL) MAVs are generally underactuated [1], [2], [3], i.e., equipped with fewer actuators than degrees-of-freedom (DOF), which results in some limitations on their performance. For example, they cannot maintain zero roll and pitch attitude during lateral translation. It is also unattainable for traditional underactuated MAVs to arbitrarily orient their fuselages to accomplish complicated grasping tasks.

In our opinion, fully actuated MAVs based on novel mechanical design should be investigated, in order to achieve complex manipulation tasks. Drawing inspiration from omnidirectional wheels, we propose a novel actuation concept for a MAV, named the Omnicopter [4]. The Omnicopter design allows for agile movements over the full 6 DOF of the robot. It has two fixed major coaxial counter-rotating propellers in the center used to provide thrust and adjust the yaw angle, and three adjustable

Yangbo Long · David J. Cappelleri
Stevens Institute of Technology,
1 Castle Point on Hudson, Hoboken, NJ 07030, USA
e-mail: {ylong1,dcappell}@stevens.edu

V. Kumar et al. (Eds.): *Adv. in Mech., Rob. & Des. Educ. & Res.*, MMS 14, pp. 215–226.
DOI: 10.1007/978-3-319-00398-6_16   © Springer International Publishing Switzerland 2013

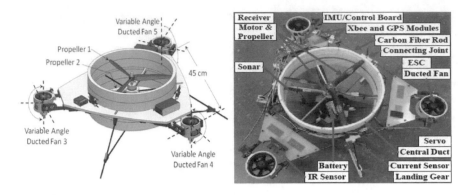

Fig. 1 Omnicopter MAV schematic (left) and prototype (right)

Table 1 Prototype components

Components	Model/Material	Quantity	Unit Price ($)
	0.125" carbon fiber rods	-	3.70
Frame	custom ABS connecting joints	-	23.95
	Depron 9 mm thick foam	-	40.00[a]
Propeller	10x7 3-blade Master Airscrew	2	7.68
Motor	BP-U2212/10 brushless outrunner	2	29.95
Ducted Fan	AEO 55 mm EDF	3	15.90
ESC	Turnigy Plush 25 Amp	5	25.99
IMU/Control Board	ArduPilot Mega	1	150.00
Transceiver Module	XBee	2	22.00
Receiver	Spektrum AR8000	1	129.99
Battery	Thunderpower 3s 2700 mAh	1	64.99
GPS	Media Tek MT3329	1	29.99
IR Sensor[b]	Sharp GP2Y0A02YK0F	4	14.95
Sonar	MaxSonar MB1200	1	44.95
Servo	Hitec HS-5055MG	3	17.99
		Total	$898.25

[a] 6 sheets spanning 13"x39".
[b] Attached only one in Fig. 1 (right).

angle ducted fans located in three places surrounding the airframe to control its roll and pitch and provide lateral forces. The Omnicopter has two configurations: C1: Fixed 90° ducted fan angles with varying rotor speeds; C2: Variable ducted fan angles and variable rotor speeds. A schematic of the Omnicopter MAV is shown in Fig. 1 (left).

An Omnicopter prototype has been constructed, as shown in Fig. 1 (right). The prototype weighs 2 lbs 3.5 oz. with an available payload at 80% power of approximately 2 lbs 6 oz ($\sim$1 kg). It was initially configured for remote control with a Spektrum AR8000 8-Channel DSMX Receiver and DX8 8-channel transmitter. Custom mounts for each of the ducted fans were 3D printed out of ABS plastic. Major components of the prototype and associated costs are listed in Table 1.

2 Omnicopter VS Other MAVs

The family of micro rotary-wing unmanned aerial vehicles (UAVs) includes ducted fan UAVs [5], [6], conventional helicopters [7], [8], tricopters [9], [10] and quadrotors [11], [12]. In the following, we briefly describe the advantages and disadvantages of each configuration and show that the Omnicopter is a promising alternative to existing VTOL MAVs.

2.1 Ducted Fan UAV (1 Rotor)

The ducted fan VTOL UAV, like iSTAR [13], is comprised of an outer duct enclosing a single propeller, fixed stators and movable vanes operated by actuators, performing thrust vectoring. It utilizes an airfoil-shaped duct to provide augmented lift. Due to small gaps between the tips of the fan blades and the wall of the duct, the loading on the blades is allowed to extend to the tips, reducing tip losses associated with free-air propellers.

However, undesirable aerodynamic characteristics are associated with these vehicles in crosswinds, namely momentum drag and asymmetric duct lift, both of which generate a positive, adverse pitching moment, preventing the vehicle from achieving steady forward flight. Unfortunately these disadvantages keep the ducted fan from being widely accepted as a reliable means of propulsion for small VTOL aircraft.

2.2 Traditional Helicopter (2 Rotors)

The conventional helicopter is characterized by a main rotor that provides thrust and a tail rotor for compensating the counter torque due to the main rotor. Blades on a helicopter are pitched in different ways to control the orientation and direction of motion using a swashplate. The main advantage of this configuration is its high maneuverability and good performance during forward flight.

Because the lateral force generated by the tail rotor is used for yaw control only, and doesn't participate to the thrust generation, the energy spent by the tail rotor can be considered as passive. Due to the complexity of the linkages and swashplate, the helicopter is more prone to mechanical faults and possible failure.

2.3 Tricopter (3 Rotors)

The tricopter consists of two body-fixed rotors and a tail tilting rotor with fixed-angle blades. The two front rotors rotate in opposite directions, thereby eliminating counter torques. The tail rotor can be tilted laterally using a servo motor in order to provide the yawing torque. The tricopter has a simpler mechanical structure compared to the above ducted fan UAV and helicopter. Indeed, the absence of stators and movable vanes, swashplate and stabilizing bar make it more robust and easier to control.

Due to its asymmetric structure, gyroscopic effects and aerodynamic torques can not be eliminated completely. Especially, the gyroscopic effect of tilting the tail rotor induces other moments on the fuselage. Also, building a servo solution for the tail rotor is challenging.

2.4   Quadrotor/Tilting Propeller Quadrotor (4 Rotors)

The quadrotor has four rotors. The front and the rear motors rotate counterclockwise while the other two rotate clockwise. The main thrust and control torques are obtained by varying the angular speed of the four rotors, without using any servo mechanism. The quadrotor configuration considerably simplifies the vehicle design and intrinsically reduces the gyroscopic effects.

However, like previously mentioned UAV types, quadrotors are still essentially underactuated. They only possess four independent control inputs with respect to their 6 DOF in space. As a consequence, it can be proven that quadrotors are only able to independently act on their Cartesian position and yaw angle, which imposes some limitations. For example, a sensor or gripper attached to the quadrotor cannot be arbitrarily oriented during flight nor it can hover in place with any body orientation.

In [14], M. Ryll et al. proposed a quadrotor with tilting propellers. With the additional four servo mechanisms, unlike standard quadrotor, this design is fully actuated. However, the inertial/gyroscopic effects resulting from the propeller rotation are neglected when designing the controller. Although simulation results are provided to validate the effectiveness and robustness of the proposed control design, its real flight performance is speculative without actual implementation.

2.5   Omnicopter (5 Rotors)

Compared to traditional MAVs, the Omnicopter has different motion principles and control modes. The yaw movement results from different speeds of the two counter-rotating coaxial propellers. The roll and pitch motions can be generated differently under the two configurations, C1 and C2, as mentioned in Sect. 1. Vertical motion can be generated by increasing or decreasing the five propellers speeds together under C1. Full position control on axes x, y and z can be achieved under C2.

The Omnicopter allows for lateral force vectors to be applied to the airframe, which gives full control over the 6 DOF of the Omnicopter body and better ability to interact with and manipulate the environment. Different from the quadrotor design with tilting propellers, the Omnicopter hardly suffers from the inertial/gyroscopic effects due to the limited size of the ducted fans.

3 Modeling, Control and Control Allocation

As mentioned before, the Omnicopter MAV has two configurations, the fixed 90° ducted fan angle configuration and the variable angle ducted fan configuration. For the former one, the modeling and control design of the Omnicopter is similar to those of a quadrotor [12], but we need a control allocation technique to distribute the computed control inputs into five motor control inputs [15]. For the latter one, we review the corresponding modeling, control and allocation from [16] in this section.

3.1 *Dynamic Model*

Modeling the Omnicopter as a rigid-body, using Newtonian mechanics, let $I = I_x, I_y, I_z$ denote the inertial frame, and $B = B_x, B_y, B_z$ the aircraft body frame, as shown in Fig. 2. Then the dynamic model is [4], [15]

$$\begin{aligned}
\dot{\xi} &= \upsilon \\
m\dot{\upsilon} &= mge_3 + Rf \\
\dot{R} &= R\omega^\times \\
J\dot{\omega} &= -\omega^\times J\omega + \tau
\end{aligned} \tag{1}$$

where ξ, υ, m, g, R, ω, J, f and τ stand for position, velocity, mass, gravitational acceleration, rotation matrix, angular velocity, inertia matrix, and force and torque in the body coordinates generated by the actuators, respectively, $\cdot^\times$ denotes the cross product operator for a vector and $e_3 = [0 \ \ 0 \ \ 1]^T$. We can linearize the original dynamic model (1) as

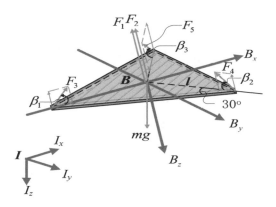

Fig. 2 Free-body diagram

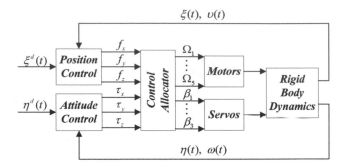

Fig. 3 Control loops for the variable angle ducted fan configuration presented in [16]

$$
\begin{aligned}
\ddot{\xi}_1 &= F_x \\
\ddot{\xi}_2 &= F_y \\
\ddot{\xi}_3 &= g + \frac{1}{m} F_z \\
\ddot{\phi} &= \tau_x \\
\ddot{\theta} &= \tau_y \\
\ddot{\psi} &= \tau_z
\end{aligned}
\tag{2}
$$

where F_x, F_y and F_z are forces in the inertial frame.

3.2 Control Design

The control block diagram is shown in Fig. 3. In the virtual control input vector $v = [F_x \ F_y \ F_z \ \tau_x \ \tau_y \ \tau_z]^T$, the virtual force inputs, v_1, v_2 and v_3, can be designed using classical PID control

$$
v_i = k_{P_i}(\xi_i^d - \xi_i) + k_{D_i}(\dot{\xi}_i^d - \dot{\xi}_i) + k_{I_i} \int_0^t (\xi_i^d - \xi_i) d\tau
\tag{3}
$$

where $i = 1,\ 2,\ 3$.

The control forces in the body frame can be computed from

$$
f = R_I^B F
\tag{4}
$$

where the rotation matrix from the inertial frame to the body frame is

$$
R_I^B = \begin{bmatrix}
c\psi c\theta & s\psi c\theta & -s\theta \\
c\psi s\theta s\phi - s\psi c\phi & s\psi s\theta s\phi + c\psi c\phi & c\theta s\phi \\
c\psi s\theta c\phi + s\psi s\phi & s\psi s\theta c\phi - s\phi c\psi & c\theta c\phi
\end{bmatrix}
\tag{5}
$$

For the attitude control inputs, v_4, v_5 and v_6, we can apply the PD control designed for the fixed angle configuration

$$v_j = k_{P_j}(\eta_j^d - \eta_j) + k_{D_j}(\dot{\eta}_j^d - \dot{\eta}_j) \tag{6}$$

where $j = 4$, 5, 6.

Therefore, we can arrive at the control inputs, $\mathbf{w}$, in the body coordinate frame

$$\mathbf{w} = [f_x \ f_y \ f_z \ \tau_x \ \tau_y \ \tau_z]^T \tag{7}$$

3.3 Control Allocation

In the case of the variable ducted fan configuration, five motor speeds (ω_1 to ω_5) and three servo motor angles (β_1 to β_3) need to be computed. The mapping between the actuator input vector $\mathbf{u} = [\omega_1^2 \ \omega_2^2 \ \omega_3^2 \ \omega_4^2 \ \omega_5^2 \ \beta_1 \ \beta_2 \ \beta_3]^T$ and the virtual control input $\mathbf{w} = [f_x \ f_y \ f_z \ \tau_x \ \tau_y \ \tau_z]^T$ is

$$\begin{aligned}
f_x &= k_{T_3}(\omega_3^2 c\beta_1 - (\omega_4^2 c\beta_2 + \omega_5^2 c\beta_3)s30°) \\
f_y &= k_{T_3}(\omega_5^2 c\beta_3 - \omega_4^2 c\beta_2)c30° \\
f_z &= -k_{T_1}\omega_1^2 - k_{T_2}\omega_2^2 - k_{T_3}(\omega_3^2 s\beta_1 + \omega_4^2 s\beta_2 + \omega_5^2 s\beta_3) \\
\tau_x &= k_{T_3}(\omega_5^2 s\beta_3 - \omega_4^2 s\beta_2)lc30° \\
\tau_y &= k_{T_3}(\omega_4^2 s\beta_2 + \omega_5^2 s\beta_3)ls30° - k_{T_3}\omega_3^2 ls\beta_1 \\
\tau_z &= k_{Q_1}\omega_1^2 - k_{Q_2}\omega_2^2
\end{aligned} \tag{8}$$

where $s = sin$ and $c = cos$, k_{T_1}, k_{T_2} and k_{T_3} are thrust factors, k_{Q_1} and k_{Q_2} are drag factors.

Solving (8) for $\mathbf{u}$, while considering actuator constraints, amounts to performing constrained nonlinear programming. Since control allocation is to be performed in real time, this may not be computationally feasible. One way to resolve this problem is to linearize (8) locally. Linearizing $\mathbf{w}(\mathbf{u})$ around $\mathbf{u}_0$ yields

$$\mathbf{w}(\mathbf{u}) = \mathbf{w}(\mathbf{u}_0) + \frac{\partial \mathbf{w}}{\partial \mathbf{u}}|_{\mathbf{u}_0}(\mathbf{u} - \mathbf{u}_0) \tag{9}$$

which leads to the linear control allocation problem

$$\bar{\mathbf{w}} = \mathbf{B}\mathbf{u} \tag{10}$$

where $\bar{\mathbf{w}} - \mathbf{w}(\mathbf{u}) - \mathbf{w}(\mathbf{u}_0) + \mathbf{B}\mathbf{u}_0$ and the effectiveness matrix $\mathbf{B} = \frac{\partial \mathbf{w}}{\partial \mathbf{u}}|_{\mathbf{u}_0}$, $\mathbf{u}_0$ is picked as the previously applied control input, $\mathbf{u}(t - \delta)$, with δ as the step size. The linearized effectiveness matrix is shown in (11).

$$
\boldsymbol{B} =
\begin{bmatrix}
0 & 0 & k_{T_3}cu_6 & -\dfrac{k_{T_3}cu_7}{2} & -\dfrac{k_{T_3}cu_8}{2}-k_{T_3}u_3su_6 & \dfrac{k_{T_3}u_4su_7}{2} & \dfrac{k_{T_3}u_5su_8}{2} \\
0 & 0 & 0 & -\dfrac{\sqrt{3}k_{T_3}cu_7}{2} & \dfrac{\sqrt{3}k_{T_3}ca_8}{2} & 0 & \dfrac{\sqrt{3}k_{T_3}u_4su_7}{2} & -\dfrac{\sqrt{3}k_{T_3}u_5su_8}{2} \\
-k_{T_1}-k_{T_2} & -k_{T_3}su_6 & -k_{T_3}su_7 & -k_{T_3}su_8 & -k_{T_3}u_3cu_6 & -k_{T_3}u_4cu_7 & -k_{T_3}u_5cu_8 \\
0 & 0 & 0 & -\dfrac{\sqrt{3}lk_{T_3}sa_7}{2} & \dfrac{\sqrt{3}lk_{T_3}su_8}{2} & 0 & -\dfrac{\sqrt{3}lk_{T_3}u_4cu_7}{2} & \dfrac{\sqrt{3}lk_{T_3}u_5cu_8}{2} \\
0 & 0 & -lk_{T_3}su_6 & \dfrac{lk_{T_3}su_7}{2} & \dfrac{lk_{T_3}su_8}{2} & -lk_{T_3}u_3cu_6 & \dfrac{lk_{T_3}u_4cu_7}{2} & \dfrac{lk_{T_3}u_5cu_8}{2} \\
k_{Q_1} & -k_{Q_2} & 0 & 0 & 0 & 0 & 0 & 0
\end{bmatrix}_{u_0}
$$

(11)

Then pseudoinverse based methods can be applied to solve the problem.

4   Simulation and Experimental Results

Based on the discussions in the former sections, simulations and experiments are carried out to demonstrate the capability of the Omnicopter.

4.1   Simulation Results

Simulations are done for both the fixed 90° ducted fan angle configuration and the variable angle ducted fan configuration. The Omnicopter was tasked to perform a circular path following task in the two configurations, respectively, which is shown in Fig. 4. To make the simulations more realistic, random white noise with zero mean and 0.01 variance have been added to the position and velocity measurements, and the attitude and angular velocity have been corrupted by noise with 0.001 variance.

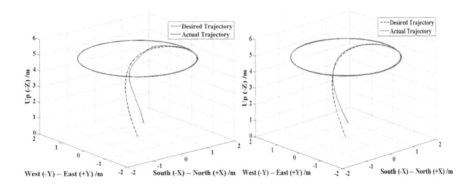

Fig. 4 3D circular path tracking performance: Fixed ducted fan configuration (left), and Variable ducted fan configuration (right)

Fig. 5 - 6 compare the position tracking performance and the attitude performance, respectively, of the two configurations. We can easily find that it has better performance in the variable angle ducted fan configuration. The Omnicopter can even keep a zero attitude when tracking the circular path in the variable angle configuration, which is impossible in the fixed angle configuration.

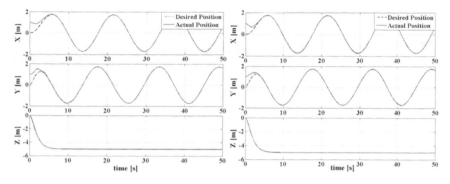

Fig. 5 Position performance: Fixed ducted fan configuration (left), and Variable ducted fan configuration (right)

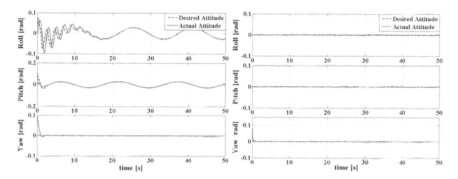

Fig. 6 Attitude performance: Fixed ducted fan configuration (left), and Variable ducted fan configuration (right)

4.2 Experimental Results

Here we present and compare the real-time experimental results [16]. Fig. 7, 8 and 9 show the experimental results for hovering performance in the fixed 90° ducted fan configuration and lateral translation performance in the variable angle ducted fan configuration. The top of the figures show attitude stabilization performance in the fixed 90° ducted fan configuration during hovering, and the bottom ones show lateral translation performance in the variable angle ducted fan configuration for the roll, pitch and yaw angles of the Omnicopter. In the current stage, the attitude control is performed autonomously on-board the MAV while the x, y and z positions, as well as the ducted fan angles of the Omnicopter, are controlled manually using a remote controller.

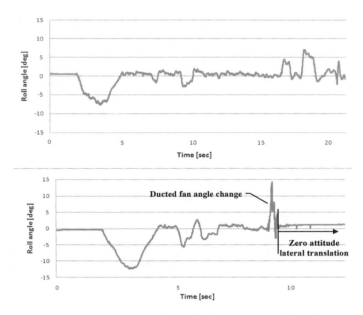

Fig. 7 Roll angle performance: Hovering in the fixed ducted fan configuration (top), and Zero attitude lateral translation in the variable ducted fan configuration (bottom)

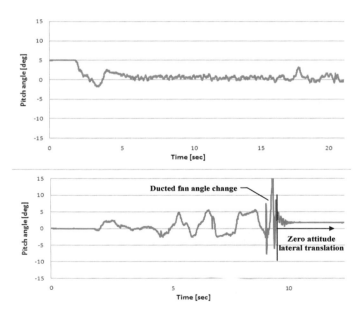

Fig. 8 Pitch angle performance: Hovering in the fixed ducted fan configuration (top), and Zero attitude lateral translation in the variable ducted fan configuration (bottom)

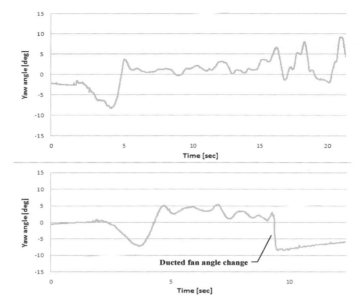

Fig. 9 Yaw angle performance: Hovering in the fixed ducted fan configuration (top), and Zero attitude lateral translation in the variable ducted fan configuration (bottom)

5   Conclusions

In this paper, we studied a new configuration of rotary wing MAV called the Omnicopter. Its novel actuation concept allows the three ducted fans to actively rotate about the assembling axes connecting them to the Omnicopter main body, which makes it possible to gain full controllability over its 6 DOF. A validation of the control performance of the Omnicopter is provided by means of simulations and experiments.

Future works will investigate the implementation of fully autonomous flights by introducing a camera-based localization system in an indoor environment, as well as GPS-guided outdoor flights. Additionally, linear and nonlinear optimization techniques will be implemented on-board for the variable angle ducted fan configuration.

References

1. Bouabdallah, S., Noth, A., Siegwart, R.: PID vs LQ Control Techniques Applied to an Indoor Micro Quadrotor. In: Proc. IEEE International Conference on Intelligent Robots and Systems, pp. 2451–2456 (2004)
2. Schafroth, D., Bermes, C., Bouabdallah, S., Siegwart, R.: Modeling, System Identification and Robust Control of a Coaxial Micro Helicopter. Control Engineering Practice 18, 700–711 (2010)

3. Plinval, H., Morin, P., Mouyon, P., Hamel, T.: Visual Servoing for Underactuated VTOL UAVs: a Linear, Homography-Based Approach. In: 2011 IEEE International Conference on Robotics and Automation (2011)
4. Long, Y., Lyttle, S., Pagano, N., Cappelleri, D.: Design and Quaternion-Based Attitude Control of the Omnicopter MAV Using Feedback Linearization. In: ASME International Design Engineering Technical Conference (2012)
5. Graf, W.: Effects of Duct Lip Shaping and Various Control Devices on the Hover and Forward Flight Performance of Ducted Fan UAVs. Master Thesis, Virginia Polytechnic Institute and State University (2005)
6. Zhao, H.: Development of a Dynamic Model of a Ducted Fan VTOL UAV. Master Thesis, RMIT University (2009)
7. Ahmed, B., Pota, H., Garratt, M.: Flight Control of a Rotary Wing UAV using Backstepping. Int. J. Robust Nonlinear Control 20, 639–658 (2010)
8. Pounds, P., Bersak, D., Dollar, A.: Stability of Small-scale UAV Helicopters and Quadrotors with Added Payload Mass under PID Control. Auton. Robot. 33, 129–142 (2012)
9. Rongier, P., Lavarec, E., Pierrot, F.: Kinematic and Dynamic Modeling and Control of a 3-Rotor Aircraft. In: Proc. IEEE International Conference on Robotics and Automation, pp. 2606–2611 (2005)
10. Escareno, J., Sanchez, A., Garcia, O., Lozano, R.: Triple Tilting Rotor Mini-UAV: Modeling and Embedded Control of the Attitude. In: American Control Conference (2008)
11. Hoffmann, G.M., Huang, H., Waslander, S.L., Tomlin, C.J.: Precision Flight Control for a Multi-vehicle Quadrotor Helicopter Testbed. Control Engineering Practice 19, 1023–1036 (2011)
12. Michael, N., Mellinger, D., Lindsey, Q., Kumar, V.: The GRASP Multiple Micro UAV Testbed. IEEE Robotics and Automation Magazine (2010)
13. iSTAR, http://defense-update.com/products/i/istar-uav.htm
14. Ryll, M., Bulthoff, H., Giordano, P.: Modeling and Control of a Quadrotor UAV with Tilting Propellers. In: IEEE International Conference on Robotics and Automation (2012)
15. Long, Y., Lyttle, S., Cappelleri, D.: Linear Control Techniques Applied to the Omnicopter MAV in Fixed Vertical Ducted Fan Angle Configuration. In: ASME International Design Engineering Technical Conference (2012)
16. Long, Y., Cappelleri, D.: Linear Control Design, Allocation, and Implementation for the Omnicopter MAV. In: IEEE International Conference on Robotics and Automation (2013)

17 Articulated Wheeled Vehicles: Back to the Future?

Xiaobo Zhou, Aliakbar Alamdari, and Venkat Krovi

Abstract. Articulated Wheeled Vehicles (AWVs) are a class of wheeled locomotion systems where the chassis is connected to a set of ground-contact wheels via actively- or passively-controlled articulations, which can regulate wheel placement with respect to chassis during locomotion. The ensuing leg-wheeled systems exploit the reconfigurability and redundancy to realize significant benefits (improved stability, obstacle surmounting capability, enhanced robustness) over both traditional wheeled-and/or legged-systems in a range of uneven-terrain locomotion applications. This article examines the history of such articulated-wheeled-vehicles leading up to the current day, while placing in context the pioneering and seminal contributions of Professor Kenneth Waldron and his students. Subsequently, we outline our own research efforts on variants of AWVs, including the creation of a systematic computational screw-theoretic framework to model, analyze, optimize and control such systems.

1 Introduction

We seek to investigate the design, modeling, analysis and implementation of multiple variants/exemplars from the broad class of locomotion systems termed Articulated Wheeled Vehicles (AWVs), originally explored by Professor Kenneth Waldron and his students [32, 33].

The characteristic feature is the attachment of the multiple wheels to a common chassis via articulated chains, which facilitates (active or passive) repositioning of the wheels with respect to chassis during locomotion. Such AWVs can provide significantly superior locomotion performance (such as uneven-terrain obstacle surmounting capability and improved suspension characteristics). Equally importantly,

Xiaobo Zhou · Aliakbar Alamdari · Venkat Krovi
Department of Mechanical and Aerospace Engineering,
State University of New York at Buffalo, Buffalo, NY 14260
e-mail: vkrovi@buffalo.edu

V. Kumar et al. (Eds.): *Adv. in Mech., Rob. & Des. Educ. & Res.*, MMS 14, pp. 227–238.
DOI: 10.1007/978-3-319-00398-6_17   © Springer International Publishing Switzerland 2013

the reconfigurability and redundancy inherent in such systems can be exploited to enhance vehicle performance characteristics (such as efficiency, stability, traction). The combination of these benefits is extremely valuable in a variety of application settings from material handling on the shop floor to challenging uneven-terrain exploration.

However, articulated wheeled vehicles are highly-constrained systems subjected to both holonomic constraints (due to the multiple closed-loops) and non-holonomic constraints (due to wheel/ground contacts). Violation of these constraints e.g. typically in terms of slipping and skidding at the ground-wheel contacts, results both in energy-dissipation and estimation-uncertainty. Hence considerable research has focused on both *enhanced-suspension-design (kinematic and kinetostatic)*, to avoid constraint violation without either sacrificing payload capacity or increasing power-consumption; and *active-coordinated-control* for enhancing mobility, stability and traction.

In this article we will first present some background on such articulated-wheeled-robots followed by an abbreviated history of literature leading up to the current day. Subsequently, we highlight aspects of our own work, inspired by and building upon the contributions of the Waldron group. We focus in on the computational/algorithmic implementation of screw-theoretic framework, that aids the modeling, analysis, refinement and control of such systems. Finally, we conclude with a discussion of the promise and potential of Articulated-Wheeled-Vehicle paradigm, which we are seeking to systematically exploit by various design- and control-related research efforts.

2 Background

Wheeled Mobile Platforms/Vehicles, comprise of a platform supported by multiple wheels which allow for relative motions between the platform and the ground. Wheeled vehicles have traditionally offered simplicity of mechanical construction and control, very favorable payload-to-weight ratio, excellent load and tractive-force distribution, enhanced stability and energy-efficiency, making them the architecture of choice for most man-made terrestrial locomotion systems.

However, despite their incredible versatility, disk wheels impose severe constraints on the design and control of the wheeled vehicle to which they may be attached. Multiple disk-wheels cannot be arbitrarily attached to a single common platform/chassis without over-constraining the system. Kinematic overconstraint (as often seen in various machinery) occurs due to the lack of compatibility between the instantaneous motions of all moving parts. However, unlike in traditional machines, the violation of the wheel-ground contact-constraints (enforced only by force-closure) is possible – and gives rise to undesirable kinematic wheel-slip (skidding/slipping/scrubbing) seen in poorly-designed wheeled vehicles. Such wheel-slip is deleterious both from the perspective of reduced efficiency (power is wasted by scrubbing) and poor performance (degraded odometric localization, uncontrollable and unpredictable stick-slip behavior).

Kinematic overconstraint has traditionally been relieved by the addition of mechanical compliance (in the form of bushings and couplings) in order to mitigate the undesired stick-slip behavior at the wheel-ground contact. A case can be made for the systematic and careful introduction of additional mechanical compliance – in the form of small articulated subchains with passive (springs/dampers) or semi-active (adjustable spring-dampers) or active (motorized) actuation. The resulting articulated leg-wheel systems form multiple closed-kinematic loops with the ground that serve to constrain and redirect the effective forces and motions on the chassis.

Thus, viewing wheeled vehicles as yet another class of a parallel-kinematic chains (with multiple articulated leg-wheel branches attached to a common chassis) allows the systematic application of the rich theory of articulated multibody systems to design, analyze, simulate and control of the ensuing systems. The nature and number of both the added wheels, together with the intermediate articulations, has a significant influence on the mobility, maneuverability, controllability, stability and efficiency of the wheeled vehicle.

From a design perspective, there is enormous diversity at various levels within selection of: (i) the individual components, like the wheels (disk wheels vs Mecanum wheels) and the articulations (lower-pair revolutes/prismatics vs higher-pair cam joints); (ii) the topology/number of joints of each subchain; and (iii) the number of sub-chains/type-of-attachment to the chassis. The suitable selection of topology, dimensions and actuation of the individual sub-chains together with the selection of the number and attachment location to the common chassis creates enormous choice.

From the control perspective, the control and reconfiguration of the collaborating leg-wheel subsystems to regulate the mobility and maneuverability of the chassis offers other challenges. System configurations must be chosen in order to: (i) minimize singular configurations of the system; (ii) enhance mutual cooperation (motions and forces) during task performance; and (iii) improve robustness to local controller lapses and environmental disturbances. Significant freedom for implementation is also available by virtue of the reconfigurability and the ability to trade-off passive-equilibration versus active-actuation.

A systematic design, analysis and control framework that builds upon individual component capability to examine system-level behavior is desirable. We are currently examining one such computational differential-geometry framework [18, 19] that builds upon the rich articulated multibody literature and provides the tools to characterize, analyze, and validate seemingly disparate articulated-wheeled locomotion systems in a unified manner.

3 Literature

3.1 Planar Locomotion/ Payload Transport

Traditionally, wheeled mobile robots were considered to operate on planar surfaces, allowing the wheels to be modeled as thin disk-wheels with a holonomic rolling-without-slip constraint in the forward direction and offering the non-holonomic no side-slip constraint in the lateral direction. Alternate designs of wheels (such as the Ball Wheel [34] or the Swedish Mecanum wheel) or tracks avoid limitations imposed by the non-holonomic constraints, but possess other design and control limitations.

Thus, multi-disc-wheel platforms have many advantages but arbitrarily attaching and actuating these disk-wheels to a common chassis creates challenges. The overall chassis/platform motion-constraints are the union of all the individual constraints from each wheel/ground contact and hence further articulations (passive/active) are required to ensure compatibility.

In the plane, the rigid body constraint takes the form of requirement of all the moving-elements to have a common instantaneous center of rotation (ICR) which can also be visualized easily. A number of authors have surveyed the various planar-wheeled platform systems and their kinematic motion analysis in the plane [11, 8]. Kinematic compatibility is established, evaluated and maintained in terms of matching of the Instantaneous Center of Rotations (ICR) of the disk-wheels and the chassis/platform, as illustrated in Fig. 1.

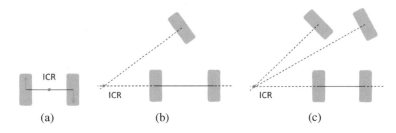

Fig. 1 Planar ICR based analysis for (a) differential drive; (b) tricycle; and (c) Ackerman steering

Campion *et al.* [13] present a systematic, general and unifying approach for derivation of kinematic- and dynamic-models of planar wheeled vehicle, with arbitrary number of various types of planar articulated-leg-wheel chains attached to a common chassis. They classified the ensuing planar composite-wheeled-vehicle systems into five generic model-classes; and for each model-class systematically address performance-related questions on system-level mobility and maneuverability, model-reducibility and controllability, and selective actuation.

In a variant on this theme, many researchers have contemplated means for coupling together multiple differentially-driven wheeled bases with rigid axles to form composite Multi-Degree-of-Freedom (MDOF) Vehicles [11]. Even operating on a perfect plane, two mobile robots with such rigid axles, cannot be rigidly coupled to each other without losing some or all of their mobility. The research challenges entailed include: passively relieving the kinematic-constraints between the multiple self-contained individual modules so as to permit arbitrarily long trains of such vehicles to be created; and coordinating the multiple degrees of freedom within this large assemblage to realize enhanced performance. Hence examples of such composite systems always feature some sort of passive-compliant-linkage examples include designs like the OmniMate/CLAPPER where differentially-driven mobile robots are attached to each other by a compliant link; or appropriately designed coupling within modular snake-robot systems like ACM-III [24].

In our own work, we examine the potential for further generalizing and extending this work to create planar Composite Wheeled Vehicles that exploit active-or passive-articulated subchains for relaxing the rigid body constraints between the various axles by introducing further articulations. The resulting articulations endow the composite vehicle with: (i) ability to accommodate changes in the relative configuration; (ii) redundant sensing for localizing the modules; and (iii) redundant actuation method for moving the common object to compensate for disturbances in the motions of the base [8, 10, 31, 18, 25, 30, 36, 37, 38, 29].

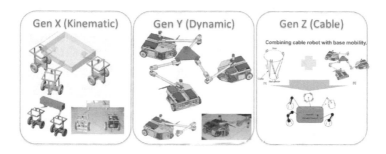

Fig. 2 Cooperative payload transport by mobile robot collectives in the ARMLAB [1]

3.2 Transitioning to the Spatial Case

The three-dimensional nature of the robot-motion with the varied ground-wheel-contact, creates many challenges. Various authors had noted that proper vehicle kinematic design, with adequate and appropriate inter-vehicle reconfiguration degrees-of-freedom, is required to permit adaptation to uneven surfaces, and allow for slip-free rolling of the wheels. Srinivasan and Nanua [27] demonstrated that a Variable Length Axle (a prismatic joint connecting two wheels joined on an axle) could accommodate the varying wheel/ground contact points wheeled vehicle systems moving on uneven terrain and help eliminate kinematic slip. From an implementation perspective, Choi and Sreenivasan [15] explored creation of Vehicles with Slip-free Motion Capability (VSMC) that facilitates passive-accommodation of the varied ground-wheel contact within the articulated chassis. Alternately, Chakraborty and Ghosal [14] introduced a Passive Variable Camber (PVC) an extra degree of freedom (DOF) at the wheel/axle joint that permits the wheel to tilt laterally relative to the axle and thereby allowing for the effective wheel/ground contact points to change without any prismatic joints. Auchter *et al.* [9] examined the performance (reduced-wheel slip and improved adaptation) of a 3-wheeled vehicle equipped with a Passive Variable Camber rear-axle to uneven terrain in simulation.

For rougher terrain applications, significantly more freedom becomes necessary between the chassis and ground contact. Hence numerous groups have examined creating hybrid-articulated leg-wheel subsystem-designs for terrestrial wheeled-locomotion systems to aid operation on rough unprepared surfaces. The leg-wheel subsystem designs consist of articulated linkages with multiple lower pair joints (revolute/prismatic) between the wheel and the chassis. This addition of individual-articulations (or even small articular-sub-systems) increases the degrees-of-freedom and provides for greater redundancy and reconfigurability. However, it also creates a need for controlling these additional degrees of freedom either passively via springs/dampers or actively with actuation. Numerous variants of such designs are possible depending upon the type, number, sequencing and nature of actuation (active/passive) of the joints. Examples range from the Mars Rover [21] and Shrimp [26] with rocker bogie suspensions, the WAAV [28] and Nomad [35] with articulated frames; to systems like the WorkPartner [22] and ALDURO [23] with powered legs and active/passive wheels. Such systems have found numerous applications in a wide variety of arenas such as exploration of extra-terrestrial [16, 21], extreme terrestrial [22, 35], and disaster environments. While high mobility, obstacle-surmounting capability and maneuverability are the obvious major requirements, additional criteria such as robustness, reliability and efficiency are also extremely desirable.

(a)   (b)   (c)

(d)   (e)   (f)

Fig. 3 Examples of leg-wheel systems: (a) ATHLETE [2]; (b) Shrimp [3]; (c) WorkPartner [4]; (d) ALDURO [5]; (e) Nomad [6]; (f) SRR [7]

4   Our Work

4.1   Analysis Framework

From an analysis and control perspective, it must be noted that much of the ICR based analysis-framework begins to break down even in the presence of the slightest notion of the ground-planarity, small-ruts and obstacles or the wheel-circularity assumptions (as may be expected in an industrial setting). In addition, the treatment of truly spatial wheeled-systems for traversing the uneven terrain required the modeling and analysis of spatial-chains. Thus an extension of the planar ICR theory to the instantaneous-screw-axes (ISA) theory proves critical for treatment of spatial cases but there is relatively limited work in this regard [12, 27]. Sreenivasan and Nanua [27] explored first- and second-order kinematic characteristics of wheeled vehicles on uneven terrain in order to determine vehicle mobility using screw-theory. Bruyninckx and Schutter [12] examined a description of the statics and velocity kinematics of serial, parallel and mobile robots based on the fundamental concepts of twists/wrenches and reciprocity and proposed a unified treatment of serial, parallel and mobile robot kinematics.

Inspired by these efforts, in our work we examined a systematic and symbolic rapid computational formulation of kinematic models for the general class of AWV. Further, automating this process by using the symbolic toolbox in MATLAB, facilitates the rapid modeling and analysis of any given design of an AWV [19]. Twist- and wrench-based approaches had previously been used to analyze motion and force capabilities of in-parallel articulated-mechanical-systems (such as parallel manipulators or multi-fingered grasping) but never for articulated-systems with rolling wheel-ground contacts. Our contribution lay in extending the twist- and wrench-based modeling framework to such articulated wheeled robotic systems (and subsuming the extant specialized approaches for ordinary wheeled robots [12]). Automating this process by use of symbolic analysis methods facilitates the rapid analysis of various AWV designs. The resulting modeling and analysis framework is well suited for both the design and control of such AWVs.

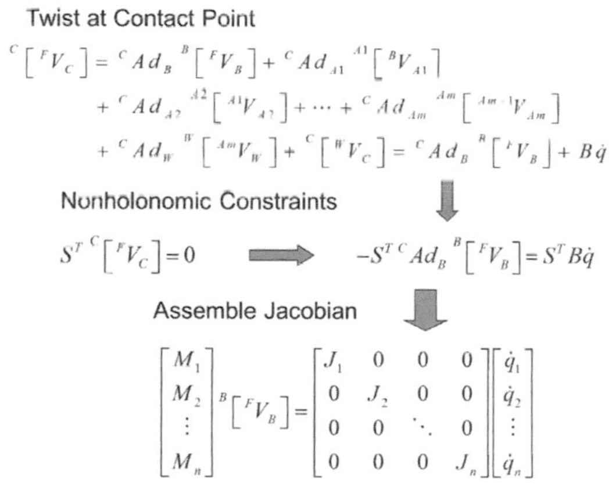

Twist at Contact Point

$$^{C}\left[^{F}V_{C}\right] = \,^{C}Ad_{B}\,^{B}\left[^{F}V_{B}\right] + \,^{C}Ad_{A1}\,^{A1}\left[^{B}V_{A1}\right]$$

$$+\,^{C}Ad_{A2}\,^{A2}\left[^{A1}V_{A2}\right] + \cdots + \,^{C}Ad_{Am}\,^{Am}\left[^{Am-1}V_{Am}\right]$$

$$+\,^{C}Ad_{W}\,^{W}\left[^{Am}V_{W}\right] + \,^{C}\left[^{W}V_{C}\right] = \,^{C}Ad_{B}\,^{B}\left[^{F}V_{B}\right] + B\dot{q}$$

Nonholonomic Constraints

$$S^{T}\,^{C}\left[^{F}V_{C}\right] = 0 \quad\Longrightarrow\quad -S^{T}\,^{C}Ad_{B}\,^{B}\left[^{F}V_{B}\right] = S^{T}B\dot{q}$$

Assemble Jacobian

$$\begin{bmatrix} M_{1} \\ M_{2} \\ \vdots \\ M_{n} \end{bmatrix} {}^{B}\left[^{F}V_{B}\right] = \begin{bmatrix} J_{1} & 0 & 0 & 0 \\ 0 & J_{2} & 0 & 0 \\ 0 & 0 & \ddots & 0 \\ 0 & 0 & 0 & J_{n} \end{bmatrix} \begin{bmatrix} \dot{q}_{1} \\ \dot{q}_{2} \\ \vdots \\ \dot{q}_{n} \end{bmatrix}$$

Fig. 4 Computational/algorithmic implementation of screw-theoretic modeling approach [19]

4.2   Passive Articulated Wheeled Vehicle Systems

The main advantages of passive AWVs are in terms of power consumption, payload capacity, and controller design. In almost all these cases, it is the addition of
articulated mechanical suspension design that endows them with their superior locomotion capabilities. For example, the JPL planetary rovers [21] and the Shrimp
rover [16] have shown enhanced terrain adaptability changing their configuration to
match the varying terrain topology by virtue of their rocker-bogie or four-bar based
suspensions. The suspension-designs typically feature multiple-closed kinematic-
loops, are designed to have few degrees of freedom, and exploit structural equilibra-
tion to passively support the weight of the system. However, in almost all cases no
systematic effort to design the articulated leg-wheel system is ever considered other
than to perform multiple parametric-studies with high-fidelity simulation runs.

In general, any such design process must take into account innumerable, often
equally important and competing considerations such as the loss of stability, tip-over
stability and ground traction for the task of locomotion on uneven terrain. However,
in our work, we will specifically focus our attention on two major complemen-
tary/conflicting design criteria large-workspace and stiff-suspension with minimal-
actuation in evaluating candidate designs for such articulated leg-wheel systems.
Our emphasis was on realizing the ability to surmount relatively-large obstacles
using minimal joint actuation within the subsystems while locomoting on uneven
terrain. Suitable selection of various kinematic parameters, such as the link-lengths
and initial configuration, as well as static parameters such as spring constants and
their preloads was critical. A novel kinetostatic design-customization framework is
employed for matching of desired kinematic and static specifications. Significant reductions in overall actuation requirements were achieved by judicious combination
of structural-equilibration, spring-assists and actuation [25].

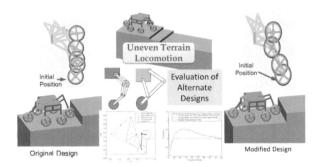

Fig. 5 Kinetostatic design of articulated leg-wheel system [25]

4.3   Fully Actuated Articulated Wheeled Systems

Fully-actuated articulated wheeled systems possess significant potential for recon-figurability (principally due to the absence of closed-loops of passive AWVs). Fur-ther, the redundant-actuation endows the system capability to optimize secondary performance indices (such as stability or traction) in addition to realizing the pri-mary locomotion tasks. On the other hand, more actuators add extra weight and control complexity necessary to resolve the redundancy in actuation. Thus the re-configurability and redundancy of fully-actuated AWVs needs to be unlocked by careful modeling, analysis and control [20, 28].

In our own work, we examined a similar over-actuated AWV design called the Reconfigurable Omnidirectional Articulated Mobile Robot [17, 18, 19]. Mul-tiple variants of kinematic control schemes were developed to ensure kinematic-constraint consistency and resolving the redundancies inherent in such articulated wheeled robots. Two generations of hardware-in-the-loop prototypes were devel-oped and simulations and real-time experiments varied out to validate localization dynamic control, and reconfiguration planning algorithms. Planned future work includes expanding our modeling framework and control scheme into 3D AWVs moving on uneven terrain.

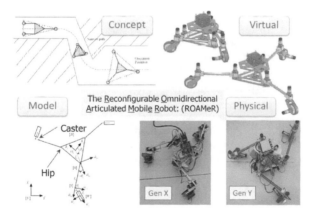

Fig. 6 The Reconfigurable Omnidirectional Articulated Mobile Robot (ROAMeR) [19]

5   Discussion

The Articulated-Wheeled-Vehicle paradigm offers remarkable and diverse opportu-nities for creation of very mobile and maneuverable terrestrial locomotion systems. However, the capabilities of articulated-wheeled locomotion systems to manipulate the chassis/payload (to improve obstacle surmounting capabilities and reduce actuation requirements) needs to be carefully unlocked by both design and control. From a design perspective, the selection of the topology, dimensions and finally configuration of the highly reconfigurable leg-wheel system plays a critical role in determining the performance. The subsequent selection of the type and number of such individual modules, together with the location and nature of attachment to the common payload determines the topology and parameters of the overall system. From the control perspective, the control and reconfiguration of the collaborating leg-wheel subsystems to regulate the mobility and maneuverability of the chassis offers opportunities and challenges. Significant freedom for implementation is also available by virtue of the reconfigurability and the ability to trade-off passive-equilibration versus active-actuation.

A systematic design, analysis and control framework that builds upon individual component capability to examine system-level behavior is desirable. We are currently examining one such screw-theoretic framework, built upon a theoretically sound articulated multibody background and implementation within an algorithmic/computational differential geometric formulation. It allows for flexible, modular and reconfigurable interchanges of component- and system-level constraints, while permitting integration into an operational framework. Quantitative measures of system-level cooperation (such as system manipulability, load-distribution and stability) aid the design-refinement, control and evaluation efforts. The ability to design, analyze and deploy under-actuated, exactly-actuated and redundantly-actuated articulated-wheeled systems using this framework can now be systematically exploited to enhance locomotion capabilities of such systems.

Acknowledgements. This work was supported in part by the National Science Foundation Award CNS-1135660.

References

1. http://mechatronics.eng.buffalo.edu
2. http://www.nasa.gov/multimedia/imagegallery/
   image_feature_748.html
3. http://www.bluebotics.com/mobile-robotics/shrimp-3
4. http://autsys.aalto.fi/en/WorkPartner/Media
5. http://www.uni-due.de/alduro/index_en.shtml
6. http://www.frc.ri.cmu.edu/projects/meteorobot/
   Nomad/Nomad.html
7. http://www-robotics.jpl.nasa.gov/systems/
   systemImages.cfm?System=6
8. Abou-Samah, M., Tang, C., Bhatt, R., Krovi, V.: A kinematically compatible framework for cooperative payload transport by nonholonomic mobile manipulators. Autonomous Robots 21(3), 227–242 (2006)
9. Auchter, J., Moore, C.A., Ghosal, A.: A Novel Kinematic Model for Rough Terrain Robots. In: Ao, S.-I., Rieger, B., Chen, S.-S. (eds.) Advances in Computational Algorithms and Data Analysis. LNEE, vol. 14, pp. 215–234. Springer, Netherlands (2009)
10. Bhatt, R.M., Tang, C.P., Krovi, V.N.: Formation optimization for a fleet of wheeled mobile robots a geometric approach. Robotics and Autonomous Systems 57(1), 102–120 (2009)

11. Borenstein, J., Everett, H.R., Feng, L.: Navigating Mobile Robots: Sensors and Techniques (1996)
12. Bruyninckx, H., Schutter, J.D.: Unified kinetostatics for serial, parallel and mobile robots (1998)
13. Campion, G., Bastin, G., Dandrea-Novel, B.: Structural properties and classification of kinematic and dynamic models of wheeled mobile robots. IEEE Transactions on Robotics and Automation 12(1), 47–62 (1996)
14. Chakraborty, N., Ghosal, A.: Kinematics of wheeled mobile robots on uneven terrain. Mechanism and Machine Theory 39(12), 1273–1287 (2004)
15. Choi, B.J., Sreenivasan, S.V.: Gross motion characteristics of articulated mobile robots with pure rolling capability on smooth uneven surfaces. IEEE Transactions on Robotics and Automation 15(2), 340–343 (1999)
16. Estier, T., Crausaz, Y., Merminod, B., Lauria, M., Piguet, R., Siegwart, R.: An innovative space rover with extended climbing abilities. In: Proceedings of the Space and Robotics
17. Fu, Q.: Kinematics of articulated wheeled robots: Exploiting reconfigurability and redundancy. M.s. thesis (2008)
18. Fu, Q., Krovi, V.: Articulated wheeled robots: Exploiting reconfigurability and redundancy. In: ASME 2008 Dynamic Systems and Control Conference, DSCC 2008 (2008)
19. Fu, Q., Zhou, X., Krovi, V.: The reconfigurable omnidirectional articulated mobile robot (ROAMeR). In: Khatib, O., Kumar, V., Sukhatme, G. (eds.) Experimental Robotics. STAR, vol. 79, pp. 871–882. Springer, Heidelberg (2012)
20. Grand, C., BenAmar, F., Plumet, F., Bidaud, P.: Decoupled control of posture and trajectory of the hybrid wheel-legged robot hylos. In: Proceedings of the 2004 IEEE International Conference on Robotics and Automation, ICRA 2004, vol. 5, pp. 5111–5116 (2004)
21. Hacot, H.: Analysis and traction control of a rocker-bogie planetary rover. M.s. thesis, MIT (1998)
22. Halme, A., Leppanen, I., Suomela, J., Ylonen, S., Kettunen, I.: Workpartner: Interactive human-like service robot for outdoor applications. The International Journal of Robotics Research 22(7-8), 627–640 (2003)
23. Hiller, M., Germann, D.: Manoeuvrability of the legged and wheeled vehicle alduro in uneven terrain with consideration of nonholonomic constraints. In: Proceedings of 2002 International Symposium on Mechatronics (ISOM 2002) (2002)
24. Hirose, S.: Biologically inspired robots: snake-like locomotors and manipulators. Oxford University Press (1993)
25. Jun, S.K., White, G.D., Krovi, V.N.: Kinetostatic design considerations for an articulated leg-wheel locomotion subsystem. Journal of Dynamic Systems, Measurement, and Control 128(1), 112–121 (2006)
26. Siegwart, R., Lamon, P., Estier, T., Lauria, M., Piguet, R.: Innovative design for wheeled locomotion in rough terrain. Robotics and Autonomous Systems 40(23), 151–162 (2002)
27. Sreenivasan, S.V., Nanua, P.: Kinematic geometry of wheeled vehicle systems. Journal of Mechanical Design 121(1), 50–56 (1999)
28. Sreenivasan, S.V., Waldron, K.J.: Displacement analysis of an actively articulated wheeled vehicle configuration with extensions to motion planning on uneven terrain. Journal of Mechanical Design 118(2), 312–317 (1996)
29. Tang, C.P., Bhatt, R., Abou-Samah, M., Krovi, V.: Screw-theoretic analysis framework for cooperative payload transport by mobile manipulator collectives. IEEE/ASME Transactions on Mechatronics 11(2), 169–178 (2006), doi:10.1109/TMECH.2006.871092
30. Tang, C.P., Krovi, V.N.: Manipulability-based configuration evaluation of cooperative payload transport by mobile manipulator collectives. Robotica 25(01), 29–42 (2007)

31. Tang, C.P., Miller, P.T., Krovi, V.N., Ryu, J.C., Agrawal, S.K.: Differential-flatness-based planning and control of a wheeled mobile manipulator: Theory and experiment. IEEE/ASME Transactions on Mechatronics 16(4), 768–773 (2011)
32. Waldron, K., McGhee, R.: The adaptive suspension vehicle. IEEE Control Systems Magazine 6(6), 7–12 (1986)
33. Waldron, K.J.: Terrain adaptive vehicles. Journal of Mechanical Design 117(B), 107–112 (1995)
34. West, M., Asada, H.: Design of ball wheel mechanisms for omnidirectional vehicles with full mobility and invariant kinematics. Journal of Mechanical Design 119(2), 153–161 (1997), doi:10.1115/1.2826230
35. Wettergreen, D., Bualat, M., Christian, D., Schwehr, K., Thomas, H., Tucker, D., Zbinden, E.: Operating Nomad during the Atacama Desert Trek, ch. 14, pp. 82–89. Springer, London (1998)
36. White, G.D., Bhatt, R.M., Krovi, V.N.: Dynamic redundancy resolution in a nonholonomic wheeled mobile manipulator. Robotica 25(02), 147–156 (2007)
37. White, G.D., Bhatt, R.M., Tang, C.P., Krovi, V.N.: Experimental evaluation of dynamic redundancy resolution in a nonholonomic wheeled mobile manipulator. IEEE/ASME Transactions on Mechatronics 14(3), 349–357 (2009)
38. Zhou, X., Tang, C.P., Krovi, V.: Cooperating mobile cable robots: Screw theoretic analysis. In: Milutinović, D., Rosen, J. (eds.) Redundancy in Robot Manipulators. LNEE, vol. 57, pp. 109–123. Springer, Heidelberg (2013)

18   A Robotic Mobile Platform for Application in Automotive Production Environment

Alberto Rovetta

Abstract. Industrial robotics is one of the booming sectors, which will have increasing importance in the future. ARIAL laboratory has realized a platform fully sensorized and able to map the environment and autonomously to navigate in it. The present work aims at describing the main features of "Andrea's" mobile platform, whose purpose is to cooperate with workers in the assembly line of automotive industries. The system integration has been set up by using ROS (Robot Operating System), a new but already cutting-edge technology for robotic systems management.

1   Introduction

POLIMI "Andrea's" platform, shown in figure 1, born as one of the main parts of Locobot project. Locobot (Low Cost Robot Co-workers) is a European project which aims to create an industrial mobile robot able to work on the assembly line and interact with workers in complete safety. The robot is composed of a robotic arm placed on a mobile autonomous platform. The purpose is to realize a machine which can navigate in full autonomy in an industrial environment, cooperatively interacting with the human workers. "Andrea's" meaning is "Autonomous Navigation with Dexterity and Robotic Environmental Actions System".

The "Andrea's" platform aims at playing three roles with a single robot, which must be able to navigate, perform pick and place tasks and cooperate with human workers without harming them.

Alberto Rovetta
Politecnico di Milano, Dipartimento di Meccanica, via Lamasa 1, 20156 Milano
e-mail: alberto.rovetta@polimi.it
         http://robotica.mecc.polimi.it

V. Kumar et al. (Eds.): *Adv. in Mech., Rob. & Des. Educ. & Res.*, MMS 14, pp. 239–244.
DOI: 10.1007/978-3-319-00398-6_18   © Springer International Publishing Switzerland 2013

2   Description of the Platform

2.1   Mecanum Wheel

One of the most attractive features of the "Andrea's" platform is represented by the system of movement, able to move the platform both longitudinally and laterally, or in rotation, without any steering. That is possible thanks to the special wheels adopted for this robot which presents on each extremity two sets of independent wheel complexes. The latter is made of an electrical motor and a gear head, connected to a mecanum type wheel and a wheel hub which links all components together and supports their weights and dynamics. To ensure an extreme freedom and ease of movement a vectorial motion wheel were adopted: that special type of wheel is characterized by a series of free spinning rolls, located on its peripherals, placed at a 45° angle in respect to the wheels rotational axis. That system guarantees that the vehicles direction and sense of movement can be controlled without the need of a complex steering system.

2.2   Safety System

During the working cycle in industrial environments unexpected events can always occur. The first objective is the safety of all people within the industry, which is why a redundant number of safety measures has been adopted.

To this end "Andrea's" platform has also been equipped with two mechanical switches for emergencies, which are able to stop immediately the worm screw brushless motors, by simultaneously opening the circuit that connects the power supply to the drives, and so cutting the current directed to the engines.

To ensure that workers in assembly line can work and move freely and safely, another system has been implemented that allows the "Andrea's" platform to detect obstacles, mobile or not, on its way and behave accordingly. If the object can represent a danger, the robot slows down and eventually it stops in case of proximity to the obstacle, automatically resuming the path in case the obstacle has been removed. This system is based on information sent by eight infrared sensors and eight sonar sensors, arranged two on each side of the robot, for a total of sixteen sensors.

2.3   Interaction with Workers and Control System

Interaction with workers is very easy thanks to an intuitive Graphical User Interface. "Andrea's" platform can be controlled in many different ways like directly by clicking on appropriate buttons on the screen with the touch screen or using a remote control by PC or even by Ipad and Iphone. The operator can also send commands to mobile platform through a Microsoft Kinect device, through which it is possible to detect the gestures indicated by worker. The robot is therefore able to recognize the commands, interpret them and move accordingly.

Another example of a control system developed for that robot is the voice command, by means of which it is possible to remotely control the movements of "Andrea's" platform.

2.4   Autonomous Navigation System

All systems have been implemented by using an innovative software called *Robotic Operating System (ROS)*. The latter provides libraries and tools to help software developers in creating robot applications. To make the control software architecture completely modular, an independent ROS node was created for each hardware component and for each control function. Each node is a standalone Linux executable, which can communicate with other nodes through message-passing on ROS topics and by calling other nodes' ROS services.

One of the most interesting stacks is represented by the Navigation System which is composed of two main parts, a *Coarse Navigation layer* and a *Fine Navigation layer*. *Coarse Navigation* basically uses a SLAM approach enabling the robot to perform a 3D mapping, using a tilting LIDAR, to localize itself within the map and to achieve autonomously the wanted position. Andrea's platform is able to calculate best trajectory which allows to reach the destination in the shortest possible time and avoiding any obstacle, expected and unexpected, or hazard to workers along its path. The purpose of *Fine Navigation*, on the other hand, is that one of positioning the robot with high precision, using a webcam facing the ceiling. By means of an algorithm that compares the images of the ceiling it is possible to reach the final position with an accuracy of about 10mm. The use of this double layer architecture enables the robot to perform basic navigation and obstacle avoidance very quickly and without excessive computational load thanks to Coarse Navigation layer, while the precise positioning on the working station is performed by the Fine Navigation layer.

2.5   Self-Balancing and Docking System

One of main purposes of "Andrea's" platform is to carry some loads placed on top of the structure by its robotic arm. During the working cycle of the Locoarm overturning moments can be generated and situations of serious danger might arise. In order to avoid any unexpected events and to balance the robot, four stabilizer robotic legs have been installed on the platform. Each system is composed by an electric linear actuators, controlled by a weight control loop, and one load-cell to measure the vertical force imposed to each leg.

To facilitate the mechanical interface of the robotic arm with the platform, an innovative robotic table has been studied and implemented. The connection with the platform has been realized through a simple docking system. The latter represents an important news, since it allows with great ease to interface the platform with different robotic arms, depending on the task assigned to the robot, and this makes the system really flexible.

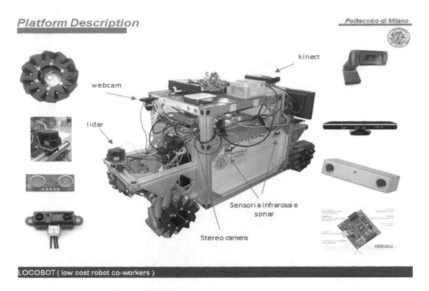

Fig. 1 POLIMI "Andrea's" platform

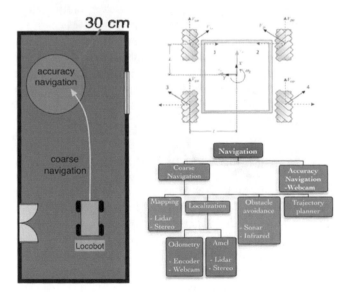

Fig. 2 Typical mecanum wheels movements and POLIMI platform navigation system

3   Conclusion

"Andrea's" Platform is a revolutionary robot equipped with a large number of security systems, able to map and to know the surrounding environment and to move independently and in complete safety in it. Equipped with a complete set of control systems that vary from voice control, to remote one and to teleoperation. "Andrea's Platform" can even receive commands from Ipad and Iphone. The purpose of the ARIAL lab is to create a machine more and more secure, that is able to easily interact with people and that has to become more and more intelligent, so that it can collaborate with workers and make their work more simple and therefore better their life.

In conclusion, "Andrea's" platform is a highly flexible solution to many problems that affect the industrial production and could be a turning point if adopted along the assembly line, greatly easing the task and the life of the workers.

References

[1] http://www.locobot.eu/
[2] Quigley, M., Gerkey, B., Conley, K., Faust, J., Foote, T., Leibs, J., Berger, E., Wheeler, R., Ng, A.: ROS: an open-source Robot Operating System, Computer Science Department, Stanford University, Stanford, CA, Willow Garage, Menlo Park, CA, Computer Science Department, University of Southern California
[3] Hanai, R., Oya, R., Izawa, T., Inaba, M.: Motion Generation for Human-Robot Collaborative Pick and Place based on Non-obstructing Strategy. Graduate School of Information Science and Technology, The University of Tokyo, 7-3-1 Hongo, Bunkyo-Ku, Tokyo, Japan
[4] Shi, E., Wang, Z., Huang, X., Huang, Y.: Study on AGV Posture Estimating Based on Distributed Kalman Fusion for Multi-Sensor. In: Inst. of Machinery Autom. Xi'an Univ. of Technol., Xi'an (2009)
[5] Bogue, R.: DaimlerChrysler installs new robot-based flexible assembly line. Industrial Robot: An International Journal 35(1), 16–18 (2008)
[6] Gamini Dissanayake, M.W.M., Newman, P., Clark, S., Durrant-Whyte, H.F., Csorba, M.: A Solution to the Simultaneous Localization and Map Building (SLAM) Problem. IEEE Trans. Robotics and Automation 17(3), 229–241 (2001)
[7] Tlane, N., de Villiers, M.: Kinematics and Dinamics Modelling of a Mecanum Wheeled Mo-bile Platform. Council for Scientific and industrial Research, Pretoria, RSA (2008)
[8] Young, K.D., Utkin, V.I., Ozgiiner, U.: A Control Engineer's Guide to Sliding Mode. YKK Syst., Mountain View, CA (1996)
[9] Dellaert, F., Fox, D., Burgard, W., Thrun, S.: Monte Carlo Localization for Mobile Robots. Computer Science Department, Carnegie Mellon University, Pittsburg PA 15213 (1999)
[10] Campbell, J., Sukthankar, R., Nourbakhash, I., Pahwa, A.: A Robust Visual Odometry and Precipice Detection System Using Consumer-grade Monocular Vision. Intel Research Pittsburgh, PA USA, Carniege Mellon University, Pittsburgh, PA USA, NASA Ames Research Center Moffett Field, CA USA (April 2005)

[11] Ohno, T., Ohya, A., Yuta, S.: Autonomus Navigation for Mobile Robots Reffering Prerecorded Image Sequence, Intelligent Robot Laboratory, Institute of Information Science and Eletronics, University of Tsukuba, Japan (1996)

[12] Bay, H., Tuytelaars, T., van Gool, L.: SURF: Speeded Up Robust Features. ETH Zurich

[13] Marder-Eppstein, E., Berger, E., Foote, T., Gerkey, B., Konolige, K.: The Office Marathon: Robust Navigation Indoor Office Environment. Willow Garage Inc., USA

[14] Savant Automation Inc., History of Automatic Guided Vehicle Systems

[15] Bresenham, J.: Algorithm for computer control of a digital plotter. Ibm. Systems J. 4(1), 25–30 (1965)

[16] Fox, D., Burgard, W., Thrun, S.: The dynamic window approach to collision avoidance. IEEE Robotics and Automation Magazine 4(1), 22–33 (1997)

[17] Latombe, J.: Robot Motion Planning. Kluwer Academic Publishers, Boston (1991) ISBN 0-7923-9206-X

[18] Khatib, O.: Real-time obstacle avoidance for robot manipulator and mobile robots. The International Journal of Robotics Research 5(1), 90–98 (1986)

[19] Khatib, M., Chatila, R.: An extended potential field approach for mobile robot sensor-based motions. In: Proceedings of International Conference on Intelligent Autonomous Systems, IAS 4 (1995)

[20] Borenstein, J., Koren, Y.: The vector field histogram - fast obstacle avoidance for mobile robots. IEEE Transaction on Robotics and Automation 7(3), 278–288 (1991)

[21] Latombe, J.: Robot Motion Planning. Kluwer Academic Publishers, Boston (1991) ISBN 0-7923-9206-X

[22] Quigley, M., Conley, K., Gerkey, B., Faust, J., Foote, T.B., Leibs, J., Wheeler, R., Ng, A.Y.: ROS: an open-source Robot Operating System. In: International Conference on Robotics and Automation. Open-Source Software Workshop (2009)

[23] Thrun, S., Burgard, W., Fox, D.: Probabilistic ROBOTICS. The Mit Press (2005)

[24] Fox, D.: KLD-sampling: Adaptive particle filters. In: Dietterich, T.G., Becker, S., Ghahramani, Z. (eds.) Proceedings of Advances in Neural Information Processing Systems (NIPS). MIT Press, Cambridge (2001)

19 Electro-Elastic Model for Dielectric Elastomers

Rocco Vertechy, Giovanni Berselli, and Vincenzo Parenti Castelli

Abstract. A continuum model is described for the study of the electro-elastic finite deformations of dielectric elastomers. The model: i) derives directly from a global energy balance; ii) does not require the postulation of any force or stress-tensor of electrical origin; iii) only requires the knowledge of permittivity and shear moduli of the considered material; and iv) is presented in Lagrangian form which is suited for the implementation in multi-physic finite element simulation environments.

1 Introduction

Dielectric Elastomers (DEs) are incompressible solids which exhibit non-linear elastic finite deformations and linear strain-independent dielectric properties. DEs can deform in response to an applied electric field, and can alter the electric field and/or potential in response to an undergone deformation. Thanks to this Electro-Mechanical (EM) coupling, DEs are currently being investigated as transduction materials for solid-state actuators, sensors and energy harvesters that are lightweight, energy-efficient, shock-resistant and low-cost [1].

Recently, a number of continuum EM models have been proposed that can be used for the accurate analysis and optimization of DE-based devices. An extensive review is provided in [2], which also presents a continuum thermo-EM model for general isotropic modified-entropic hyperelastic dielectrics.

Rocco Vertechy
Scuola Superiore Sant'Anna, Pisa, Italy
e-mail: r.vertechy@sssup.it

Giovanni Berselli
University of Modena and Reggio Emilia, Italy
e-mail: giovanni.berselli@unimore.it

Vincenzo Parenti Castelli
University of Bologna, Italy
e-mail: vincenzo.parenti@unibo.it

V. Kumar et al. (Eds.): *Adv. in Mech., Rob. & Des. Educ. & Res.*, MMS 14, pp. 245–250.
DOI: 10.1007/978-3-319-00398-6_19   © Springer International Publishing Switzerland 2013

This paper describes a reduced continuum finite-deformation EM model which is suited for the analysis and finite element simulation of DEs. The validity of the considered model has already been tested in an number practical case studies [3].

Section 2 provides the statement of the problem; Section 3 defines the total EM energy of a general system comprising elastic dielectrics and conductors; Section 4 derives the balance equations, boundary conditions and constitutive relations for the considered general EM system, Section 5 specifies the constitutive relations holding for typical DE materials.

2   Problem Definition

Referring to Fig. 1, consider a closed and electrically isolated EM system $\mathcal{B}$, which comprises dielectric and conducting bodies (electrodes) that move and deform in free space under the action of externally applied loads of electro-mechanical origin. For every motion and deformation of $\mathcal{B}$: 1) no mass can enter or leave the boundary of $\mathcal{B}$; 2) energy can cross the boundary of $\mathcal{B}$ in the form of electrical and mechanical work; 3) no interaction occurs between the electrical charges that lie within $\mathcal{B}$ and those outside (i.e. the boundary of $\mathcal{B}$ is either electrically shielded from its exterior or has an infinite extent).

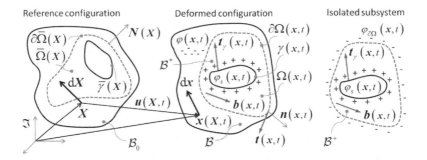

Fig. 1 EM system: reference and actual configuration, and isolated subsystem

Regarding kinematics, define with $\Im$ a fixed frame with respect to which the motions and deformations of $\mathcal{B}$ are measured (with $\mathcal{B}$ specifically indicating the current deformed configuration), and identify with $\mathcal{B}_0$ the reference (stress-free) configuration. For any arbitrary time instant $t \geq 0$, consider a general material point P of $\mathcal{B}$ and indicate with X and $x(X,t)$ (where $x(X, 0) \equiv X$) the position vectors expressing the location occupied by P when the EM system is in $\mathcal{B}_0$ and $\mathcal{B}$ respectively. Then, with reference to Fig. 1, the following definitions hold

$$u(X,t) = x(X,t) - X, \ \mathbf{F}(X,t) = \partial x/\partial X = \mathrm{GRAD}(x), \ J(X,t) = \det \mathbf{F}, \tag{1}$$

where: u is the displacement field; $\mathbf{F}$ is the deformation gradient; J is the Jacobian determinant.

Regarding EM system loadings, the physical space contained within B features: a distribution of electric charges (namely free and injected electrons or ions), with densities $\varphi(x,t)$ and $\varphi_\gamma(x,t)$, respectively defined per unit of deformed volume dv of B and per unit of deformed area ds of the physical surface $\gamma(t)$ (for instance a conducting electrode); a distribution of matter with mass density, $\rho(x,t)$, defined per unit volume dv. The same physical space is also subjected to: purely mechanical loads represented by a body force field (for instance the gravity field), $b(x,t)$, defined per unit volume dv; and a traction vector, $t_\gamma(x,t)$, defined per unit area ds of $\gamma(t)$ (for instance a body boundary).

Beside the displacement field, the other variables that complete the description of the state of B are the electric potential $\phi(x,t)$, the electric displacement vector $D(x,t)$, and the electric field $E(x,t)$ such that

$$E = -\partial\phi/\partial x = -\mathrm{grad}\phi. \tag{2}$$

3   Conservation of Energy

Consider an arbitrary but closed subsystem of B, hereafter called B^*, which (for every time instant $t > 0$) is identified by the volume $\Omega(t)$ and bounded by the closed surface $\partial\Omega(t)$ with unit normal $n(x,t)$. In Fig. 1, one of these possible subsystems is indicated with a blue dash-dotted line.

Irrespective of the specific response of the substances contained therein, the evolution of B^* is governed by a balance of EM energy. Differently from B, B^* is not electrically isolated, and thus interactions may exist between the electrical charges that lie within $\Omega(t)$ and those outside. According to potential theory [2], and as represented on the right side of Fig. 1, B^* is equivalent to an identical electrically isolated subsystem B^+ which has the boundary $\partial\Omega(t)$ covered by a single layer of charges with surface density

$$\varphi_{\partial\Omega} = -\mathbf{D} \cdot \mathbf{n}. \tag{3}$$

Thus, the conservation of total EM energy for the arbitrary subsystem B^* reads as

$$\mathrm{d}(\mathcal{K} + \mathcal{W})/\mathrm{d}t = \mathcal{P}_{me} + \mathcal{P}_{el}, \tag{4}$$

where $\mathcal{P}_{me}(t)$ and $\mathcal{P}_{el}(t)$ are the external mechanical and electrical powers entering in B^+ (that is in B^*) from the outside of its boundary $\partial\Omega(t)$, namely

$$\mathcal{P}_{me} = \int_{\Omega(t)-\gamma(t)} \boldsymbol{b} \cdot \dot{\boldsymbol{u}} \mathrm{dv} + \int_{\gamma(t)} \boldsymbol{t}_\gamma \cdot \dot{\boldsymbol{u}} \mathrm{ds} + \int_{\partial\Omega(t)-\gamma(t)} \boldsymbol{t} \cdot \dot{\boldsymbol{u}} \mathrm{ds} , \qquad (5.1)$$

$$\mathcal{P}_{el} = \int_{\Omega(t)-\gamma(t)} \phi \mathrm{d}\left(\varphi \mathrm{dv}\right)/\mathrm{dt} + \int_{\gamma(t)} \phi \mathrm{d}\left(\varphi_\gamma \mathrm{ds}\right)/\mathrm{dt} + \int_{\partial\Omega(t)-\gamma(t)} \phi \mathrm{d}\left(\varphi_{\partial\Omega} \mathrm{ds}\right)/\mathrm{dt} , \qquad (5.2)$$

with $\dot{\boldsymbol{u}}(\boldsymbol{x},t)$ being the velocity field ($\dot{\boldsymbol{u}} = d\boldsymbol{u}/dt$); whereas $\mathcal{K}(t)$ and $\mathcal{W}(t)$ are the kinetic and potential energies associated to the physical space contained in $\mathcal{B}^+$

$$\mathcal{K}(t) = \int_{\Omega(t)-\gamma(t)} 0.5\rho\dot{\boldsymbol{u}}^2 \mathrm{dv} , \quad \mathcal{W} = \int_{\Omega(t)-\gamma(t)} \left(\rho\Psi + \boldsymbol{E}\cdot\boldsymbol{D}\right)\mathrm{dv} , \qquad (5.3)$$

with $\Psi(\mathbf{F},\mathbf{E})$ being the energy density (per unit volume dv) of deformation and polarization of a given material. Note that Ψ does not include the energy required to build the electrostatic field in $\mathcal{B}^+$ (this is accounted by the term $\boldsymbol{E}\cdot\boldsymbol{D}$).

Equation (4), together with equations (5), represents the conservation of total EM energy of the arbitrary subsystem $\mathcal{B}^*$, expressed in global form and referred to the deformed configuration $\mathcal{B}$ of the overall EM system. For solids, it is generally more convenient to express the balance equations with respect to the reference configuration $\mathcal{B}_0$ (the so called Lagrangian description). This makes it possible to evaluate the considered integrals and to perform all time-derivatives with respect to fixed spatial domains; namely the arbitrary volume $\bar{\Omega} = \Omega(0)$ with boundary $\partial\bar{\Omega} = \partial\Omega(0)$, and the physical surface $\bar{\gamma} = \gamma(0)$.

Introducing from Eq. (1) the volume ratio relationship, $\mathrm{dv} = J\mathrm{dV}$, and the Nanson's formula, $\boldsymbol{n}\mathrm{ds} = \boldsymbol{N}\mathrm{dS}$ (with dV and dS indicating the infinitesimal undeformed volume and surface of $\mathcal{B}_0$, and $\boldsymbol{N}$ being the unit normal to dS), and defining the Lagrangian electric field $\bar{\boldsymbol{E}}$ from Eqs. (1) and (2) as

$$\bar{\boldsymbol{E}} = -\partial\phi/\partial\boldsymbol{X} = -\mathrm{GRAD}\phi = \mathbf{F}^T \boldsymbol{E} , \qquad (6)$$

equations (5) can then be rewritten as

$$\mathcal{P}_{me} = \int_{\bar{\Omega}-\bar{\gamma}} \bar{\boldsymbol{b}} \cdot \dot{\boldsymbol{u}} \mathrm{dV} + \int_{\bar{\gamma}} \bar{\boldsymbol{t}}_\gamma \cdot \dot{\boldsymbol{u}} \mathrm{dS} + \int_{\partial\bar{\Omega}-\bar{\gamma}} \bar{\boldsymbol{t}} \cdot \dot{\boldsymbol{u}} \mathrm{dS} , \qquad (7.1)$$

$$\mathcal{P}_{el} = \int_{\bar{\Omega}-\bar{\gamma}} \phi \mathrm{d}\left(\bar{\varphi}\mathrm{dV}\right)/\mathrm{dt} + \int_{\bar{\gamma}} \phi \mathrm{d}\left(\bar{\varphi}_\gamma \mathrm{dS}\right)/\mathrm{dt} - \int_{\partial\bar{\Omega}-\gamma} \phi \mathrm{d}\left(\bar{\boldsymbol{D}}\cdot\boldsymbol{N}\mathrm{dS}\right)/\mathrm{dt} , \qquad (7.2)$$

$$\mathcal{K} = \int_{\bar{\Omega}-\bar{\gamma}} 0.5\bar{\rho}\dot{\boldsymbol{u}}^2 \mathrm{dV} , \quad \mathcal{W} = \int_{\bar{\Omega}-\bar{\gamma}} \left(\bar{\rho}\Psi + \bar{\boldsymbol{E}}\cdot\bar{\boldsymbol{D}}\right)\mathrm{dV} , \qquad (7.3)$$

where $\bar{\boldsymbol{b}}$, $\bar{\boldsymbol{t}}$, $\bar{\boldsymbol{t}}_\gamma$, $\bar{\varphi}$, $\bar{\varphi}_\gamma$, $\bar{\rho}$ and $\bar{\boldsymbol{D}}$ are the Lagrangian variables

$$\bar{b} = Jb\,,\ \bar{t} = t\frac{\mathrm{d}s}{\mathrm{d}S}\,,\ \bar{t}_\gamma = t_\gamma \frac{\mathrm{d}s}{\mathrm{d}S}\,,\ \bar{\varphi} = J\varphi\,,\ \bar{\varphi}_\gamma = \varphi_\gamma \frac{\mathrm{d}s}{\mathrm{d}S}\,,\ \bar{\rho} = J\rho\,,\ \bar{D} = J\mathbf{F}^{-1}D\,. \qquad (8)$$

In obtaining Eq. (7.2), Eq. (3) has been used.

Resorting to the Gauss's divergence theorem along with Eqs. (1) and (6), the conservation of total EM energy of $\mathcal{B}^*$ in Lagrangian description follows as

$$\int\limits_{\bar{\Omega}-\bar{\gamma}}\left[\left(\bar{\rho}\ddot{u} - \mathrm{DIV}\left(\bar{\rho}(\partial\Psi/\partial\mathbf{F})^T\right) - \bar{b}\right)\cdot\dot{u} + \left(\bar{D} + \bar{\rho}\,\partial\Psi/\partial\bar{E}\right)\cdot\dot{\bar{E}}\right]\mathrm{d}V\ +$$

$$+\int\limits_{\partial\bar{\Omega}-\bar{\gamma}}\left[\left(\bar{\rho}(\partial\Psi/\partial\mathbf{F})^T\cdot N\right) - \bar{t}\right]\cdot\dot{u}\,\mathrm{d}S - \int\limits_{\bar{\gamma}}\left(\left[\!\left[\bar{\rho}(\partial\Psi/\partial\mathbf{F})^T\right]\!\right]\cdot N + \bar{t}_\gamma\right)\cdot\dot{u}\,\mathrm{d}S\ +\quad . \qquad (9)$$

$$-\int\limits_{\bar{\Omega}-\bar{\gamma}}\phi\frac{\mathrm{d}}{\mathrm{d}t}\left(\bar{\varphi} - \mathrm{DIV}\left(\bar{D}\right)\right)\mathrm{d}V - \int_{\bar{\gamma}}\phi\frac{\mathrm{d}}{\mathrm{d}t}\left(\bar{\varphi}_\gamma - \left[\!\left[\bar{D}\right]\!\right]\cdot N\right)\mathrm{d}S = 0$$

4   Balance Equations and Constitutive Relations

Equation (9) holds for any arbitrary volume $\bar{\Omega}$ (with boundary $\partial\bar{\Omega}$) and for any general EM process. Thus, satisfaction of Eq. (9) requires:

$$\bar{\rho}\ddot{u} = \mathrm{DIV}\left(\mathbf{P}\right) + \bar{b}\ \text{ on }\ \bar{\Omega}-\bar{\gamma}\,,\text{ and }\ \bar{t}_\gamma = -\left[\!\left[\mathbf{P}\right]\!\right]\cdot N\ \text{ on }\ \bar{\gamma}\,, \qquad (10)$$

$$\mathrm{DIV}\left(\bar{D}\right) = \bar{\varphi}\ \text{ on }\ \bar{\Omega}-\bar{\gamma}\,,\text{ and }\ \left[\!\left[\bar{D}\right]\!\right]\cdot N = \bar{\varphi}_\gamma\ \text{ on }\ \bar{\gamma}\,, \qquad (11)$$

$$\mathbf{P}^T = \bar{\rho}\,\partial\Psi/\partial\mathbf{F}\ \text{ with }\ \bar{t} = \mathbf{P}\cdot N\,, \qquad (12)$$

$$\bar{D} = -\bar{\rho}\,\partial\Psi/\partial\bar{E}\,. \qquad (13)$$

For the considered EM system, Eqs. (10) and (12) represent the Lagrangian form of the balance of linear momentum (with the second relation of Eq. (12) being the stress theorem holding in the reference configuration), whereas Eqs. (11) and (13) are the Lagrangian form of the electrostatic equations.

5   Constitutive Relations for Dielectric Elastomers

Equations (10)-(13) hold for any conservative elastic dielectric body that admits an energy density function of deformation and polarization. Particular problem solutions require specific definitions of $\Psi(\mathbf{F},E)$. A possible form for DEs is

$$\Psi = \Psi_{MR} + \Psi_{es} + \Psi_{vol}\,, \qquad (14.1)$$

$$\Psi_{MR} = \frac{c_1}{\bar{\rho}}\left[\text{trace}\left(\mathbf{FF}^T\right) - 3 \right] + \frac{c_2}{\bar{\rho}}\left[\left(\text{trace}\left(\mathbf{FF}^T\right)\right)^2 - \text{trace}\left(\left(\mathbf{FF}^T\right)^2\right) - 3 \right],$$ (14.2)

$$\Psi_{es} = -0.5\varepsilon E^2/\rho = -0.5\varepsilon J\bar{E}\cdot\left(\mathbf{F}^{-1}\mathbf{F}^{-T}\bar{E}\right)/\bar{\rho}, \quad \Psi_{vol} = -p\left(J - 1\right)/\bar{\rho}.$$ (14.3)

where Ψ_{MR} is the Mooney-Rivlin strain-energy function for hyperelastic materials (only dependent on the DE shear moduli c_1 and c_2), Ψ_{es} is a purely electrostatic energy function (only dependent on the DE permittivity ε), and Ψ_{vol} is a constraining term introduced to enforce the incompressibility condition ($J = 1$, with p being a Lagrange multiplier identifiable as a hydrostatic pressure).

With this energy density function, the constitutive relations (12) and (13), which complete the EM model for DEs together with Eqs. (10) and (11), read as

$$\mathbf{P}^T = 2\left[c_1\mathbf{1} + c_2\left(\text{trace}\left(\mathbf{FF}^T\right) - \mathbf{FF}^T\right)\right]\mathbf{F} - p\mathbf{F}^{-T} +$$
$$+ \varepsilon\left[\left(\mathbf{F}^{-T}\bar{E}\right)\otimes\left(\mathbf{F}^{-T}\bar{E}\right) - 0.5\bar{E}\cdot\left(\mathbf{F}^{-1}\mathbf{F}^{-T}\bar{E}\right)\mathbf{1}\right]\mathbf{F}^{-T},$$ (15)

$$\bar{D} = \varepsilon\mathbf{F}^{-1}\mathbf{F}^{-T}\bar{E}.$$ (16)

6   Conclusions

An electro-mechanical finite-deformation model for dielectric elastomers has been presented. The model is fully coupled and features the balance equations of linear momentum and of electrostatics, associated boundary conditions, and constitutive relations only dependent on three parameters. The model is expressed in a Lagrangian formulation which enables its direct use in finite element codes.

Acknowledgments. R.V. acknowledge the financial support from the EC, in the framework of the project "PolyWEC - New mechanisms and concepts for exploiting electroactive Polymers for Wave Energy Conversion" (FP7-ENERGY.2012.10.2.1, grant: 309139).

References

1. Carpi, F., De Rossi, D., Kornbluh, R., Pelrine, P., Sommer-Larsen, P.: Dielectric Elastomers as Electromechanical Transducers: Fundamentals, Materials, Devices, Models and Applications of an Emerging Electroactive Polymer Technology. Elsevier, Oxford (2008)
2. Vertechy, R., Berselli, G., Parenti Castelli, V., Bergamasco, M.: Continuum thermo-electro-mechanical model for electrostrictive elastomers. Journal of Intelligent Material Systems and Structures 24(6), 761–778 (2013), doi:10.1177/1045389X12455855
3. Vertechy, R., Frisoli, A., Bergamasco, M., Carpi, F., Frediani, G., De Rossi, D.: Modeling and experimental validation of buckling dielectric elastomer actuators. Smart Materials and Structures 21(9), 94005 (2012), doi:10.1088/0964-1726/21/9/094005

20 Designing Positive Tension for Wire-Actuated Parallel Manipulators

Leila Notash

Abstract. For wire-actuated parallel manipulators, the minimum 2-norm solution for the vector of wire tensions, calculated utilizing the Moore-Penrose generalized inverse of the Jacobian matrix, could result in negative tension for wires. In this paper, a methodology for generating the minimum 2-norm non-negative wire tension vector, when the null space basis of the Jacobian matrix of these manipulators is spanned by one or more vectors, is presented. Two planar parallel manipulators, a four-wire 2 degrees of freedom and a six-wire 3 degrees of freedom, are simulated to illustrate the proposed methodologies.

1 Introduction

In wire/cable-actuated manipulators, also known as wire/cable-driven manipulators, the motion of mobile platform (end effector) is constrained by the wires/cables. Because wires act in tension and cannot exert forces in both directions along their lines of action, i.e., their inputs are unidirectional and irreversible, to fully constrain an m degrees of freedom (DOF) rigid body suspended by wires, in the absence of gravity and external force/moment (wrench), the number of wires/actuators should be larger than the DOF of manipulator (Figures 1).

The failure could be defined as any event that affects the performance of manipulator such that it cannot complete its task as required. Wire-actuated manipulators could fail because of the hardware and/or software failures, including failure of a wire, sensor, actuator, or transmission mechanism; as well as computational failure. These failures could result in the loss of DOF, actuation, motion constraint, and information [1]. From the force point of view, the failure occurs if the wire does not provide the required force, e.g., when the actuator force/torque is lost partially or fully or the actuator is saturated. This could also happen when the wire is broken or slack (zero tension), wire is jammed (constant length), or its actuating mechanism malfunctions such that a different wire force is provided.

Leila Notash
Queen's University, Kingston, Canada
e-mail: notash@me.queensu.ca

V. Kumar et al. (Eds.): *Adv. in Mech., Rob. & Des. Educ. & Res.*, MMS 14, pp. 251–263.
DOI: 10.1007/978-3-319-00398-6_20 © Springer International Publishing Switzerland 2013

For a given mobile platform wrench, there are infinite solutions for the wire tension vector as $n \geq m+1$ for wire-actuated parallel manipulators. Applying the generalized inverse of the Jacobian matrix, the minimum 2-norm solution could result in negative tension for wires, which is not acceptable and is considered as "failure" in this article. The homogeneous solution is used to adjust the tension to positive values if the platform position and orientation (pose) is within the wrench closure workspace. In Section 2, methodologies for achieving positive tension, with minimum 2-norm for the wire tension vector, and for calculating the non-negative vectors that span the null space basis of the Jacobian matrix are presented. Simulation results are reported in Section 3. The article concludes with Section 4.

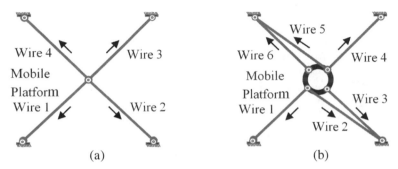

Fig. 1 Planar parallel manipulators (a) four-wire 2 DOF; and (b) six-wire 3 DOF

2   Wrench Recovery for Negative Wire Tension

For the n-wire-actuated parallel manipulators, the $n \times 1$ vector of wire forces $\boldsymbol{\tau} = \begin{bmatrix} \tau_1 & \cdots & \tau_n \end{bmatrix}^T$ is related to the $m \times 1$ vector of forces and moments (wrench) $\mathbf{F}$ applied by the platform with the $m \times n$ transposed Jacobian matrix $\mathbf{J}^T$ as

$$\mathbf{F} = \mathbf{J}^T \boldsymbol{\tau} = \begin{bmatrix} \mathbf{J}_1^T & \mathbf{J}_2^T & \cdots \mathbf{J}_i^T & \cdots & \mathbf{J}_{n-1}^T & \mathbf{J}_n^T \end{bmatrix} \boldsymbol{\tau} = \sum_{j=1}^{n} \mathbf{J}_j^T \tau_j \qquad (1)$$

where $m \leq 6$ depending on the dimension of task space. Column j of $\mathbf{J}^T$, $\mathbf{J}_j^T$, corresponds to the wrench applied on the platform by the jth wire/actuator. The solution of $\mathbf{F} = \mathbf{J}^T \boldsymbol{\tau}$ for the wire tensions is

$$\boldsymbol{\tau} = \boldsymbol{\tau}_p + \boldsymbol{\tau}_h = \mathbf{J}^{\#T} \mathbf{F} + \left(\mathbf{I} - \mathbf{J}^{\#T} \mathbf{J}^T \right) \mathbf{k} = \mathbf{J}^{\#T} \mathbf{F} + \mathbf{N} \boldsymbol{\lambda} \qquad (2)$$

where $\mathbf{J}^{\#T}$ is the Moore-Penrose generalized inverse of $\mathbf{J}^T$, $\boldsymbol{\tau}_p = \mathbf{J}^{\#T} \mathbf{F}$ is the minimum 2-norm (particular) solution, and $\boldsymbol{\tau}_h = \left(\mathbf{I} - \mathbf{J}^{\#T} \mathbf{J}^T \right) \mathbf{k}$ and $\boldsymbol{\tau}_h = \mathbf{N} \boldsymbol{\lambda}$ represent the homogenous solution. $\left(\mathbf{I} - \mathbf{J}^{\#T} \mathbf{J}^T \right) \mathbf{k}$ is the projection of the $n \times 1$ arbitrary vector $\mathbf{k}$ onto the null space of $\mathbf{J}^T$. Columns of the $n \times (n-m)$ matrix $\mathbf{N}$ correspond to the null space basis of $\mathbf{J}^T$, referred here as the null space vectors, and λ is an $(n-m)$-vector. When one or more entries of $\mathbf{\tau}_p = \mathbf{J}^{\#T}\mathbf{F}$ are negative the wire tensions could be adjusted by identifying the correctional tension $\mathbf{\tau}_h$ that would set all the wire tensions to positive values provided the manipulator pose is in the wrench closure workspace. The adjusted wire tensions should satisfy the tension limits $\tau_{\min} \le \tau_l = \tau_{pl} + \tau_{hl} = \tau_{pl} + \displaystyle\sum_{j=1}^{n-m} n_{lj}\lambda_j \le \tau_{\max}$, for $l=1,\cdots,n$, where n_{ij} corresponds to entry i of the jth null space vector.

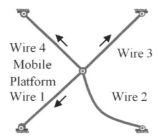

Fig. 2 Four-wire actuated parallel manipulator with slack wire 2

2.1 Conditions for Non-negative Wire Tension

When $\mathbf{F} \in \Re(\mathbf{J}^T)$, $\mathbf{F}_{\Re^\perp} = \left(\mathbf{I} - \mathbf{J}^T\mathbf{J}^{\#T}\right)\mathbf{F} = 0$, and the pose is in the wrench closure workspace of manipulator positive wire tensions could be calculated. The criteria for having non-negative wire tension could be defined based on the orthonormal bases of the null spaces of the $m \times n$ transposed Jacobian matrix $\mathbf{J}^T$ and the $m \times (n+1)$ augmented transposed Jacobian matrix $\mathbf{J}^T_{aug}$ by re-writing equation (1) as $\mathbf{J}^T\mathbf{\tau} - \mathbf{F} = 0$ and then $\mathbf{J}^T_{aug}\mathbf{\tau}_{aug} = \begin{bmatrix} \mathbf{J}^T_1 \ \mathbf{J}^T_2 \cdots \mathbf{J}^T_{n-1} \ \mathbf{J}^T_n \ -\mathbf{F} \end{bmatrix}\mathbf{\tau}_{aug} = 0$. The orthonormal basis of the null space of $\mathbf{J}^T$ is defined by $n-m$ number of n-vectors, i.e., the dimension of the null space vectors of $\mathbf{J}^T$ is $n \times 1$, while that of $\mathbf{J}^T_{aug}$ is defined by $n-m+1$ number of $(n+1)$-vectors.

The sufficient condition for rectifying the negative tension of particular solution to positive tension is the existence of a null space vector of $\mathbf{J}^T$ with all positive entries (e.g., refer to [2]). In the presence of external wrench, even if there is no null space vector of $\mathbf{J}^T$ with consistent sign, positive wire tension is feasible if there exist a null space vector of the augmented Jacobian matrix $\mathbf{J}^T_{aug}$ with non-negative values for the first n entries (corresponding to wires) and positive value for the $(n+1)$ th entry (corresponding to $\mathbf{F}$).

2.2 Methodology for Adjusting Negative Wire Tension

When the minimum norm solution results in negative tension for wire i, i.e., $\tau_{pi} < 0$, wire i could be considered as "failed" and its tension should be set to a non-negative value τ_{ci}. If wire i is left as slack (Figure 2) $\tau_{ci} = 0$. To increase the wrench capability and stiffness of the manipulators, the tension of wire i could be adjusted to a positive value, $\tau_{ci} > 0$. Rewriting equation (2) for wire i

$$\tau_{pi} + \tau_{hi} = \tau_{pi} + \sum_{j=1}^{n-m} n_{ij} \lambda_j = \tau_{ci} \tag{3}$$

Then, the platform wrench becomes

$$\mathbf{F}_f = \begin{bmatrix} \mathbf{J}_1^T & \mathbf{J}_2^T & \cdots & \mathbf{J}_i^T & \cdots & \mathbf{J}_{n-1}^T & \mathbf{J}_n^T \end{bmatrix} \boldsymbol{\tau}_f = \sum_{j=1}^{n} \mathbf{J}_j^T \tau_{pj} - \mathbf{J}_i^T (\tau_{pi} - \tau_{ci}) \tag{4}$$

where $\boldsymbol{\tau}_f = [\tau_{p1}\, \tau_{p2} \cdots \tau_{ci} \cdots \tau_{p\,n-1}\, \tau_{p\,n}]^T$ and the change in tension of wire i after adjusting its negative value is $|\tau_{pi} - \tau_{ci}|$. To provide the platform wrench $\mathbf{F}$, the remaining wires must balance the wrench corresponding to the adjusted negative wire tension. With the "correctional" force provided by the remaining wires $\boldsymbol{\tau}_{corr} = [\tau_{corr1}\, \tau_{corr2} \cdots 0 \cdots \tau_{corrn-1}\, \tau_{corrn}]^T$, the recovered wrench will be

$$\mathbf{F}_r = \mathbf{J}^T \boldsymbol{\tau}_f + \mathbf{J}_f^T \boldsymbol{\tau}_{corr} \tag{5}$$

where entry i of $\boldsymbol{\tau}_{corr}$ and column i of the Jacobian matrix $\mathbf{J}_f^T = \begin{bmatrix} \mathbf{J}_1^T & \mathbf{J}_2^T & \cdots & 0 & \cdots & \mathbf{J}_{n-1}^T & \mathbf{J}_n^T \end{bmatrix}$ are replaced by zeros. Then, the change in the platform wrench will be $\mathbf{F} - \mathbf{F}_r = \mathbf{J}^T (\boldsymbol{\tau}_p - \boldsymbol{\tau}_f) - \mathbf{J}_f^T \boldsymbol{\tau}_{corr}$. When the minimum 2-norm solution results in negative tension for k wires, after adjusting the negative tensions to positive values, the platform wrench that should be balanced by the remaining wires is $\sum_k \mathbf{J}_i^T (\tau_{pi} - \tau_{ci}) = \mathbf{J}^T (\boldsymbol{\tau}_p - \boldsymbol{\tau}_f)$, where the summation is taken over the wires with negative tension.

To fully compensate for the adjusted negative tension, i.e., for $\mathbf{F} - \mathbf{F}_r = \mathbf{0}$, the correctional force provided by the remaining wires should be [3]

$$\boldsymbol{\tau}_{corr} = \mathbf{J}_f^{\#T} \sum \mathbf{J}_i^T (\tau_{pi} - \tau_{ci}) = \mathbf{J}_f^{\#T} \mathbf{J}^T (\boldsymbol{\tau}_p - \boldsymbol{\tau}_f) \tag{6}$$

where k columns of $\mathbf{J}^T$, corresponding to the wires with negative tension, are replaced by zeros resulting in $\mathbf{J}_f^T$. Then, the overall wire force will be

$$\boldsymbol{\tau}_{tot} = \boldsymbol{\tau}_f + \boldsymbol{\tau}_{corr} = \mathbf{J}_f^{\#T}\mathbf{J}^T\boldsymbol{\tau}_p + (\mathbf{I} - \mathbf{J}_f^{\#T}\mathbf{J}^T). \tag{7}$$

These $\boldsymbol{\tau}_{corr}$ and $\boldsymbol{\tau}_{tot}$ are minimum 2-norm solutions for the chosen τ_{ci} [4]. If $\mathbf{J}_f^T$ has full row-rank, i.e., $\mathbf{F}$ belongs to the range space of $\mathbf{J}_f^T$, $\mathbf{F} \in \Re(\mathbf{J}_f^T)$, the right-generalized inverse (GI) of $\mathbf{J}_f^T$ is $\mathbf{J}_f^{\#T} = \mathbf{J}_f \left(\mathbf{J}_f^T \mathbf{J}_f\right)^{-1}$ as the vector of wire forces is physically consistent. Otherwise, the weighted left-GI is used.

When the minimum norm solution results in negative tension for k wires and the pose is in the wrench closure workspace $\tau_{pi} + \tau_{hi} = \tau_{ci} \geq \tau_{min} \geq 0$ for each of k wires, where τ_{min} is the minimum allowable tension. In the following subsections, formulations of $\tau_{ci} \geq 0$ and the non-negative null space vector(s) of $\mathbf{J}^T$ and $\mathbf{J}_{aug}^T$, when the pose is in the wrench closure workspace, are presented.

2.3   Minimum 2-Norm with Negative Tension for One Wire

When $n \geq m+1$ and the minimum 2-norm solution results in negative tension for wire i, $\tau_{pi} < 0$, considering $\tau_{pi} + \sum_{j=1}^{n-m} n_{ij}\lambda_j = \tau_{ci} \geq \tau_{min} \geq 0$, there is no condition on τ_{ci} provided the adjusted wire tensions do not exceed the maximum value, i.e., $\tau_{min} \leq \tau_l \leq \tau_{max}$, for $l = 1, \cdots, n$. Then for the chosen $\tau_{ci} \geq \tau_{min}$ value, the minimum 2-norm solutions for the correctional and overall wire tension vectors are calculated using equations (6) and (7).

It is worth mentioning that if the homogenous solution $\boldsymbol{\tau}_h$ of $\boldsymbol{\tau} = \boldsymbol{\tau}_p + \boldsymbol{\tau}_h$ is used to adjust the negative tension $\tau_{pi} < 0$ to a non-negative value, the null space vectors should be formulated such that the pertinent entry of at least one null space vector is positive while the other entries do not alter the corresponding positive particular solution to a negative value.

2.4   Minimum 2-Norm with Negative Tension for More Than One Wire

When $n \geq m+1$ and the minimum 2-norm solution results in negative tension for at least two wires, there is no condition on $\tau_{ci} \geq \tau_{min} \geq 0$ if these tensions could be taken as free (non-pivot) variables. When any of these wires with negative tension are dominating the corresponding $\tau_{ci} \geq \tau_{min} \geq 0$ is selected and τ_c of the remaining wires are calculated. The wire tensions after adjustment should not exceed the maximum value, i.e., $\tau_{min} \leq \tau_l \leq \tau_{max}$, for $l = 1, \cdots, n$. Then for the chosen/calculated $\tau_{ci} \geq \tau_{\min}$ values, the minimum 2-norm solutions for the correctional and overall wire tension vectors are calculated using equations (6) and (7).

For instance, when $n = m+1$ and the minimum norm solution results in negative tension for $k \geq 2$ wires the null space vector of $\mathbf{J}^T$ has all non-negative entries (positive entries pertinent to wires with negative τ_p) in the wrench closure workspace. If the tensions of these wires cannot be taken as the free variables then using the non-negative null space vector

$$\lambda = \max(\lambda_i, \lambda_j, \cdots) = \max\left(\frac{\tau_{ci} + |\tau_{pi}|}{n_i}, \frac{\tau_{cj} + |\tau_{pj}|}{n_j}, \cdots\right) \tag{8}$$

where $|\tau_{pi}|$ is the magnitude of τ_{pi}. The largest of $\lambda_i, \lambda_j, \ldots, \lambda$, corresponds to the dominating wire, e.g., wire i, and its τ_c is set to $\tau_{\min}$. Then, the tension of non-dominating wires, e.g., wire j, is calculated using $\tau_{cj} = \tau_{pj} + n_j\lambda > \tau_{\min}$, and the minimum 2-norm solutions for the correctional and overall wire tension vectors are calculated using equations (6) and (7) for these τ_c values.

When $n > m+1$ and the minimum norm solution produces negative tension for $k \geq 2$ wires and the tension of at least one of these wires cannot be taken as the free variable then the non-negative values for these wires cannot be selected arbitrarily. For example, when the tension of wire i is the only free variable, considering the normalized null space vectors of $\mathbf{J}^T$, the entry corresponding to the dominating wire i is always smaller than the ones relating to the remaining wires with negative tension. Hence, $\tau_{ci} = \tau_{\min}$ and τ_c of other wires are calculated.

When the platform pose is in the wrench closure workspace and $n = m+1$ a unique solution for λ with minimum 2-norm for $\boldsymbol{\tau}$ could be formulated regardless of the number of wires with negative particular solution. When $n > m+1$ the null space of $\mathbf{J}^T$ is spanned by $n-m$ vectors. If the homogenous solution $\boldsymbol{\tau}_h$ is used to adjust the negative tension $\tau_{pi} < 0$ to a non-negative value there could exist many linear combinations of these null space vectors, i.e.,

$$\tau_{hl} = \sum_{j=1}^{n-m} n_{lj}\lambda_j = n_l\lambda, \quad \text{satisfying} \quad \tau_{\min} \leq \tau_{pl} + \tau_{hl} = \tau_{pl} + \sum_{j=1}^{n-m} n_{lj}\lambda_j \leq \tau_{\max} \quad \text{for}$$

$l = 1, \cdots, n$, while the entries of null space vector $\mathbf{n}$ corresponding to negative tension should be non-zero. One of these combinations results in minimum 2-norm for $\boldsymbol{\tau}$ and $\boldsymbol{\tau}_{corr}$.

2.5 Non-negative Null Space Vectors

A non-negative null space vector $\mathbf{n}$ of $\mathbf{J}^T$, among all $\boldsymbol{\tau}_h$ of $\mathbf{J}^T \boldsymbol{\tau}_h = \mathbf{0}$, is formulated such that $\tau_{pl} + n_l \lambda \geq \tau_{cl}$, for $l = 1, \cdots, n$. Equation (1) is rearranged by augmenting $\mathbf{J}^T$ with wrench $\mathbf{F}$ to form $\mathbf{J}_{aug}^T$ using $\mathbf{J}^T \boldsymbol{\tau} - \mathbf{F} = \mathbf{0}$

$$\begin{bmatrix} \mathbf{J}^T & -\mathbf{F} \end{bmatrix} \begin{bmatrix} \boldsymbol{\tau} \\ 1 \end{bmatrix} = \mathbf{J}_{aug}^T \boldsymbol{\tau}_{aug} = \mathbf{0} \tag{9}$$

Then, the problem is redefined as identifying the non-negative solutions for the $(n+1) \times 1$ augmented tension vector $\boldsymbol{\tau}_{aug}$ while the $(n+1)$ th entry of $\boldsymbol{\tau}_{aug}$ is zero. An $(n+1) \times (n+1+m)$ matrix $^{(0)}\mathbf{D}$ could be constructed by augmenting the identity matrix $\mathbf{I}_{n+1}$ with the transpose of $\mathbf{J}_{aug}^T$

$$^{(0)}\mathbf{D} = \begin{bmatrix} \mathbf{I} & \mathbf{J}_{aug} \end{bmatrix} = \begin{bmatrix} \mathbf{I} & \begin{bmatrix} \mathbf{J}^T & -\mathbf{F} \end{bmatrix}^T \end{bmatrix} \tag{10}$$

Matrices $^{(j)}\mathbf{D}$, $j = 1, 2, \cdots$ and $j \leq m$, are formed successively by non-negative linear combination of its rows such that the entries of $(n+1+j)$ column of matrix $^{(j)}\mathbf{D}$ are zero, e.g., refer to [5]. The procedure is terminated when all the columns with index larger than $(n+1)$ are null vectors or when $j = m$. Then, the rows with zeros for entries $(n+2)$ to $(n+1+m)$ correspond to the null space vectors of $\mathbf{J}^T$ or $\mathbf{J}_{aug}^T$ and are saved as final $\mathbf{D}$.

The non-negative solution for the null space vectors of the $m \times (n+1)$ $\mathbf{J}_{aug}^T$ are obtained using the first $(n+1)$ entries of the rows of final $\mathbf{D}$ with non-zero $(n+1)$ th entry and normalizing them such that the $(n+1)$ th entry is identity. These normalized vectors are non-negative solutions for $\boldsymbol{\tau}_{aug}$, i.e., the vertices of the convex polyhedron. The first $(n+1)$ entries of the rows of final $\mathbf{D}$ with zero $(n+1)$ th entry, which are independent of the values of wrench $\mathbf{F}$, correspond to the non-negative null space vectors of $\mathbf{J}^T$, i.e., the extreme rays of the polyhedron. The non-negative null space basis vectors of $\mathbf{J}^T$ (and $\mathbf{J}_{aug}^T$) could be linearly combined to obtain the non-negative null space vector(s) with certain (or maximum) non-zero entries. When a specific entry of all null space vectors of $\mathbf{J}^T$ is zero there is no redundancy in the corresponding wire [3].

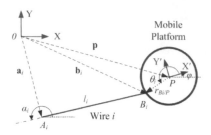

Fig. 3 Parameters of planar wire-actuated parallel manipulators

3   Case Study

For planar manipulators, the platform is connected to the base by n wires, each wire with a length of l_i and orientation of α_i (Figure 3). The attachment points of wire i to the base and platform are denoted as points A_i and B_i, respectively. The angular positions of points B_i on the platform are denoted by θ_i. The wire forces $\boldsymbol{\tau}$ are related to the wrench $\mathbf{F}$ applied by the platform as

$$\mathbf{F} = \mathbf{J}^T \boldsymbol{\tau} = \begin{bmatrix} \cos\alpha_1 & \cdots & \cos\alpha_n \\ \sin\alpha_1 & \cdots & \sin\alpha_n \\ v_1 & \cdots & v_n \end{bmatrix} \begin{bmatrix} \tau_1 \\ \vdots \\ \tau_n \end{bmatrix} \tag{11}$$

where column i of the $3{\times}n$ matrix $\mathbf{J}^T$, i.e., $\mathbf{J}_i^T$, represents the axis of wire i with the direction cosines of $\cos\alpha_i = l_{ix}/l_i$ and $\sin\alpha_i = l_{iy}/l_i$, and v_i is the moment of wire axis with respect to the origin of $\Gamma(X',Y',Z')$, point P, formulated in $\Psi(X,Y,Z)$ as

$$v_i = -\cos\alpha_i\left(b_{iy} - p_y\right) + \sin\alpha_i\left(b_{ix} - p_x\right) \tag{12}$$

3.1   Four-Wire Manipulator with 2 DOF

For a 2 DOF translational manipulator with four wires (Figure 1a), the coordinates of A_i, $i = 1, \ldots, 4$, are $(-2, -1.5)$, $(2, -1.5)$, $(2, 1.5)$ and $(-2, 1.5)$, respectively. The null space basis of the $2{\times}4$ matrix $\mathbf{J}^T$ is spanned by two $4{\times}1$ vectors.

Example 1. At the platform pose of $\mathbf{p} = [0\ 0]^T$ in the wrench closure workspace

$$\mathbf{J}^T = \begin{bmatrix} \cos\alpha_1 & \cdots & \cos\alpha_4 \\ \sin\alpha_1 & \cdots & \sin\alpha_4 \end{bmatrix} = \begin{bmatrix} -0.800 & 0.800 & 0.800 & -0.800 \\ -0.600 & -0.600 & 0.600 & 0.600 \end{bmatrix} \tag{13}$$

For $\mathbf{F} = [0\ \ 24]^T$ Newtons, the minimum 2-norm vector of wire forces is $\boldsymbol{\tau}_p = \mathbf{J}^{\#T}\mathbf{F} = [-10\ \ -10\ \ 10\ \ 10]^T$ with negative tension for wires 1 and 2 and a magnitude of $\|\tau_p\|_2 = 20$. Considering the null space vectors of $\mathbf{J}^T$, formulated using the entries corresponding to wires 1 and 2 (with negative tension) as the free variables in the reduced row echelon form of $\mathbf{J}^T$,

$$\mathbf{N} = [\mathbf{n}_1 \quad \mathbf{n}_2] = \begin{bmatrix} 0 & 1 \\ 1 & 0 \\ 0 & 1 \\ 1 & 0 \end{bmatrix} \tag{14}$$

$$\mathbf{N}_{aug}^T = [0 \quad 0 \quad 0.707 \quad 0.707 \quad 0.035] \tag{15}$$

For $\tau_{c1} = 1$ N and $\tau_{c2} = 1$ N, which is equivalent to using $c_1 = 11$ and $c_2 = 11$ in $\tau = \tau_p + c_1 \mathbf{n}_1 + c_2 \mathbf{n}_2 \geq \tau_{min}$,

$$\tau_f = [1 \quad 1 \quad 10 \quad 10]^T \tag{16}$$

$$\tau_{corr} = \mathbf{J}_f^{\#T} \sum \mathbf{J}_i^T (\tau_{pi} - \tau_{ci}) = [0 \quad 0 \quad 11 \quad 11]^T \tag{17}$$

$$\tau_{tot} = \tau_f + \tau_{corr} = [1 \quad 1 \quad 21 \quad 21]^T \tag{18}$$

which produces the original wrench, with $\|\tau_{corr}\|_2 = 15.556$, $\|\tau_{tot}\|_2 = 29.732$ and $\tau_{aug} = [1 \quad 1 \quad 21 \quad 21 \quad 1]^T$.

Example 2. At the platform pose of $\mathbf{p} = [2.0 \quad 0.0]^T$ meters, same X coordinate as that for the anchors of wires 2 and 3,

$$\mathbf{J}^T = \begin{bmatrix} \cos \alpha_1 & \cdots & \cos \alpha_4 \\ \sin \alpha_1 & \cdots & \sin \alpha_4 \end{bmatrix} = \begin{bmatrix} -0.936 & 0 & 0 & -0.936 \\ -0.351 & -1.000 & 1.000 & 0.351 \end{bmatrix} \tag{19}$$

For $\mathbf{F} = [-46.817 \quad 20.534]^T$ Newtons, the minimum 2-norm vector of wire forces is $\tau_p = \mathbf{J}^{\#T} \mathbf{F} = [21.791 \quad -9.140 \quad 9.140 \quad 28.209]^T$, with negative tension for wire 2 and a magnitude of $\|\tau_p\|_2 = 37.917$. The non-negative null space vector of $\mathbf{J}^T$, $\mathbf{n} = [0 \quad 1 \quad 1 \quad 0]^T$, has a non-zero entry corresponding to wire 2, and indicates equal antagonistic contributions of wires 2 and 3, which are collinear at this platform pose, to the homogeneous solution. Two sparse non-negative null space vectors of $\mathbf{J}_{aug}^T$ are

$$\mathbf{N}_{aug}^T = \begin{bmatrix} \mathbf{n}_{aug1}^T \\ \mathbf{n}_{aug2}^T \end{bmatrix} = \begin{bmatrix} 0 & 0 & 0.060 & 1.000 & 0.020 \\ 1.000 & 0 & 0.762 & 0 & 0.020 \end{bmatrix} \tag{20}$$

which correspond to two sparse non-negative solutions for $\mathbf{J}_{aug}^{T}\boldsymbol{\tau}_{aug} = \mathbf{0}$; $\boldsymbol{\tau}_{aug} = [0\ \ 0\ \ 2.978\ \ 50.000\ \ 1]^{T}$ and $\boldsymbol{\tau}_{aug} = [50.000\ \ 0\ \ 38.090\ \ 0\ \ 1]^{T}$, both with zero input from wire 2 for this pose and platform wrench $\mathbf{F}$.

The null space basis of $\mathbf{J}^{T}$ is spanned by two vectors

$$\mathbf{N} = [\mathbf{n}_1\ \ \mathbf{n}_2] = \begin{bmatrix} 0 & -1.000 \\ 1.000 & 0.702 \\ 1.000 & 0 \\ 0 & 1.000 \end{bmatrix} \tag{21}$$

The tension of wire 2 is set to $\tau_{c2} = \tau_{min} = 1$ N. Then

$$\boldsymbol{\tau}_f = [21.791\ \ 1\ \ 9.140\ \ 28.209]^{T} \tag{22}$$

$$\boldsymbol{\tau}_{corr} = \mathbf{J}_f^{\#T}\mathbf{J}^{T}(\boldsymbol{\tau}_p - \boldsymbol{\tau}_f) = \mathbf{J}_f^{\#T}\mathbf{J}_2^{T}(\tau_{p2} - \tau_{c2}) = [-2.856\ \ 0\ \ 8.134\ \ 2.856]^{T} \tag{23}$$

$$\boldsymbol{\tau}_{tot} = \boldsymbol{\tau}_f + \boldsymbol{\tau}_{corr} = [18.935\ \ 1\ \ 17.274\ \ 31.065]^{T} \tag{24}$$

which produces the original wrench, and $\|\boldsymbol{\tau}_{corr}\|_2 = 9.082$, $\|\boldsymbol{\tau}_{tot}\|_2 = 40.286$. This is equivalent to using $\boldsymbol{\tau} = \boldsymbol{\tau}_p + c_1\mathbf{n}_1 + c_2\mathbf{n}_2 \geq \tau_{min}$ with $c_1 = 8.134$ and $c_2 = -2.856$. It is noteworthy that this $\boldsymbol{\tau}_{tot}$ corresponds to $\boldsymbol{\tau}_{aug} = [18.935\ \ 1.000\ \ 17.274\ \ 31.065\ \ 1]^{T}$ with non-zero input from wire 2, relating to the dense non-negative null space vector of $\mathbf{J}_{aug}^{T}$, $\mathbf{n}_{aug} = [0.610\ \ 0.032\ \ 0.556\ \ 1.000\ \ 0.032]^{T}$.

3.2 Six-Wire Manipulator with 3 DOF

For a 3 DOF manipulator with six wires (Figure 1b), the coordinates of A_i, $i = 1$, ..., 6, are (−2, −1.5), (2, −1.5), (2, −1.5), (2, 1.5), (−2, 1.5) and (−2, 1.5) respectively. The position of connection point B_i on the mobile platform is set at a constant radius of $r_{Bi/P} = 0.25$ m. The angular coordinates, θ_i, $i = 1, ..., 6$, of the wire connections to the platform are, respectively, $180°$, $180°$, $0°$, $0°$, $0°$ and $180°$. The null space of the 3×6 matrix $\mathbf{J}^{T}$ is spanned by three 6×1 vectors.

Example 3. At the platform pose of $\mathbf{p} = [0.5\ \ 1.0]^{T}$ meters and $\phi = 60°$, the wire forces $\boldsymbol{\tau}$ are related to wrench $\mathbf{F}$ applied by the platform using $\mathbf{F} = \mathbf{J}^{T}\boldsymbol{\tau}$, where

$$\mathbf{J}^{T} = \begin{bmatrix} -0.721 & 0.580 & 0.452 & 0.979 & -0.994 & -0.957 \\ -0.693 & -0.815 & -0.892 & 0.202 & 0.107 & 0.289 \\ -0.069 & 0.227 & -0.209 & -0.187 & 0.229 & -0.243 \end{bmatrix} \tag{25}$$

For $\mathbf{F} = [0.784 \ -54.566 \ -0.993]^T$ applied by the mobile platform, i.e., for a force of $[0.784 \ -54.566]^T$ Newtons and a moment of -0.993 N-m about the Z direction, the minimum 2-norm vector of wire forces is $\boldsymbol{\tau}_p = \mathbf{J}^{\#T}\mathbf{F} = [22.546 \ 18.020 \ 24.689 \ -7.692 \ -1.143 \ -1.919]^T$, with negative tension for wires 4, 5 and 6 and a magnitude of $\left\|\boldsymbol{\tau}_p\right\|_2 = 38.817$. The number of wires with negative tension is $k = 3$ and $n - m = 3$. Three sparse non-negative null space vectors of $\mathbf{J}^T$ are

$$\mathbf{N}^T = \begin{bmatrix} \mathbf{n}_1^T \\ \mathbf{n}_2^T \\ \mathbf{n}_3^T \end{bmatrix} = \begin{bmatrix} 0 & 0 & 0.317 & 0.869 & 1.000 & 0 \\ 0 & 0.532 & 0 & 1.000 & 0.817 & 0.497 \\ 0.308 & 0.094 & 0 & 1.000 & 0.817 & 0 \end{bmatrix} \tag{26}$$

The five sparse non-negative null space vectors of $\mathbf{J}_{aug}^T$ are

$$\mathbf{N}_{aug}^T = \begin{bmatrix} \mathbf{n}_{aug1}^T \\ \mathbf{n}_{aug2}^T \\ \mathbf{n}_{aug3}^T \\ \mathbf{n}_{aug4}^T \\ \mathbf{n}_{aug5}^T \end{bmatrix} = \begin{bmatrix} 0 & 13.189 & 52.833 & 0 & 30.901 & 0 & 1.000 \\ 0 & 101.523 & 0 & 21.369 & 0 & 82.525 & 1.000 \\ 0 & 62.378 & 19.072 & 0 & 0 & 45.954 & 1.000 \\ 51.105 & 28.795 & 0 & 21.369 & 0 & 0 & 1.000 \\ 28.458 & 21.879 & 19.072 & 0 & 0 & 0 & 1.000 \end{bmatrix} \tag{27}$$

with a rank of 4, where columns 2 to 5 of $\mathbf{N}_{aug}$ are dependent with a rank of 3.

The entry of each null space vector $\mathbf{n}_i$, $i = 1, 2, 3$, corresponding to wire 6 is smaller than the ones relating to wires 4 and 5. As well, the entries of $\mathbf{n}_3$ and $\mathbf{n}_1$ corresponding to wires 4 and 5 are respectively the largest. The tension of wires 4, 5 and 6 are set to $\tau_c = \tau_{min} = 1$ N and

$$\boldsymbol{\tau}_f = [22.546 \ 18.020 \ 24.689 \ 1 \ 1 \ 1]^T \tag{28}$$

$$\boldsymbol{\tau}_{corr} = \mathbf{J}_f^{\#T}\mathbf{J}^T(\boldsymbol{\tau}_p - \boldsymbol{\tau}_f) = [5.432 \ 4.783 \ -5.417 \ 0 \ 0 \ 0]^T \tag{29}$$

$$\boldsymbol{\tau}_{tot} = \boldsymbol{\tau}_f + \boldsymbol{\tau}_{corr} = [27.978 \ 22.803 \ 19.272 \ 1 \ 1 \ 1]^T \tag{30}$$

which produces the original wrench, and $\left\|\boldsymbol{\tau}_{corr}\right\|_2 = 9.040$, $\left\|\boldsymbol{\tau}_{tot}\right\|_2 = 40.952$. This is equivalent to using $\boldsymbol{\tau} = \boldsymbol{\tau}_p + c_1\mathbf{n}_1 + c_2\mathbf{n}_2 + c_3\mathbf{n}_3 \geq \tau_{min}$ with $c_1 = -17.087$, $c_2 = 5.878$ and $c_3 = 17.663$.

Example 4. At the platform pose of $\mathbf{p} = [-0.5 \ -1.0]^T$ meters and $\phi = -60°$

$$\mathbf{J}^T = \begin{bmatrix} -0.887 & 0.965 & 0.993 & 0.658 & -0.513 & -0.516 \\ -0.462 & -0.263 & -0.119 & 0.753 & 0.858 & 0.857 \\ 0.250 & -0.176 & 0.200 & 0.237 & -0.004 & 0.005 \end{bmatrix} \tag{31}$$

For $\mathbf{F} = [47.873 \ 75.817 \ 19.368]^T$ applied by the mobile platform, i.e., for a force of $[47.873 \ 75.817]^T$ Newtons and a moment of 19.368 N-m about the Z direction, the minimum 2-norm vector of wire forces is

$\tau_p = \mathbf{J}^{\#T}\mathbf{F} = [\text{-6.924}\quad \text{-7.990}\quad 31.380\quad 56.606\quad 18.096\quad 18.779]^T$, with negative

tension for wires 1 and 2 ($|\tau_{p1}| < |\tau_{p2}|$) and a magnitude of $\|\tau_p\|_2 = 70.575$. Three

sparse non-negative null space vectors of $\mathbf{J}^T$ are

$$\mathbf{N}^T = \begin{bmatrix} \mathbf{n}_1^T \\ \mathbf{n}_2^T \\ \mathbf{n}_3^T \end{bmatrix} = \begin{bmatrix} 0.692 & 1.000 & 0 & 0 & 0 & 0.681 \\ 0.705 & 1.000 & 0 & 0.011 & 0.677 & 0 \\ 0.705 & 1.000 & 0.013 & 0 & 0.688 & 0 \end{bmatrix} \tag{32}$$

all with larger entries corresponding to wire 2 compared to that for wire 1. Wire 1 is the dominating wire while the number of wires with negative tension is $k = 2 < n - m = 3$. The following four sparse non-negative null space basis vectors of $\mathbf{J}_{aug}^T$ each have zero entry corresponding to wire 1, while wire 2 tension is a pivot variable in all four.

$$\mathbf{N}_{aug}^T = \begin{bmatrix} \mathbf{n}_{aug1}^T \\ \mathbf{n}_{aug2}^T \\ \mathbf{n}_{aug3}^T \\ \mathbf{n}_{aug4}^T \end{bmatrix} = \begin{bmatrix} 0 & 1.348 & 99.927 & 0 & 102.561 & 0 & 1.000 \\ 0 & 1.348 & 0 & 83.116 & 15.845 & 0 & 1.000 \\ 0 & 1.772 & 0 & 82.857 & 0 & 16.231 & 1.000 \\ 0 & 4.048 & 97.943 & 0 & 0 & 103.295 & 1.000 \end{bmatrix} \tag{33}$$

The tension of wire 1 is set to $\tau_{c1} = \tau_{min} = 1$ N and then the tension of wire 2 is calculates as $\tau_{c2} = 3.462$ N, and

$$\tau_f = [1\quad 3.462\quad 31.380\quad 56.606\quad 18.096\quad 18.779]^T \tag{34}$$

$$\tau_{corr} = \mathbf{J}_f^{\#T}\mathbf{J}^T(\tau_p - \tau_f) = [0\quad 0\quad \text{-0.000}\quad \text{-0.000}\quad \text{-0.000}\quad 7.795]^T \tag{35}$$

$$\tau_{tot} = \tau_f + \tau_{corr} = [1\quad 3.462\ 31.380\ 56.606\ 18.096\ 26.573]^T \tag{36}$$

which produces the original wrench, and $\|\tau_{corr}\|_2 = 7.795$, $\|\tau_{tot}\|_2 = 72.357$. This is equivalent to using $\tau = \tau_p + c_1\mathbf{n}_1 + c_2\mathbf{n}_2 + c_3\mathbf{n}_3 \geq \tau_{min}$ with $c_1 = 11.452$ and $c_2 = c_3 = 0$.

4   Conclusion

Using the generalized inverse of the Jacobian matrix of wire-actuated parallel manipulators, the minimum 2-norm solution for the vector of wire tensions could result in negative tension for one or more wires. The negative tensions are generally adjusted to positive values utilizing the null space vectors of the transposed Jacobian matrix. In this paper, a methodology for adjusting the negative tension of the minimum 2-norm solution using the generalized inverse of the transposed Jacobian matrix, when the null space of the transposed Jacobian matrix is spanned by one or more vectors, were presented. The proposed methodology produces the minimum 2-norm non-negative solution for the vector of wire tensions and was implemented on a four-wire 2 DOF and a six-wire 3 DOF planar manipulators.

References

1. Notash, L., Huang, L.: On the design of fault tolerant parallel manipulators. Mech. and Machine Theory 38(1), 85–101 (2003)
2. Roberts, R.G., Graham, T., Lippitt, T.: On the inverse kinematics, statics, and fault tolerance of cable-suspended robots. J. Robotic Systems 15(10), 581–597 (1998)
3. Notash, L.: Failure recovery for wrench capability of wire-actuated parallel manipulators. Robotica 30(6), 941–950 (2012)
4. Notash, L.: Wrench recovery for wire-actuated parallel manipulators. In: Proc. of 19th CISM-IFToMM Symp. Robot Design, Dynamics, and Control (RoManSy) (2012)
5. Schuster, R., Schuster, S.: Refined algorithm and computer program for calculating all non-negative fluxes admissible in steady states of biochemical reaction systems with or without some flux rates fixed. CABIOS 9(1), 79–85 (1993)

21 Control of Humanoid Hopping Based on a SLIP Model

Patrick M. Wensing and David E. Orin

Abstract. Humanoid robots are poised to play an ever-increasing role in society over the coming decades. The structural similarity of these robots to humans makes them natural candidates for applications such as elder care or search and rescue in spaces designed for human occupancy. These robots currently, however, do not have the capability for fast dynamic movements which may be required to quickly recover balance or to traverse challenging terrains. Control of a basic dynamic movement, hopping, is studied here through simulation experiments on a 26 degree of freedom humanoid model. Center of mass trajectories are planned with a spring-loaded inverted pendulum (SLIP) model and are tracked with a task-space controller. Unauthored arm movements emerge from the task-space approach to produce continuous dynamic hopping at 1.5 m/s.

Fig. 1 A combination of SLIP model planning and task-space control allows a continuous dynamic hop to be controlled at real-time rates. The structure of the Task-Space Controller allows unauthored arm action to emerge which prevents extra torso pitching during leg thrust and during positioning of the feet in flight.

Patrick M. Wensing · David E. Orin
The Ohio State University
Dept. of Electrical and Computer Engineering
Columbus, OH 43210
e-mail: {wensing.2,orin.1}@osu.edu

V. Kumar et al. (Eds.): *Adv. in Mech., Rob. & Des. Educ. & Res.*, MMS 14, pp. 265–274.
DOI: 10.1007/978-3-319-00398-6_21   © Springer International Publishing Switzerland 2013

1   Introduction

With the abundance of promising recent work in humanoid robots, these systems are becoming ever closer to operating alongside humans in the home and in the workplace. Mechanical improvements to many of the state-of-the-art humanoids [5, 9, 16, 17] are continually occurring to push the potential applications that they may provide. Aside from mechanical improvements, control of these systems continues to advance as well, for instance, providing intuitive human to robot interactions [1], stable locomotion over mildly uneven terrain [14], and balance recovery from environmental disturbances [3].

Despite these efforts, humanoid robots still have little capability for fast dynamic movements, such as hopping or jumping. These types of movements require coordinated interactions between many degrees of freedom in order to manage the rapid interchanges of kinetic and potential energies throughout stages of stance and flight. Stable performance of these movements is further complicated by short periods of stance, during which large ground forces on the system must be managed within their frictional and unidirectional limits to provide corrective interactions.

As a basic dynamic movement, hopping provides a platform to evaluate control approaches for dynamic motion without the need to address the more complex limb phasings found in derivative movements such as running. Future humanoids operating in challenging environments will require aggressive movements such as a hop to clear obstacles or to traverse areas with widely separated footholds. Hop control has been explored previously in bipeds, where specialized compliant actuators [7] were used to store and return energy to the system during stance. Here, instead, compliant dynamics are mimicked in the humanoid through the use of a physics-based spring-loaded inverted pendulum (SLIP) model to generate reference dynamics for the humanoid center of mass (CoM). This approach enables continuous forward hopping, and manages joint coordination through the use of a task-space controller to select joint torques. Results are shown for full 3D hopping in simulation with a 26 degree of freedom (DoF) humanoid. A snapshot of this model mid-flight is shown in Fig. 1. The fluid resultant hopping motion features arm swing during stance and flight as a natural coordination strategy from the task-space control approach, and emerges without manual authoring.

The remainder of this paper is laid out as follows. Notation to describe the dynamics of the humanoid model and the SLIP template are developed briefly in Section 2. Section 3 presents the Task-Space Controller at a high-level and describes the generation of SLIP-based CoM reference trajectories. Results are presented in Section 4 with a summary provided in Section 5.

2   Simulation and Template Models

The humanoid shown in Fig. 1 is a 26-DoF system that is modeled after a 6 foot (1.83 m), 160 pound (72.6 kg) male. Further details on the model are given in [18]. The configuration of the system can be described by $q = [\, q_b^T \;\; q_a^T \,]^T$, where

$q_b \in SE(3)$ is the unactuated position and orientation of the torso (referred to as the floating base) and q_a denotes the configuration of the actuated joints. The joint rate and acceleration vectors, $\dot{q} \in \mathbb{R}^{26}$ and $\ddot{q} \in \mathbb{R}^{26}$, are partitioned similarly. The standard dynamic equations of motion are:

$$H(q)\ddot{q} + C(q,\dot{q})\dot{q} + G(q) = S_a^T \tau + J_s(q)^T F_s \tag{1}$$

where H, $C\dot{q}$, and G are the familiar mass matrix, velocity product terms, and gravitational terms, respectively. Here F_s collects ground reaction forces (GRFs) for appendages in support, and J_s is a combined support Jacobian. The matrix $S_a = [0_{20 \times 6} \ 1_{20 \times 20}]$ is a selection matrix for the actuated joints and $\tau \in \mathbb{R}^{20}$ is the joint torque vector. The full 3D dynamics of the humanoid are simulated with the DynaMechs [13] simulation package. This simulator employs a penalty-based contact model which includes compliance and damping in the normal and tangential directions at each planar contact point. No force feedback is provided to the controller.

To generate dynamic hopping, the approach presented in Section 3 will seek to mimic the CoM dynamics described by a SLIP model. This template model for locomotion, shown in Fig. 2, has been shown to describe the CoM dynamics incredibly well for high-speed forward locomotion in a wide array of insects and animals [2, 4, 8]. Species as diverse as crabs to kangaroos bounce in a dynamically similar fashion at high-speeds, and demonstrate similar effective leg stiffnesses relative to their size and weight [2]. In biological systems, the selection of an effectively compliant gait at high speeds instead of a stiff legged gait (which is employed at lower speeds) is due in part to energetic savings that are enabled by the passive compliance of muscles, tendons, and ligaments [4]. Although we assume no joint compliance for the humanoid model used here, it is envisioned that the addition of passive and variable compliance will continue to be an active area of actuation research [6, 10], enabling future humanoids to perform these types of movements with efficiency and power.

The SLIP model assumes a linear leg spring during stance, with a rest length equal to the touchdown length of the spring. Stance terminates when the leg spring again reaches its rest length, and is followed by ballistic CoM dynamics in flight.

Fig. 2 SLIP stance model for forward locomotion. The position is given relative to an anchor location as (x, z). The model includes a Hookean leg spring with spring constant k. During flight, a ballistic model is assumed for the point mass.

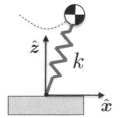

This point-mass model implicitly assumes massless legs that can be arbitrarily repositioned in flight for preparation of the upcoming touchdown.

3   Prioritized Task-Space Control

Task-space (also called operational-space) control provides a convenient framework to allow for control of the salient characteristics of a behavior or movement, without requiring large amounts of motion detail in the high-dimensional configuration space of the humanoid. For instance, natural task spaces such as the CoM and configurations of the feet can be used to generate dynamic walking [12] with minimal required hand authoring. The role of task-space control within the control system used here is shown in Fig. 3. Roughly, the task-space control problem is to select joint torques to reproduce some commanded task dynamics as closely as possible. While task-space control for a manipulator is a well studied problem, the underactuation of a humanoid in flight and stance introduces additional constraints on the solution to the problem. In this work, a state machine (consisting simply of flight and stance states) is used to inform the Task-Space Controller of these constraints, and to manage tracking of the tasks (CoM, feet, posture, etc.).

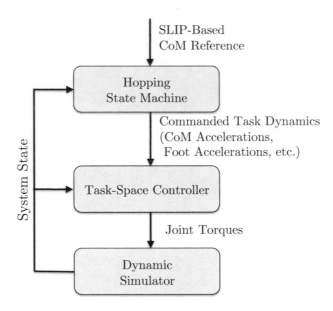

Fig. 3 Overall system block diagram. SLIP-based reference trajectories and hand authored foot trajectories are tracked by the Task-Space Controller. The Hopping State Machine monitors if the system is in flight or stance, and turns off CoM control during flight.

In our recent work [18] we proposed a conic-optimization-based solution to the task-space control problem that uses numerical optimization software to select physically feasible contact forces and joint torques. Roughly, the optimization-based approach enacts task-space control while simultaneously ensuring that the classic force distribution problem (FDP) [11] remains feasible to generate the task dynamics with forces under the feet. To ensure solvability of the FDP, the contact wrench acting on each of the N_S support feet is broken up into forces $\boldsymbol{f}_{s_{ij}} \in \mathbb{R}^3$ which act at each of the N_{P_i} contact vertices for foot i. Then, given a commanded task acceleration $\dot{\boldsymbol{v}}_{t,c}$, an optimization problem described by (2)-(4) can be solved to select ground forces, joint torques, and joint accelerations that are consistent with the system dynamics.

$$\min_{\ddot{q},\tau,\boldsymbol{f}_{s_{ij}}} \frac{1}{2}\|\boldsymbol{J}_t\,\ddot{q} + \dot{\boldsymbol{J}}_t\,\dot{q} - \dot{\boldsymbol{v}}_{t,c}\|^2 \tag{2}$$

$$\text{subject to } \boldsymbol{H}\,\ddot{q} + \boldsymbol{C}\,\dot{q} + \boldsymbol{G} = \boldsymbol{S}_a^T\,\tau + \sum_{i=1}^{N_S}\sum_{j=1}^{N_{P_i}} \boldsymbol{J}_{s_{ij}}^T \boldsymbol{f}_{s_{ij}} \tag{3}$$

$$\boldsymbol{f}_{s_{ij}} \in \mathscr{C}_i \quad \forall i \in \{1,\ldots,N_S\}, j \in \{1,\ldots,N_{P_i}\}. \tag{4}$$

Here $\boldsymbol{J}_t$ is a task Jacobian, $\boldsymbol{J}_{s_{ij}}$ is a Jacobian for contact vertex j of foot i, and $\mathscr{C}_i$ is a friction cone for foot i. $\boldsymbol{J}_t$ may be a Jacobian for a stack of tasks and may include, for instance, foot and CoM Jacobians within its rows. In this optimization, (2) enforces optimal tracking of the task dynamics, while (3) and (4) ensure that the the optimal task dynamics are physically realizable.

If a strict hierarchy of importance exists amongst the tasks, then a Prioritized Task-Space Control (PTSC) problem exists. The optimization problem above can be solved first to optimize tracking of the highest-priority task alone, and then subsequently to optimize tracking of the lower-priority tasks. These subsequent optimizations require additional constraints to be added to the problem as described in [18]. This formulation can also be used to regulate angular momentum, as described in [18], even though angular momentum is not amenable to a task Jacobian.

3.1 SLIP-Based CoM Reference Trajectories

SLIP-based CoM reference trajectories are used to generate the commanded CoM accelerations. First, a periodic trajectory of the SLIP model, through stance and flight, is found off-line. This off-line process tunes the SLIP touchdown angle, maximum CoM height during flight, and effective leg stiffness to obtain a periodic gait with user-specified stance and flight times. An example periodic trajectory is shown in Fig. 4. We note that choosing too high of an effective leg stiffness causes the system to slow down from one step to the next, while too low of an effective leg stiffness causes the system to speed up. This is shown in Fig. 5 for a variation of 30 percent above and below the stiffness for periodic locomotion. All cases shown use the touchdown angle and top-of-flight height shown in Fig. 4.

The periodic SLIP trajectory is followed on-line through a simple PD control law to select the commanded CoM acceleration (which composes three of the

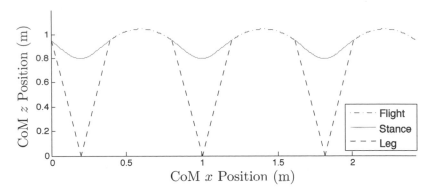

Fig. 4 SLIP-based CoM reference trajectory for 1.5 m/s forward hopping. Touchdown and liftoff angles are symmetric, which holds for every 1 step periodic trajectory of the SLIP model.

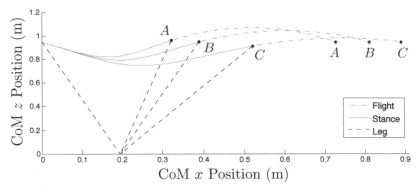

Fig. 5 Comparison of CoM trajectories for varying leg stiffness. Trajectory B uses the leg stiffness found to generate periodic CoM motion, while A and C employ stiffnesses that are 30% greater and 30% less, respectively.

components of $\dot{v}_{t,c}$). With the CoM position given as p_G, the commanded acceleration $\ddot{p}_{G,c}$ is specified as

$$\ddot{p}_{G,c} = \ddot{p}_{G,d} + K_D(\dot{p}_{G,d} - \dot{p}_G) + K_P(p_{G,d} - p_G) \tag{5}$$

where $(p_{G,d}, \dot{p}_{G,d}, \ddot{p}_{G,d})$ are the desired values from the SLIP based trajectory. The lateral position of the CoM is commanded to remain at a fixed initial position. We note that it is possible to employ a series of PD setpoints, as in [18] to achieve a standing jump. However, continuous hopping requires more careful design of CoM trajectories. The biological grounding of the SLIP model makes it a natural choice to generate these trajectories, and does so with little required hand authoring.

3.2 State-Based Control Summary

Periods of stance and flight require different task-space dynamics to be controlled. During periods of stance, CoM control is active, and the feet are constrained to not accelerate (linearly or rotationally). Feet are chosen as a first priority (within PTSC), and the CoM as a second priority. Given its many DoFs, the system is redundant to achieve these tasks. Thus, we add additional pose tasks for each joint to promote the return to a natural posture [18]. In addition, the net system angular momentum (as expressed at the CoM) is regulated to zero in the forward and vertical directions to promote balance [15]. These additional tasks more than exhaust the redundancy available after tracking the CoM and feet. As a result, task-weightings are placed on each degree of freedom when evaluating the error norm in (2). For instance, it is important to maintain upright torso posture during any movement, yet the specific motion of the arms is largely unimportant. As a result, task weightings assigned to torso posture are approximately seven times higher than those on the arm DoFs.

In flight, the CoM follows a ballistic trajectory and is not controlled, while the feet remain a high priority in preparation for the upcoming touchdown. Foot trajectories are manually designed relative to the CoM and are commanded to accelerate with PD laws similar to (5). Due to conservation of angular momentum in flight, the configuration of the system at landing is sensitive to flight foot trajectories. That is, any change in angular momentum of the legs during flight must be countered by opposite angular momentum changes in the upper body. The practical implication of this fact is that leg transfer trajectories that are performed with the feet closer to the body result in less pitched-forward torso posture at touchdown.

4 Results

The control approach described enables continuous forward hopping at 1.5 m/s. Snapshots from this motion in simulation[1] are shown in Fig. 6. Fig. 7 shows the close tracking of the SLIP reference velocities despite impact disturbances that degrade the tracking at touchdown. We note that the gait shown here uses a nondimensional leg stiffness [2] of 19.6, which normalizes the effective leg stiffness

Fig. 6 Simulation snapshots for periodic hopping at 1.5 m/s. Arm swing motions are a natural coordination that emerge due to the reduced task weighting of the arm joints in the PTSC.

---

[1] A video of this hopping motion is provided at
http://go.osu.edu/Wensing_Orin_Waldron2013

for systems of different size and weight. This effective leg stiffness is within one standard deviation of average stiffnesses observed in biological hoppers such as kangaroos [2].

We note that the arm swing trajectories that emerge as a coordination strategy from the PTSC allow for tighter regulation of the torso posture as shown in Fig. 8. Arms swing backwards during flight to offset the change in angular momentum of the legs. Without their influence, the torso pitches forward additionally prior to touchdown. While not shown here, the removal of foot lift during flight foot positioning has similar negative influence on the torso during flight.

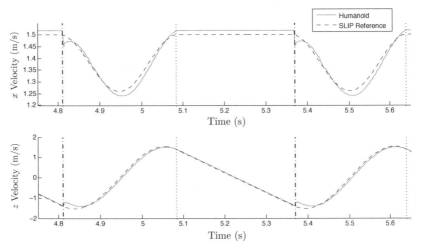

Fig. 7 CoM velocity tracking for 1.5 m/s forward hopping. Vertical bars indicate transitions between stance and flight states.

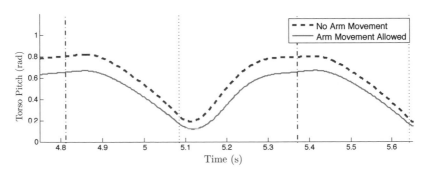

Fig. 8 The flexibility to allow arm movement with PTSC allows for tighter regulation of the torso pitch during stance and flight. The torso pose controller here employs a zero setpoint of nominally upright.

By varying the leg stiffness and other SLIP template characteristics, other more dynamic (and less biologically grounded) hopping gaits may be generated. As shown in the video accompaniment to this work, gaits with additional foot clearance can be generated through specification of periodic CoM trajectories with lower effective leg stiffness. A second gait showcased in the accompanying video employs an effective leg stiffness that is approximately one-half of that required for the gait in Fig. 4. A larger touchdown angle, coupled with this decreased stiffness allows for more vertical CoM variation in stance, and provides longer flight times with higher maximum heights.

5   Summary

The spring-loaded inverted pendulum model has been shown to be an enabling template for the generation of continuous dynamic hopping. The ability of the SLIP model to capture the salient dynamics of a periodic hop enables the control approach here to be applied with little more than the authoring of a set of foot trajectories. A task-space control approach, which enforces feasibility of the force distribution problem at each instant, effectively manages system balance through prioritization of balance tasks and weighted pose tracking.

While the approach here has shown positive results for hopping, the methods presented should enable control of other dynamic movements with significant contributions from out-of-plane effects. Simple template models of dynamic locomotion that capture out-of-plane effects, such as the 3D-SLIP model, will be studied in future work to enable a richer set of dynamic movement capabilities for humanoid robots.

Acknowledgments. This paper is dedicated to Professor Ken Waldron, who was a great inspiration to our work in legged locomotion at Ohio State. His leadership in several research projects sponsored by DARPA and NSF has had a significant impact on our program in robotics.

This work was supported by a National Science Foundation Graduate Research Fellowship to Patrick Wensing, and by Grant No. CNS-0960061 from the NSF with a subaward to The Ohio State University.

References

1. Arumbakkam, A., Yoshikawa, T., Dariush, B., Fujimura, K.: A multi-modal architecture for human robot communication. In: IEEE-RAS Int. Conf. on Humanoid Robots, pp. 639–646 (2010)
2. Blickhan, R., Full, R.: Similarity in multilegged locomotion: Bouncing like a monopode. Journal of Comparative Physiology A 173, 509–517 (1993)
3. Cho, B.K., Park, S.S., Oh, J.H.: Stabilization of a hopping humanoid robot for a push. In: IEEE-RAS Int. Conf. on Humanoid Robots, pp. 60–65 (2010)
4. Farley, C.T., Glasheen, J., McMahon, T.A.: Running springs: speed and animal size. Journal of Experimental Biology 185(1), 71–86 (1993)

5. Gouaillier, D., Collette, C., Kilner, C.: Omni-directional closed-loop walk for NAO. In: IEEE-RAS Int. Conf. on Humanoid Robots, pp. 448–454 (2010)
6. Grebenstein, M., Albu-Schaffer, A., Bahls, T., Chalon, M., Eiberger, O., Friedl, W., Gruber, R., Haddadin, S., Hagn, U., Haslinger, R., Hoppner, H., Jorg, S., Nickl, M., Nothhelfer, A., Petit, F., Reill, J., Seitz, N., Wimbock, T., Wolf, S., Wusthoff, T., Hirzinger, G.: The DLR hand arm system. In: IEEE Int. Conf. on Robotics and Automation, pp. 3175–3182 (2011)
7. Hester, M., Wensing, P.M., Schmiedeler, J.P., Orin, D.E.: Fuzzy control of vertical jumping with a planar biped. In: Proc. of the ASME Int. Design Engineering Technical Conferences, Montreal, Canada, pp. DETC2010-28857:1–28857:9 (2010)
8. Holmes, P., Full, R.J., Koditschek, D., Guckenheimer, J.: The dynamics of legged locomotion: Models, analyses, and challenges. SIAM Rev. 48(2), 207–304 (2006)
9. Kaneko, K., Kanehiro, F., Morisawa, M., Akachi, K., Miyamori, G., Hayashi, A., Kanehira, N.: Humanoid robot HRP-4 - humanoid robotics platform with lightweight and slim body. In: IEEE/RSJ Int. Conf. on Intelligent Robots and Systems, pp. 4400–4407 (2011)
10. Knox, B.T., Schmiedeler, J.P.: A unidirectional series-elastic actuator design using a spiral torsion spring. Journal of Mechanical Design 131(12), 125001:1–125001:5 (2009)
11. Kumar, V., Waldron, K.: Force distribution in closed kinematic chains. IEEE Journal of Robotics and Automation 4(6), 657–664 (1988)
12. de Lasa, M., Mordatch, I., Hertzmann, A.: Feature-based locomotion controllers. In: ACM SIGGRAPH 2010 Papers, vol. 29, p. 131:1–131:10. ACM, New York (2010)
13. McMillan, S., Orin, D.E., McGhee, R.B.: DynaMechs: an object oriented software package for efficient dynamic simulation of underwater robotic vehicles. In: Underwater Robotic Vehicles: Design and Control, pp. 73–98. TSI Press, Albuquerque (1995)
14. Morisawa, M., Kanehiro, F., Kaneko, K., Kajita, S., Yokoi, K.: Reactive biped walking control for a collision of a swinging foot on uneven terrain. In: IEEE-RAS Int. Conf. on Humanoid Robots, pp. 768–773 (2011)
15. Orin, D.E., Goswami, A.: Centroidal momentum matrix of a humanoid robot: Structure and properties. In: IEEE/RSJ Int. Conf. on Intelligent Robots and Systems, pp. 653–659 (2008)
16. Park, I.W., Kim, J.Y., Lee, J., Oh, J.H.: Mechanical design of humanoid robot platform KHR-3 (KAIST humanoid robot 3: HUBO). In: IEEE-RAS Int. Conf. on Humanoid Robots, pp. 321–326 (2005)
17. Sakagami, Y., Watanabe, R., Aoyama, C., Matsunaga, S., Higaki, N., Fujimura, K.: The intelligent ASIMO: system overview and integration. In: IEEE/RSJ Int. Conf. on Intelligent Robots and Systems, vol. 3, pp. 2478–2483 (2002)
18. Wensing, P.M., Orin, D.E.: Generation of dynamic humanoid behaviors through task-space control with conic optimization. In: Proc. of the IEEE Int. Conf. on Robotics and Automation (to appear, 2013)

22 Duty Factor and Leg Stiffness Models for the Design of Running Bipeds

Muhammad E. Abdallah and Kenneth J. Waldron

Abstract. Supporting the design process for running biped robots, analytical models are presented for two aspects of running: the duty factor (DF) of the gait, and the stiffness value of the leg. For a given running speed, an optimal DF exists that minimizes the energy expenditure. We present a formula for the optimal DF based on a model of the energetics, and the results are compared to both human data and simulation results. In addition, a model is presented for the stiffness value of the leg as a function of the physical properties, speed, and DF. The *Gait Resonance Point* is proposed as a design target for compliant running. At this point, the gait matches the spring resonance and the stiffness value becomes independent of the DF.

1 Introduction

As the field of running robotics continues to progress, a growing need exists for more rigorous tools to support the design process. How can a running biped robot be designed to not only run stably but to also satisfy performance and gait specifications? Accordingly, this work presents design rules and models for two aspects of running: the duty factor (DF) of the gait and the stiffness value of the leg.

The DF is the fraction of a stride period a specific leg spends in contact with the ground. It carries significant implications for the energy consumption and ground impact force of a gait. Despite these consequences, no documented consideration for the DF, to our knowledge, has been given for the design of any of the existing running bipeds. We show here that an optimal DF exists such that the energy expenditure is minimized. We present an analytical formula for the optimal DF applicable to any speed or design for the robot.

Muhammad E. Abdallah
General Motors R&D, Warren, MI 48090
e-mail: muhammad.abdallah@gm.com

Kenneth J. Waldron
Stanford University, Stanford, CA 94305
e-mail: kwaldron@stanford.edu

V. Kumar et al. (Eds.): *Adv. in Mech., Rob. & Des. Educ. & Res.*, MMS 14, pp. 275–293.
DOI: 10.1007/978-3-319-00398-6_22   © Springer International Publishing Switzerland 2013

It has already been shown that an optimal DF exists, where the energy consumption for a desired speed is minimized. Alexander calculates the energy costs of a model for human running, and he shows the existence of this optimal DF [6]. He displays the results at two discrete speeds given human parameters. What we need, however, is a formula that allows us to compute the optimal DF regardless of the speed or parameters. Minetti provided such a formula for the internal work, but it is not applicable to the total work of the system [20]. Nishii does present a formula for the total work as a function of the DF; however, he models quasi-static locomotion [23].

On the other hand, the use of springs for running has received more attention in the literature. The presence of compliance in the leg of a running biped is invaluable as an energy storage and thrust mechanism. It has been proven essential to the running of biological systems [5, 13]. It has also been instrumental in the more dynamically successful robotic runners [26, 22, 3, 10, 24, 12]. Other robots have achieved running without passive compliance. Their implementation, however, is either the dynamically limited *ZMP*-controlled running [1, 21, 17, 15], or has not been disclosed [9].

Surprisingly, little theory exists for selecting the stiffness value according to gait specifications. Raibert shares the effective stiffnesses for his hoppers but not the selection process [26]. Rad et al. designed their stiffness to maximize the active energy input during stance [25]. Schmeideler and Waldron selected their stiffness value according to biomimetic models [28]. Thompson and Raibert introduced a first-pass approximation of setting the contact time equal to half the harmonic cycle; then they manually adjusted the stiffness to produce the desired results [30]. Ahmadi and Buehler began with an analytical model of purely vertical hopping. They then incorporated model-specific empirical formulas to relate the stiffness to the speed [2]. None of these studies provides a generalizable design process or laws for selecting the stiffness of an arbitrary biped for an arbitrary speed.

We derive here an analytical formula for the leg stiffness according to the physical properties of the system and the speed characteristics of the gait. A process to translate the stiffness value from the basic mass-spring model to the virtual leg of a general biped is presented. In addition, we present the *Gait Resonance Point* as a design target for compliant biped running. At this point, the stiffness value becomes independent of the DF. It also lends itself to simple design rules governing the stiffness value.

We will first address the DF formula by analyzing the energy cost of running. Then, we will turn to the stiffness analysis and present the relevant models.

2 The Energy Cost of Running

The change in kinetic energy (KE) of a multibody system can be partitioned into two segments. The first is defined as the *external work*; it consists of the KE of the center of mass (CM) in a ground reference frame. The second is defined as the *internal work*, and it consists of the KE of the limbs in the CM reference frame [11, 20].

This framework allows for a simple characterization of the energy expenditure. The external work itself can be further decomposed. Contrasting the top-of-flight and the bottom-of-contact instants reveals two components for the external work. First, the external work must provide the change in height, i.e. the hopping height energy. Second, it must provide the change in KE as the system accelerates from its minimum horizontal speed at the nadir to its maximum horizontal speed at the apex.

Similarly, the internal work can be characterized by two components. First, it must provide the energy to swing the non-support leg, work conducted primarily by the respective hip. Second, it must provide the work conducted by all the other joints, primarily to maintain posture.

Of these four total components, two dominate the energy cost: the hopping height energy and the swing energy. As will be shown, the work needed for these two components scales by the square of the DF. Hence, we will consider these two components only. Neglecting the other components will compromise the absolute measure of the energy expenditure but not the location of the minima, since these two functions so greatly dominate the rate-of-change.

These two functions capture the tradeoff inherent in the DF. Consider the case of running with a smaller DF, i.e. a longer flight time. This long flight time entails a large hopping height to maintain the air time. At the same time, it increases the stride period and thus reduces the swing energy required. This tradeoff determines the optimal DF.

2.1 The Energy Components

For the following energy calculations, we will consider the net absolute work performed over a step. This assumes fully active actuation. A step includes a contact phase and the subsequent flight phase, i.e. half the stride. Fig. 1 shows a schematic of the step. A symmetrical contact phase is assumed here. This symmetry is inherent in spring-operated runners [27, 29], and serves as a simplification for more general systems. θ_o and l_o represent the angle and length, respectively, of the virtual leg at touchdown.

Given a change in height of h, the net absolute work performed for the hopping height is $2mgh$. Neglecting the change in height during contact, the hopping height energy as a function of the liftoff vertical velocity (v_z) is

$$E_{hh} = mv_z^2. \tag{1}$$

For the swing energy derivation, we will model the swing leg as a lumped-mass pendulum. It has a length of l_s, a mass of m_l, a swing angle of α, and a hip torque of τ_{sw}. The linearized dynamics follow, where ω_n is the pendulum natural frequency.

$$\ddot{\alpha} + \omega_n^2 \alpha = \frac{1}{m_l l_s^2} \tau_{sw} \tag{2}$$

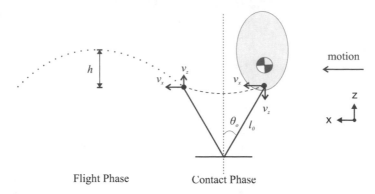

Fig. 1 The behavior of the contact and flight phases for running. v_x and v_z represent the two velocity components at both touchdown and liftoff.

Given a sinusoidal input, $\tau_{sw} = \tau_o \cos(\omega t)$, the solution becomes

$$\alpha(t) = \theta_o \cos(\omega t), \tag{3}$$

where ω is the stride frequency. We are concerned here with the steady-state, forced solution. From (2) and (3), one can solve for the maximum torque:

$$\tau_o = \theta_o m_l l_s^2 \left(\omega^2 - \omega_n^2\right). \tag{4}$$

The change in energy for the swing equals the work done over the period, T, such that

$$\Delta E = \int_0^T \tau_{sw} d\alpha. \tag{5}$$

Substituting from (3) and (4) and applying a double-angle trigonometric identity,

$$\Delta E = \theta_o^2 m_l l_s^2 \omega (\omega^2 - \omega_n^2) \int_0^T \frac{1}{2} \sin(2\omega t) dt. \tag{6}$$

Performing the integration,

$$\Delta E = -\frac{\theta_o^2 m_l l_s^2 (\omega^2 - \omega_n^2)}{4} \left(\cos(2\omega T) - 1\right). \tag{7}$$

Then, simplifying once again with a double-angle identity,

$$\Delta E = \frac{\theta_o^2 m_l l_s^2 (\omega^2 - \omega_n^2)}{2} \sin^2(\omega T). \tag{8}$$

Due to the symmetry of the gait, the net absolute energy expenditure is twice the value at mid-swing. This gives us our final expression for the swing energy.

$$E_{sw} = \theta_o^2 m_l l_s^2 (\omega^2 - \omega_n^2) \tag{9}$$

2.2 Introducing the DF

The total energy cost is the sum of both components, $E_{tot} = E_{hh} + E_{sw}$. We will now express this result as a function of the DF.

The time duration for the contact and flight phases, t_c and t_f respectively, can be determined from the kinematics of Fig. 1. Given a vertical liftoff velocity of v_z, the flight time is:

$$t_f = \frac{2v_z}{g}, \tag{10}$$

where g is the gravitational acceleration. Approximating the horizontal velocity as a constant v_x, the contact time is:

$$t_c = \frac{2l_o \sin(\theta_o)}{v_x}. \tag{11}$$

The DF, β, is defined as the ratio of a leg's contact time to the stride period. Assuming symmetric right and left steps,

$$\beta = \frac{t_c}{2(t_c + t_f)}. \tag{12}$$

To determine E_{tot} as a function of the DF, we need to solve for ω and v_z. The stride frequency, ω, follows.

$$\omega = \frac{2\pi}{2(t_c + t_f)} \tag{13}$$

$$-\frac{2\pi}{t_c}\beta \tag{14}$$

To determine v_z, solve for (12), (11), and (10).

$$v_z = \frac{g l_o \sin(\theta_o)}{v_x}\left(\frac{1}{2\beta} - 1\right) \tag{15}$$

Substituting for ω and v_z, the final energy cost formula can now be determined:

$$E_{tot} = c_1\left(\frac{1}{2\beta} - 1\right)^2 + c_2\beta^2 - c_3, \tag{16}$$

where c_i are constant functions of the parameters.

$$c_1 = m\left(\frac{g l_o \sin(\theta_o)}{v_x}\right)^2 \tag{17}$$

$$c_2 = \theta_0^2 m_l l_s^2 \left(\frac{\pi v_x}{l_o \sin(\theta_o)}\right)^2 \tag{18}$$

$$c_3 = \theta_0^2 m_l l_s g. \tag{19}$$

This formula quantifies the energy tradeoff associated with the DF. The energy expenditure at higher DF's is dominated by the swing energy; while at lower DF's it is dominated by the hopping height energy. Fig. 2 displays the energy cost for an average human running at 4 m/s, exhibiting the minimum energy behavior. Throughout this work, the parameters used reflect an average human with a mass of 73 kg and a leg-length of 0.8 m. The full list of parameters is available in Table 1.

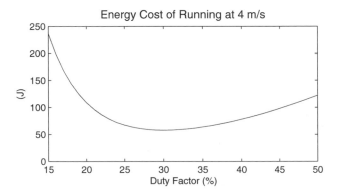

Fig. 2 Plot of the energy cost formula (16), shown for an average human running at 4 m/s. The minima depends on the running speed and physical parameters of the system.

Table 1 The Physical Parameters Based on Anthropomorphic Data

Parameter	Symbol	Value
total mass	m	73 kg
leg mass	m_l	12.5 kg
initial leg length	l_o	0.8 m
swing leg length	l_s	0.4 m
initial leg angle	θ_o	0.4 rads

3   The Optimal Duty Factor

To find the minima of the energy cost, we solve for $\frac{d}{d\beta} E_{tot} = 0$. This results in our final characteristic equation, the *Optimal DF Formula*.

$$\left(\frac{4c_2}{c_1} \right) \beta^4 + 2\beta - 1 = 0 \tag{20}$$

The optimal DF is a function of the physical properties of the system and the speed desired. The results of this formula for an average human are shown in Fig. 3. Note, closed-form solutions exist for fourth-order, or quartic, equations. These solutions, however, are prohibitively complex.

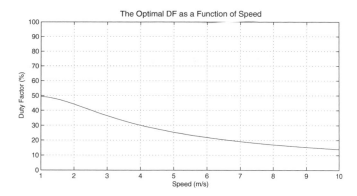

Fig. 3 Optimal DF for an average human running, based on (20). The results suggest decreasing DF's at higher speeds.

This formula was validated in two ways. First, it was compared to data of human running. As a widely accepted premise, biological systems adapt their gaits to minimize energy consumption [6, 16]. Minetti experimentally determined the DF of humans running at a range of speeds [20]. He used an average sample size of over 30 runners, and his data is referenced here.

Second, the results were validated in simulation. Using the dynamic model described in the Appendix, the biped was tested at two select speeds. For each speed, a range of DF's was applied and the energy consumption tabulated. The model approximated an average human running at speeds of 3.5 m/s and 4.0 m/s respectively. Fig. 4 displays the energy curves for each simulation run.

Both the human data and the simulation model exhibited close correspondence with the Optimal DF Formula. Results of the formula nearly fell within a standard deviation of the average human DF's. This close correspondence occurred despite neglecting the error between mechanical and metabolic work and despite not knowing the actual physical properties of the runners. The final results are shown in Fig. 5, comparing the Optimal DF Formula with the human data and simulation results. The formula was computed here with the same average human parameters used throughout.

It is well known that running animals passively conserve energy using internal springs [5, 13]. We assume that the proportion of energy conserved in both oscillations (hopping and swinging) is similar. To design a robot with passive springs, simply scale the constants in (20) by the proportion of energy expected to be actively supplied: c_1 for hopping and c_2 for swinging.

We now turn to the stiffness analysis by first presenting the mass-spring model.

4   The SLIP Model for Stiffness Analysis

A simple yet effective model for compliant running is the basic mass-spring, also known as the Spring Loaded Inverted Pendulum (SLIP) model. It consists of a

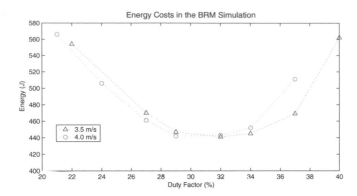

Fig. 4 Energy costs at constant speeds for the simulation. The 4.0 m/s run reached a minimum at approximately 30% DF. The 3.5 m/s run reached a minimum at approximately 32% DF.

massless, spring-operated leg attached to the center of mass of the body. It has been widely used in the study of both biological and robotic systems, and its applicability has been well documented [8, 19, 29]. Fig. 6 shows the model.

While running at steady-state, the SLIP model must exhibit a symmetric contact phase [29]. This symmetry is reflected in both the configuration and velocities as shown in Fig. 6.

McMahon and Cheng utilized this symmetry to analyze the dynamics of running [19]. They conducted iterative numerical integrations of the SLIP dynamical equations to determine the parameters that satisfy this symmetry. They then presented empirical formulas that fit the data within errors of 0.5%.

Using their empirical formulas, we can display the spring natural frequency of the system, ω_n, as a function of the speed and DF. The results for an average human are shown in Fig. 7.

We will use McMahon's data as a benchmark for our approximations. His formulas would suffice for determining the needed SLIP stiffness if not for two factors. First, the formulas are very complex. Twenty-four unique parameters are needed to compute the stiffness at any one liftoff speed (v_z). More importantly, his formulas are only valid for a limited range of conditions.

The next section will present an alternative, analytical formula for the stiffness and compare it to McMahon's results.

5 The SLIP Governing Equation

To derive an analytical model for the SLIP, consider the equation of motion for the vertical direction. The free-body diagram is shown in Fig. 6. R is the ground reaction force, which equals the spring force. k and l_o are the stiffness and initial length of the spring.

$$-mg - k(l - l_o)\cos(\theta) = m\ddot{z} \tag{21}$$

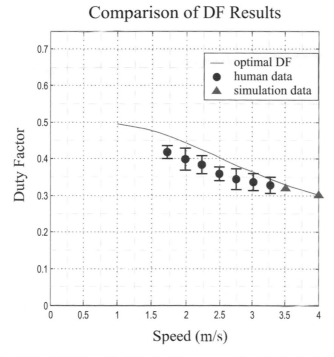

Fig. 5 The Optimal DF Formula (20) closely corresponds to both simulation results and natural human DF's. The experimental human data is provided by Minetti and displayed as a mean ± S.D. [20]

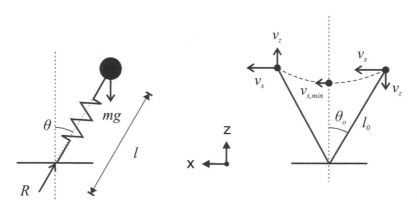

Fig. 6 The SLIP model. The free body diagram is shown to the left. The symmetry of the contact phase is shown to the right.

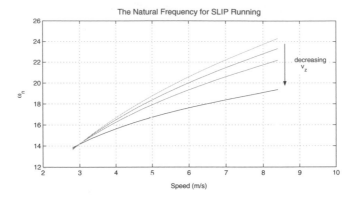

Fig. 7 The natural frequency required for steady-state running of the SLIP model, based on the empirical formulas of McMahon and Cheng [19]. Each curve represents a liftoff velocity, v_z. From top to bottom, $v_z = 0.84, 0.70, 0.56$, and 0.28 m/s. Note, the DF changes inversely with v_z.

Given an origin at the ground contact, the relation between the spring length and height is $z = l\cos(\theta)$. Substituting for l in (21) results in

$$m\ddot{z} + kz + (mg - kl_o\cos(\theta)) = 0. \tag{22}$$

We will replace $\cos(\theta)$ with its average value, denoted as $\overline{c\theta}$. This results in a single-variable, linear equation of motion.

$$m\ddot{z} + kz + (mg - kl_o\overline{c\theta}) = 0. \tag{23}$$

It can be shown that the geometric average of $\cos(\theta)$ is

$$\overline{c\theta} = \frac{\sin(\theta_o)}{\theta_o}. \tag{24}$$

Introducing the following change of variable for z, (23) can be expressed as the familiar harmonic oscillator.

$$y = z + \frac{mg}{k} - l_o\overline{c\theta} \tag{25}$$

$$\ddot{y} + \omega_n^2 y = 0 \tag{26}$$

ω_n here is the spring natural frequency, $\sqrt{k/m}$.

This equation can be solved by introducing the following boundary conditions. These conditions are based on the symmetry of the contact phase, starting at mid-stance and ending at liftoff. t_{ms} represents half the contact time.

$$\dot{y}(0) = 0 \tag{27}$$

$$\dot{y}(t_{ms}) = v_z \tag{28}$$

$$y(t_{ms}) = l_o(c\theta_o - \overline{c\theta}) + \frac{mg}{k} \tag{29}$$

For brevity, $\cos(\theta_o)$ is expressed in shorthand as $c\theta_o$. (29) derives from the condition that the final height is $l_o c\theta_o$.

We will introduce the *contact frequency*, ω_c, as the frequency of the contact phase. Its period is twice the contact time.

$$\omega_c = \frac{\pi}{t_c}$$

$$\omega_c = \frac{\pi v_x}{2l_o \sin(\theta_o)} \tag{30}$$

Solving for the boundary conditions and (30) gives us our final formula relating the spring frequency to the system and gait parameters. We refer to it as the *SLIP Governing Equation*.

$$\left(g - \omega_n^2 l_o(\overline{c\theta} - c\theta_o)\right) \tan\left(\frac{\pi}{2}\frac{\omega_n}{\omega_c}\right) + v_z\omega_n = 0 \tag{31}$$

The stiffness a biped needs to run is a function of both physical properties and gait characteristics. It is a function of the mass, leg length, and initial leg angle of the system. It is also a function of the desired speed and DF of the gait, since the DF is inversely proportional to v_z. The SLIP Governing Equation provides an analytical tool for determining the stiffness value given those parameters. No closed form solution is provided; however, numerical solutions are easily computed.

The equation predicts the spring frequency with good accuracy. Compared to McMahon's data, it produced errors less than 8%. A side-by-side comparison is shown in Fig. 8.

6   Applying the SLIP to a Telescoping Biped

We are now capable of solid predictions for the theoretical SLIP model, but how applicable is the SLIP to a real system? This section analyzes the dynamics of a telescoping biped in relation to the SLIP. It presents a model of the stiffness for the virtual leg of a *general* biped.

It turns out that an actual robot requires smaller stiffness values than the SLIP model predicts. The difference is due to several ideal assumptions in the SLIP, particularly the massless legs and the hip-centered body CM. We will now analyze the dynamics of a general biped—one with massive legs and a body CM displaced from the hip—to determine the required stiffness in relation to the SLIP.

The free body diagram for the biped at touchdown is shown in Fig. 9. We will analyze the torso to obtain a generalized version of the SLIP equation of motion presented in (23).

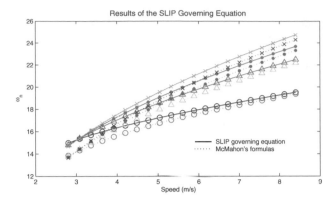

Fig. 8 The SLIP Governing Equation accurately predicts the spring frequency needed. Its results are compared here to McMahon's empirical formulas. The errors are less than 8%.

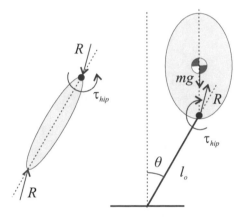

Fig. 9 Free body diagram of a general biped at touchdown. The support leg is shown to the left, and the remaining system is shown to the right.

Consider the dynamics of the support leg. During running, the reaction forces on the leg are of such magnitude that they dominate the gravitational and inertial forces seen by the leg. In the absence of any hip torques, this produces reaction forces at the two ends that are approximately equal and opposite, akin to a static two-force member. This phenomenon underlies the well-established behavior exhibited by running animals: their ground reaction forces point along the axis of their legs [4, 14, 7]. Further documentation of this phenomenon has been presented by Abdallah [1].

Consider the leg reaction force, R, on the torso. As just described, R acts principally in the direction of the leg axis. Given the displaced CM a general biped carries, however, a torque is now needed at the hip (τ_{hip}) to maintain the torso upright. This torque causes a deviation in R from the leg axis—a deviation that is consistently towards the vertical. Swinging through the contact phase, R will thus span a tighter range about the vertical than the corresponding range of $2\theta_o$ spanned by the leg. Hence, a more appropriate determination for the average cosine of (24) is

$$\overline{c\theta} = 1. \tag{32}$$

In addition, it is apparent from the analysis that the mass this force faces does not include the mass of the support leg. Hence, the effective mass for the natural frequency includes the body and all other limbs to the exclusion of one leg.

The process for computing the leg stiffness for a general biped can be summarized in two steps:

1. Solve for the spring frequency from the SLIP Governing Equation (31) using $\overline{c\theta} = 1$.
2. Translate the frequency into a stiffness using the effective mass described above: $k = \omega_n^2(m - m_l)$.

To solve for the frequency in the first step, one can either select v_z directly or according to the desired DF. In the latter case, (15) provides the solution for v_z as a function of the DF. The Optimal DF Formula (20) provides a compelling choice for the DF in this case. Examples of spring implementations are available elsewhere for both telescoping [26, 25] and articulated [28, 22] legs.

7 Simulation Results

This stiffness model for a general biped was tested and verified in simulation. A fully dynamic simulation of a planar telescoping biped was created. The prismatic legs were loaded with passive springs, and the stiffness value was determined using the two-step process of the previous section. A full description of the model and simulation environment is available in the Appendix.

The model was tested at two speeds. First, it was tested at a target speed of 3.5 m/s, with a vertical liftoff velocity of 1.0 m/s. Given these velocities and the model properties, the theoretical stiffness computed to 13.1 kN/m. This stiffness value was applied to the model and the results proved successful. The model settled at a steady-state run with velocities of $v_x = 3.45$ m/s and $v_z = 1.1$ m/s. Fig. 10 shows the velocities of the overall CM for the run. Note, a pretension of 2 cm was needed in the spring at each step to compensate for the impact energy losses.

The model was also tested at a second speed of 3.0 m/s, with the same vertical velocity of 1.0 m/s. The theoretical stiffness was computed as 11.4 kN/m. The simulation validated the prediction once again with the model reaching steady-state at a speed of 3.0 m/s and a vertical speed of 1.1 m/s.

8 The Gait Resonance Point

The SLIP stiffness curves display an interesting behavior: the stiffness values tend to converge at a select speed. This phenomenon is observable in Fig. 7. Although the

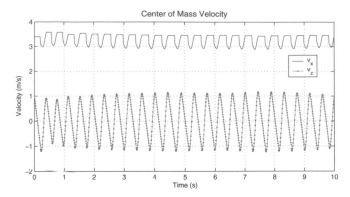

Fig. 10 The design process was implemented to select a SLIP stiffness and then translate it to the telescoping biped. The stiffness was designed to achieve a speed of 3.5 m/s with a liftoff vertical velocity of 1.0 m/s. The model in Fig. 14 was tested in simulation, and the above results verify the accuracy of the design models.

stiffness is normally a function of both the speed and the DF, it becomes effectively constant at this speed, independent of the DF.

This behavior consistently demonstrated itself in McMahon's formulas, even with changes to the parameters. The SLIP Governing Equation (31) allows us to solve for this point. It represents the solution of the equation at its singularity, where the *tan* term approaches infinity and its coefficient approaches zero. This translates into the following two conditions.

$$\frac{\omega_n}{\omega_c} = 1 \tag{33}$$

$$g - \omega_n^2 l_o(\overline{c\theta} - c\theta_o) = 0 \tag{34}$$

Since the contact frequency matches the spring natural frequency at this point, we refer to it as the *Gait Resonance Point*.

These conditions lead to simple design rules for the GRP. Solving for (30), (33), and (34), the following rules can be derived for the spring frequency and the velocity of the GRP. The approximation of $\overline{c\theta} = 1$ was applied.

$$\omega_n = \omega_c \tag{35}$$

$$v_x = \frac{2}{\pi} \sqrt{g l_0(1 + c\theta_o)} \tag{36}$$

A comparison of the analytical GRP with the frequency curves is shown in Fig. 11. A large portion of the error in the prediction is due to using the approximation $\overline{c\theta} = 1$. The approximation reduces the accuracy with respect to the SLIP model but increases the applicability to a general biped.

Thompson and Raibert suggested (35) as a first-pass rule for determining the stiffness value of their running biped [30]. We show that this rule is accurate at a

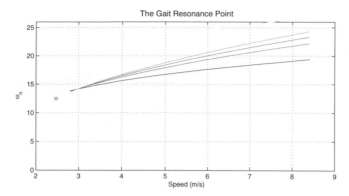

Fig. 11 The *Gait Resonance Point* targets the intersection point—at which the stiffness becomes effectively independent of the DF. The GRP is shown as the star, and the frequency curves are the same curves from Fig. 7.

select speed, as determined by (36). It also serves as a good initial guess for the numerical solutions of the SLIP Governing Equation.

The GRP offers a design target with two advantages. First, the stiffness is effectively independent of the DF; hence, we can implement different DF's without changing the stiffness. Second, the stiffness is determined by simple design rules, where the spring frequency equals the contact frequency. The main disadvantage of the GRP is in the restrictiveness of the condition. Running at a specific speed entails a required leg-length, as expressed by (36). Fig. 12 plots the velocities and corresponding leg-lengths for the GRP.

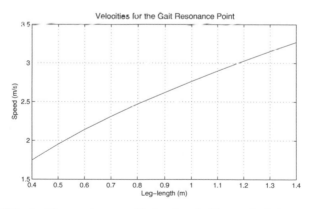

Fig. 12 The GRP velocities require a specific leg-length. These calculations were based on a leg angle of 0.4 radians.

If the GRP does not fit the desired design specifications, one can return to the SLIP Governing Equation to determine the needed stiffness. A DF needs to be selected in this case, and the aforementioned Optimal DF Formula can provide a basis for this selection.

The next section provides a physical interpretation of the GRP behavior.

8.1 Interpretation

As seen in (23), the vertical motion of the SLIP models a simple harmonic oscillator during contact. Due to its symmetric boundary velocities, the contact phase will trace an arbitrary, even sinusoidal range. This range will complete a full half-cycle only if the initial height matches the equilibrium height. It can be shown that this condition is satisfied only at the GRP.

Running at the GRP will therefore model a true harmonic oscillation, in which the contact phase matches a complete half-cycle of the mass-spring. This explains the two featured advantages of the GRP. First, the spring resonant frequency (ω_n) equals the contact frequency (ω_c). Second, the stiffness value becomes independent of the DF. The DF determines the amplitude of the oscillation, but the frequency of a harmonic oscillator is independent of the amplitude.

Fig. 13 shows the general case, where the contact phase traces an arbitrary even range. A change in the DF (through v_z) changes the amplitude of the oscillation. Since the contact oscillation does not match a full harmonic, a change in amplitude produces a change in the resonant frequency as well. You can see this effect in the figure. The GRP represents the exception, where a change in DF (and hence the amplitude) does not change the frequency.

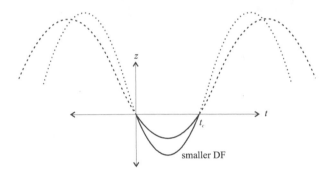

Fig. 13 Traces of height during the contact phase are shown as the solid lines. The sinusoidal curves fitting those traces are shown as the dotted lines. In this general case, changing the DF results in a change of both amplitude and frequency. At the GRP, the contact phase matches a full half-cycle of the harmonic; hence, a change in amplitude (due to the DF) does not result in a change of frequency.

9   Conclusion

This work answers the following basic question: *I want to design a biped robot to run at a target speed. What DF should I design it for? And what stiffness value does it need?*

In designing a running biped, selecting a stiffness to meet a desired gait specification is not trivial. It depends on the speed, configuration, and DF in highly-nonlinear systems. The DF itself warrants careful consideration, given its effect on the energy consumption and impact forces of the gait.

It is one thing to design a compliant robot that can run stably; it is another thing to design it for set gait specifications. This work presents formulas and design laws that supplement the design process for such specifications.

Acknowledgements. The authors wish to acknowledge the financial support of the National Science Foundation, grant number IIS-0535226, during the course of this work.

Appendix: The Simulation Model

The simulation system modeled a telescoping legged biped, shown in Fig. 14. The model consists of three rigid bodies, representing the torso and two legs, connected by a revolute joint at the hip. The feet are point-masses connected by prismatic joints to the legs.

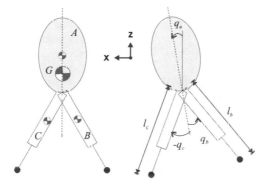

Fig. 14 The biped model used for the dynamic simulation

A fully dynamic simulation was created. The foot-ground contact was modeled as a rigid, inelastic contact using motion constraints. Friction was assumed sufficient to avoid slip. For the DF tests, the model implemented the control strategy of Abdallah and Waldron for running [1]. For the stiffness tests, the prismatic joints were loaded with purely passive springs, where the springs were slightly pretensioned at each step to compensate for the impact losses. The physical parameters modeled an average human as shown in Table 2.

The equations of motion were generated in Autolev™, a symbolic manipulator for dynamics applications [18]. A C program was developed to determine the appropriate constraint equations, solve for the equations of motion, and perform the integration using a variable-step integrator. Further details on the simulation system are available in [1].

Table 2 Model Parameters

	distance from hip to CM (m)	mass (kg)	inertia (kg m^2)
torso	0.34	48.3	8.12
leg	0.384	11.5	1.03
foot	0.8 (nominal)	1.0	0

References

1. Abdallah, M.: Mechanics motivated control and design of biped running. Ph.D. dissertation, Stanford University, Stanford, CA (June 2007)
2. Ahmadi, M., Buehler, M.: Stable control of a simulated one-legged running robot with hip and leg compliance. IEEE Trans. on Robotics and Automation 13, 96–104 (1997)
3. Ahmadi, M., Buehler, M.: The ARL Monopod II running robot: control and energetics. In: IEEE Intl. Conf. on Robotics and Automation, Detroit, MI, pp. 1689–1694 (May 1999)
4. Alexander, R.: Mechanics and scaling of terrestrial locomotion. In: Pedley, T.J. (ed.) Scale Effects in Animal Locomotion, ch. 6, pp. 93–110. Academic Press, London (1977)
5. Alexander, R.: Elastic Mechanisms in Animal Movement. Cambridge University Press, Cambridge (1988)
6. Alexander, R.: A model of bipedal locomotion on compliant legs. Phil. Trans. of the Royal Society of London 338, 189–198 (1992)
7. Biewener, A.A.: Animal Locomotion. Oxford University Press, Oxford (2003)
8. Blickhan, R.: The spring-mass model for running and hopping. Journal of Biomechanics 22(11/12), 1217–1227 (1989)
9. Boston Dynamics, Cheetah, http://www.bostondynamics.com/robot_cheetah.html
10. Brown, B., Zeglin, G.: The bow leg hopping robot. In: IEEE Intl. Conf. on Robotics and Automation, Leuven, Belgium, pp. 781–786 (May 1998)
11. Cavagna, G., Kaneko, M.: Mechanical work and efficiency in level walking and running. Journal Physiology 268, 467–481 (1977)
12. Cotton, S., Olaru, I., Bellman, M., Ven, T., Godowski, J., Pratt, J.: Fastrunner: A fast, efficient and robust bipedal robot. In: IEEE Intl. Conf. on Robotics and Automation, St. Paul, MN (May 2012)
13. Farley, C.T., Glasheen, J., McMahon, T.A.: Running springs: speed and animal size. Journal of Experimental Biology 185, 71–86 (1993)
14. Full, R., Blickhan, R., Ting, L.: Leg design in hexapedal runners. Journal of Experimental Biology 158, 369–390 (1991)
15. Hirai, K., Hirose, M., Haikawa, Y., Takenaka, T.: The development of the Honda humanoid robot. In: IEEE Intl. Conf. on Robotics and Automation, Leuven, Belgium, pp. 1321–1326 (May 1998)
16. Hoyt, D., Taylor, R.: Gaits and the energetics of locomotion in horses. Nature 292, 239–240 (1981)
17. Kajita, S., Nagasaki, T., Yokoi, K., Kaneko, K., Tanie, K.: Running pattern generation for a humanoid robot. In: IEEE Intl. Conf. on Robotics and Automation, Washington, DC (May 2002)
18. Kane, T., Levinson, D.: Dynamics Online: Theory and Implementation with AUTOLEV. Online Dynamics, Inc., Sunnyvale (1996)

19. McMahon, T.A., Cheng, G.C.: The mechanics of running: How does stiffness couple with speed? Journal of Biomechanics 23(suppl. 1), 65–78 (1990)
20. Minetti, A.E.: A model equation for the prediction of mechanical internal work of terrestrial locomotion. Journal of Biomechanics 31, 463–468 (1998)
21. Nagasaka, K., Kuroki, Y., Suzuki, S., Itoh, Y., Yamaguchi, J.: Integrated motion control for walking, jumping, and running on a small bipedal entertainment robot. In: IEEE Intl. Conf. on Robotics and Automation, New Orleans, LA (April 2004)
22. Nichol, J., Palmer, L., Waldron, K.: Design of a leg system for quadraped gallop. In: Huang, T. (ed.) Proceedings of the 11th Congress in Mechanism and Machine Science. China Machinery Press, Tianjin (2003)
23. Nishii, J.: An analytical study of the cost of transport for legged locomotion. In: Proceedings of the 2nd Intl. Symposium on Adaptive Motion of Animals and Machines, Kyoto, Japan (March 2003)
24. Park, H., Sreenath, K., Hurst, J., Grizzle, J.: Identification of a bipedal robot with a compliant drivetrain: Parameter estimation for control design. Control Systems Mag. 31 (April 2011)
25. Rad, H., Gregorio, P., Buehler, M.: Design, modeling and control of a hopping robot. In: IEEE/RSJ Conf. on Intelligent Systems and Robots, Yokohama, Japan, pp. 1778–1785 (July 1993)
26. Raibert, M.: Legged Robots that Balance. MIT Press, Cambridge (1986)
27. Raibert, M.: Running with symmetry. The Intl. Journal of Robotics Research 5(4) (1986)
28. Schmiedeler, J.P., Waldron, K.J.: Leg stiffness and articulated leg design for dynamic locomotion. In: Design Engineering Techinal Conf.s and Computers and Information in Engineering Conf. ASME, Montreal (2002)
29. Schwind, W.: Spring loaded inverted pendulum running: a plant model. Ph.D. dissertation, University of Michigan, Ann Arbor, MI (1998)
30. Thompson, C.M., Raibert, M.H.: Passive dynamic running. In: Hayward, V., Khatib, O. (eds.) Experimental Robotics I. LNCIS, vol. 139, pp. 74–83. Springer, Heidelberg (1989)

23 Unfinished Business: Impulsive Models of Quadrupedal Running Gaits

James P. Schmiedeler and Lawrence Funke

Abstract. Ken Waldron has significantly advanced the understanding of the dynamics of high-speed quadrupedal locomotion through his research work in both modeling and experimentation. This paper revisits an impulsive model of quadrupedal running gaits that Waldron developed and seeks to find all feasible steady-state gait solutions for it. Prior work had reported only single solutions to the nonlinear systems of equations defining each gait. Using the Bertini software to implement a homotopy continuation method, all solutions were found for the trot, pace, bound, half-bound, and canter gaits of a biologically sized quadruped moving with a fixed stride period in the presence of drag. New solutions were identified for the trot and pace, and differences from previously reported solutions for the bound and half-bound were found. The approach has not yet been successful in comprehensively solving the transverse and rotary gallop systems of equations, so that remains a topic of ongoing research. In general, however, surprisingly few physically meaningful solutions were found for any gait despite the large numbers of possible solutions to consider.

1 Introduction

During one of their first research meetings as advisor and graduate student early in the fall of 1996, Ken Waldron handed Jim Schmiedeler a paper he had written with Sunil Agrawal on an impulsive model of quadruped running [1]. The paper argues that the stance phase represents a small enough percentage of the stride period at high speeds that an impulsive model can capture the key dynamics with adequate accuracy. Agrawal and Waldron [1] had found steady-state solutions for several gaits using graphical techniques with their model, and Waldron's brief

James P. Schmiedeler · Lawrence Funke
University of Notre Dame, Department of Aerospace and Mechanical Engineering, Notre Dame, In, 46556
e-mail: {schmiedeler.4,lfunke}@nd.edu

V. Kumar et al. (Eds.): *Adv. in Mech., Rob. & Des. Educ. & Res.*, MMS 14, pp. 295–304.
DOI: 10.1007/978-3-319-00398-6_23   © Springer International Publishing Switzerland 2013

instructions to Schmiedeler at the onset of his graduate program were to adapt the model to investigate how drag on a running quadruped affected the phasing of the footfalls in a gallop. Waldron hand wrote some notes to start the effort, which outlined how Schmiedeler could formulate a system of equations enforcing steady-state conditions and then solve them numerically. This task would consume the next 15 months of Schmiedeler's research effort and culminate in his M.S. thesis [4] and his first journal publication [6].

Early in 1997, Schmiedeler had made enough progress coding a Newton's Method approach to solving the system of nonlinear equations characterizing the transverse gallop to realize that he was not making any progress toward finding physically meaningful solutions. Waldron's suggestion to address this issue was to pursue a continuation method that would allow one to start with a known solution to a simpler system and then track that solution as a homotopy was used to transition from the simpler system to the full galloping model system. Schmiedeler promptly began reading Wampler, Morgan, and Sommese's paper on the topic in the *ASME Journal of Mechanical Design* [7], which had been published during Waldron's tenure as editor of JMD. A couple of weeks passed before the next group research meeting was held with Waldron, Shankar Venkataraman, Po-Hua Yang, Chris Hubert, David Orin, and a couple of Orin's graduate students. At his turn going around the table in Waldron's office at the corner of Robinson Lab, Schmiedeler reported proudly that he had solved the five-position Burmester problem in the previous two weeks. Waldron replied with appropriate skepticism to ask why Schmiedeler had been pursuing this problem that was completely unrelated to his research and already entirely solved. The explanation that this was in fact an example from the continuation methods paper in JMD that Schmiedeler had used to verify his code was working was accepted with laughter, but probably should have been the start of his update rather than an addendum to it.

Regrettably, the story in many ways stops there because Schmiedeler never was able to parlay his success in solving the five-position Burmester problem via continuation methods into success solving his galloping equations in a similar manner. Instead, he ultimately found solutions using his original Newton's Method approach and a kind of brute force homotopy. Through a large number of code executions, he happened onto an initial guess for the solution to the system in the absence of drag that enabled convergence in this simpler case and then slowly increased the drag and updated the initial guesses with prior solutions. This strategy worked for both the rotary and transverse gallops [4] and ultimately for the trot, pace, bound, canter, and half-bound in subsequent analysis of the vertical excursion of the mass center in these gaits as part of Schmiedeler's dissertation work [3]. It was always, however, an inferior approach to Waldron's suggestion of continuation because it provided only one solution among the many possible and was thus highly dependent on the initial guess. The present paper is an effort to correct this inadequacy of Schmiedeler's research under Waldron by finding all of the solutions for the various dynamic quadrupedal gaits of the impulsive model using the rigorous approach of continuation that Waldron had recommended early on in the research.

2 Numerical Continuation

Recently, the numerical continuation methods described in [7] have been made readily available to researchers in a free software package known as Bertini [2]. This software takes the variables and equations of a polynomial system as input and employs a homotopy continuation method to solve for all solutions of the system. If all of the variables are declared in the same group, as they are in this work, the total number of possible solution paths to follow for a system can be found by taking the degrees of each equation and multiplying them together. For example, in the system of Eqs. 1 and 2 to be solved such that f and g are zero, the total number of possible solution paths to follow is 4.

$$f = x^2 - 1 \tag{1}$$
$$g = y^2 - 1 \tag{2}$$

In this case, each path yields a finite solution, and they are clearly $(1,1)$, $(-1,1)$, $(1,-1)$, and $(-1,-1)$. In more general cases, the system is not this trivial, some solutions will be imaginary, and not all possible solution paths will yield finite solutions. The systems described in the following sections range from tens of possible solution paths to over a million.

3 Quadruped Model

The model employed here is identical to that reported in Schmiedeler's dissertation [3], and a schematic is shown in Fig. 1. The trunk is a laterally symmetric rigid body with the center of mass located forward of the geometric center. The legs are massless, and the shoulder joints, hip joints, and center of mass all lie in a plane. The shoulder and hip joints are modeled as revolutes that allow rotation only in the sagittal plane. The legs are assumed to act as pure thrust generators, with the line of action of the leg impulse passing from the point of foot contact through the corresponding shoulder/hip joint. (Lateral impulses, and accordingly lateral body translations, are neglected.) Therefore, the configuration of the massless legs is immaterial, which is why they are drawn as straight lines in Fig. 1 as opposed to articulated chains. The length of these lines is the optimal working length of the legs for providing instantaneous thrust, and the legs are constrained to operate at this length for the infinitesimal stance phase. The geometric and inertial parameters (estimated from [5]) of the model are provided in Table 1.

For all of the gaits modeled, the changes in trunk orientation are assumed to be small enough to validate the small angle approximation for the roll, pitch, and yaw angles, and the yaw angle at the initial footfall is taken to be zero since the focus is on straight-line motion. Symmetry is assumed for the two front leg impulses and the two hind leg impulses. A constant drag force acting through the mass center in the direction opposite forward progression represents all energy losses in that direction. All solutions reported here were generated for a drag force of 47 N, a stride period of 0.4 s, and a forward speed of 3.28 $\frac{m}{s}$ for comparison with results in Schmiedeler's

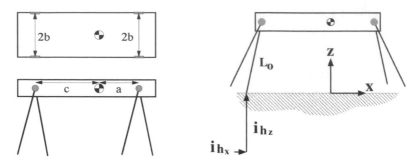

Fig. 1 Schematic views of the quadruped model from the top and side showing body dimensions, the fixed coordinate frame, and a left hind leg impulse

Table 1 Geometric and inertial parameters of the quadruped model

Parameter	Value	Units	Description
a	0.17	m	Longitudinal distance, mass center to shoulders
c	0.31	m	Longitudinal distance, mass center to hips
b	0.08	m	Lateral distance, mass center to shoulders/hips
L_o	0.7	m	Optimal working length of legs
m	33	kg	Trunk mass
I_x	0.69	$kg \cdot m^2$	Roll moment of inertia
I_y	3.37	$kg \cdot m^2$	Pitch moment of inertia
I_z	2.94	$kg \cdot m^2$	Yaw moment of inertia

dissertation [3], but the methodology could certainly be applied to examine varying drag, stride period, and/or speed. Steady-state motion is imposed by constraining the height of the mass center, the roll, pitch, and yaw angles and angular velocities, and the longitudinal and vertical velocities to all be the same at the beginning and end of a stride cycle, as marked by the initial and subsequent footfall of the left front leg.

4 Running Gait Results

Because each running gait is characterized by different combinations of symmetries, the system of equations defining each gait is unique, and the differences are identified below. The actual equations are omitted for brevity since they are already documented in Schmiedeler's dissertation [3]. The variables common across gaits are listed in Table 2. Only a subset of all possible gaits are examined here because, regrettably, the systems of equations for the transverse and rotary gallops have yet to be solved satisfactorily using Bertini.

Table 2 Variables appearing in the systems of equations defining the various quadrupedal running gaits

Variable	Description
t_2	Time of second footfall
t_3	Time of third footfall
z_1	Height of mass center at start of gait cycle
u_{zo}	Vertical velocity of mass center at start of gait cycle
d_f	Longitudinal distance from mass center to front footfall
d_h	Longitudinal distance from mass center to hind footfall
θ_{x1}	Roll angle at start of gait cycle
θ_{y1}	Pitch angle at start of gait cycle
ω_{xo}	Roll angular velocity at start of gait cycle
ω_{yo}	Pitch angular velocity at start of gait cycle
i_{fx}	Horizontal component of front impulses
i_{hx}	Horizontal component of hind impulses
i_{fz}	Vertical component of front impulses
i_{hz}	Vertical component of hind impulses

4.1 Trot

With the legs operating in diagonal pairs, the timings of the footfalls in the trot are necessarily separated by one half the stride period to achieve the imposed symmetry. There are only 5 equations in the system to solve for five variables: z_1, i_{fx}, i_{fz}, i_{hx}, and i_{hz}. For implementation in Bertini, though, two additional equations are required to eliminate square roots from the original equations and achieve a polynomial system. This is accomplished by defining new variables $I_f = \sqrt{i_{fx}^2 + i_{fz}^2}$ and $I_h = \sqrt{i_{hx}^2 + i_{hz}^2}$ and including Eqs. 3 and 4 in the system.

$$0 = I_f^2 - i_{fx}^2 - i_{fz}^2 \tag{3}$$
$$0 = I_h^2 - i_{hx}^2 - i_{hz}^2 \tag{4}$$

For the trot system of seven equations, the total number of possible solution paths is 48, but only 24 of the solutions were finite, 18 real and finite, and 14 unique. Of the unique solutions, ten were not physically possible due to either the initial mass center height or the vertical component of an impulse being negative. Of the four remaining, two were bipedal gaits - one employing only the hind legs and the other only the front legs. Of the two meaningful solutions, one matches that reported in Schmiedeler's dissertation [3], and the other is a new solution shown in Table 3. This new solution has the somewhat unusual characteristic of the hind legs acting as brakes to oppose the forward thrust provided by the front legs, so although physically possible, it would be an unnatural choice. Therefore, no truly meaningful additional solutions were found for the trot using Bertini.

Table 3 Trot solutions

Parameters	New Solution	Previous Solution	Units
z_1	0.653	0.691	m
i_{fx}	17.512	9.407	$N \cdot s$
i_{hx}	-8.112	-0.007	$N \cdot s$
i_{fz}	41.370	41.646	$N \cdot s$
i_{hz}	23.376	23.100	$N \cdot s$

4.2 Pace

Similar to the trot, the pace has the lateral footfalls separated by one half the stride period and is characterized by 5 equations in the same 5 variables. Within Bertini, Eqs. 3 and 4 were included to yield a system with 96 possible solution paths. Finite solutions were found for 26 of these paths, with 24 being real and finite, 12 of those being unique, and again only 4 of those being physically possible. Two were again bipedal solutions in which impulses were zero at either the front or hind legs. Of the remaining two solutions, one matches the previously reported solution, and the other is a new solution listed in Table 4 that is similar to the new trot solution in that the horizontal component of the hind leg impulse is negative. Like the trot, then, no truly meaningful new solutions were found for the pace.

Table 4 Pace solutions

Parameters	New Solution	Previous Solution	Units
z_1	0.648	0.723	m
i_{fx}	21.884	5.985	$N \cdot s$
i_{hx}	-12.484	3.415	$N \cdot s$
i_{fz}	41.227	41.227	$N \cdot s$
i_{hz}	23.519	23.519	$N \cdot s$

4.3 Bound

With the front and hind legs operating in pairs, the bound is slightly different in that the system contains 9 equations, but 11 variables: t_3, z_1, u_{zo}, d_f, d_h, θ_{y1}, ω_{yo}, i_{fx}, i_{fz}, i_{hx}, and i_{hz}. As in Schmiedeler's dissertation [3], the approach here is to select values for i_{fx} and i_{fz} to enable solutions for the remaining variables. Once again, Eq. 4 was used, but specifying i_{fx} and i_{fz} eliminated any need for Eq. 3. The resulting system of 10 equations has 288 possible solution paths to follow. Using the selected values of i_{fx} and i_{fz} from Schmiedeler's dissertation yields 8 finite solutions, 4 of which were also real and unique. Two of these were not physically possible because the timing of the hind leg footfalls did not occur within the gait cycle. The other two were unique, but very similar - identical to three decimal places and to the solution previously reported [3]. When somewhat different values of i_{fx} and i_{fz} were selected and the system was solved, two physically meaningful solutions were found again, but they differed from each other in their values for t_3, u_{zo}, and ω_{yo}. Table 5 lists

the solutions for one such case in which the impulse values were taken from the
bounding solution found for the half-bound system of equations as discussed in the
following section. This suggests that there are likely two unique physically meaningful solutions for the bound, but the difference between them can be negligible
depending on the specified values of the front leg impulses.

Table 5 Bound solutions

Parameters	Solution 1	Solution 2	Units
t_3	0.377	0.023	s
z_1	0.689	0.689	m
u_{zo}	-0.598	-1.879	$\frac{m}{s}$
d_f	0.170	0.170	m
d_h	0.566	0.566	m
θ_{y1}	-0.067	-0.067	rad
ω_{yo}	3.887	0.236	$\frac{rad}{s}$
i_{fx}	0.000	0.000	$N \cdot s$
i_{hx}	9.400	9.400	$N \cdot s$
i_{fz}	40.868	40.868	$N \cdot s$
i_{hz}	23.878	23.878	$N \cdot s$

4.4 Half-Bound

The half bound is similar to the bound except that the unique footfalls of the two
front legs yield a solvable system containing 11 equations in 11 variables (t_2, t_3, z_1,
u_{zo}, d_h, θ_{x1}, θ_{y1}, ω_{xo}, ω_{yo}, i_{fz}, and i_{hz}) provided i_{fx} is assumed to be zero, as in [3].
Including Eq. 4 yields a system of 12 equations with 2,304 possible solution paths
to follow. Only 66 were finite, and only 16 of those were also real, all of which were
unique. Thirteen of the real solutions were not physically possible because the second or third footfall in each occurred outside the gait cycle. Of the remaining three,
two were actually bound solutions with t_2 equal to 0. The remaining valid solution is
listed in Table 6 and is new in the sense that it does not match the solution reported
in Schmiedeler's dissertation [3]. This was a surprise to the authors, and a definitive
explanation for why the previous solution was not found using Bertini remains elu-
sive. The current hypothesis is that the previous solution did not actually satisfy all
of the equations to the necessary precision. For this numerical method, a non-zero
residual within one order of magnitude of the number of decimal places to which the
solution is reported is anticipated. In this case, the previous solution was considered
accurate to 15 decimal places, so the residual for each equation was expected to be
on the order of 10^{-14} or less, yet one of the equations yielded a residual on the order
of 10^{-13}. Therefore, the current hypothesis is that the previously reported solution
is very close to being a solution to the system, but in fact falls just short of satisfying
all the equations in the system. Therefore, Bertini did not find this previous solution
but found a significantly different solution that does satisfy all of the equations to
within the expected residual.

Table 6 Half-bound solution

Parameters	Value	Units
t_2	0.161	s
t_3	0.280	s
z_1	0.728	m
u_{zo}	-0.443	$\frac{m}{s}$
d_h	0.565	m
θ_{x1}	-0.227	rad
θ_{y1}	0.056	rad
ω_{xo}	-1.896	$\frac{rad}{s}$
ω_{yo}	2.052	$\frac{rad}{s}$
i_{fz}	40.668	$N \cdot s$
i_{hz}	24.078	$N \cdot s$

4.5 Canter

In the canter, one diagonal pair of legs operates together, while the legs in the other diagonal pair have unique footfalls. This yields a canter system consisting of 13 equations in 13 variables (t_2, t_3, z_1, u_{zo}, d_f, d_h, θ_{x1}, ω_{xo}, ω_{yo}, i_{fx}, i_{fz}, i_{hx}, and i_{hz}) provided that the initial yaw angular velocity and pitch angle are both assumed to be zero, as in [3]. Equations 3 and 4 are included for a system of 15 equations with 122,880 possible solution paths. Using the resources of Notre Dame's Center for Research Computing (CRC), Bertini ran for approximately 50 hours to track all of these paths. Results showed that 201 of the solutions were finite and 53 of those were real, with all of them being unique. Of these, only three were physically possible. Two of the three actually represented pronk gaits because the footfall times were all 0, and the final solution matched the one previously reported [3]. Therefore, no new solutions were found for the canter gait.

4.6 Gallop

Following the assumptions in Schmiedeler's dissertation [3], the rotary gallop is defined by 14 equations in 14 variables and the transverse gallop by 15 equations in 15 variables. Both systems require the use of Eqs. 3 and 4, which brings the total system sizes to 16 and 17 equations, respectively. Like the canter, the rotary gallop system has 122,880 possible solution paths, and the transverse gallop system has 10 times as many possible solution paths. No solutions are reported here for either gallop system due to difficulties in their implementation in Bertini. After more than 4 times as much computation time as for the canter (8 days 16 hours), Bertini did not yield any physically possible solutions for the rotary gallop. After 15 days of computation, the Notre Dame CRC kills any job still running, and in that time, Bertini failed to completely solve the transverse gallop system.

To address these issues, the authors pursued the grouping of variables within Bertini to reduce the number of solution paths to track in an effort to reduce the computation time. Within the software, a single variable group can be declared for all variables, a different group for each variable, or a combination resulting in more than one group with one or more variables in each. For all of the results reported in the previous sections, a single variable group that contained all of the variables was used. In general, the use of too many or too few groups is equally undesirable. Consider a system containing the equation $xy - 1 = 0$. If x and y are listed in the same variable group, Bertini considers this equation to have degree 2 despite the equation being linear in each variable. If x and y are declared in separate variable groups, though, the equation is considered to have degree 1. Since the number of possible solution paths is determined by multiplying the degrees of the equations together, the manner in which the variables are declared is important. By dividing the variables in both gallop systems into 3 groups, one can prevent any two variables that multiply each other from appearing in the same group, while also minimizing the total number of variable groups. This reduces the number of possible solution paths in the transverse gallop system to 325,328. The rotary gallop system was divided into the same 3 variable groups, with the obvious difference that the one variable that appears in the transverse but not the rotary gallop was removed. This surprisingly resulted in an increase in the number of possible solution paths to 179,668. When these updated systems were run using the Notre Dame CRC, the opposite behavior was observed. Bertini failed to completely solve the rotary gallop system within 15 days, so the job was killed. For the transverse gallop system, solutions were found after a computation time of approximately 8 days; however, none of these solutions were physically possible. Therefore, the gallop solutions remain a topic of ongoing research, likely requiring greater expertise with Bertini through collaboration with either Andrew Sommese or Charles Wampler.

5 Conclusions

The authors are genuinely surprised by the results presented in this paper. Specifically, they anticipated that many more physically possible, if not meaningful, solutions would be found at least for the more complex gaits of the half-bound and canter. Given the nonlinear nature of the governing equations, it was unexpected that only one or two feasible solutions would be found for each gait. With the transverse and rotary gallop systems still unsolved, there is certainly the possibility that these will yield more solutions, but one cannot anticipate that with confidence based on the results for the simpler gaits presented here. It is also possible that the selected level of drag and/or value of the stride period placed unforeseen limits on the systems such that more physically meaningful solutions could be found by selecting significantly different values for these parameters.

References

1. Agrawal, S.K., Waldron, K.J.: Impulsive model for a quadruped running machine. In: Proceedings of the Winter Annual Meeting of the ASME, vol. 11, pp. 139–148 (1989)
2. Bates, D.J., Hauenstein, J.D., Sommese, A.J., Wampler, C.W.: Bertini: Software for Numerical Algebraic Geometry, http://www.nd.edu/~sommese/bertini (cited February 11, 2013)
3. Schmiedeler, J.: The Mechanics of and Robotic Design of Quadrupedal Galloping. PhD thesis, The Ohio State University (2001)
4. Schmiedeler, J.: The Effect of Drag on Footfall Phasing in Quadruped Galloping. MS thesis, The Ohio State University (1998)
5. Schmiedeler, J.P., Siston, R., Waldron, K.J.: The significance of leg mass in modeling quadrupedal running gaits. In: Bianchi, G., Guinot, J.C., Rzymkowski, C. (eds.) RO-MANSY 14: Theory and Practice of Robots and Manipulators, pp. 481–488. Springer (2002)
6. Schmiedeler, J.P., Waldron, K.J.: The mechanics of quadrupedal galloping and the future of legged vehicles. International Journal of Robotics Research 18, 1224–1234 (1999)
7. Wampler, C.W., Morgan, A.P., Sommese, A.J.: Numerical continuation methods for solving polynomial systems arising in kinematics. Journal of Mechanical Design 112, 59–68 (1990)

24 Automatic Full Body Inverse Dynamic Analysis Based on Personalized Body Model and MoCap Data

M.J. Tsai, Allen Lee, and H.W. Lee

Abstract. A dual mode 3D body scanning/motion capturing system has been developed for creating personalized 3D body model as well as capturing body motion. The 3D body scanned data of an actor is arranged into a customized and structured digital body model. The body geometric parameters such as the centroid, moment of inertia, and principle axes can be accurately calculated from body segments. In addition, kinematic parameters such as joint angles, velocities, and accelerations of body segments can be computed from the motion data. Combining both the body geometric and kinematic parameters, free body diagrams are employed to balance the joint forces and moments of each body segments using New-Euler method. Finally, a full body inverse dynamics algorithm is applied to analyze the captured motion data performed by the actor. The resultant joint forces/moments are compared to those from the literatures. The whole process is completed in an automatic way without manual intervention.

Keywords: Full body Inverse dynamics, Joint force, 3D body model.

1 Introduction

Currently, commercial available 3D body scanners cannot capture body motion; whereas body motion capturing (mocap) systems cannot scan the body. Typically, in a mocap system, a skeleton or a fictional body model is created by a CAD system to analyze the body motion. Users should try very hard to register the

M.J. Tsai · Allen Lee · H.W. Lee
Department of Mechanical Engineering, National Cheng Kung University
Tainan, Taiwan
e-mail: {mjtsai,n16991473,hwlee}@mail.ncku.edu.tw

V. Kumar et al. (Eds.): *Adv. in Mech., Rob. & Des. Educ. & Res.*, MMS 14, pp. 305–322.
DOI: 10.1007/978-3-319-00398-6_24   © Springer International Publishing Switzerland 2013

markers that are worn by a real body onto the virtual model. The tracing markers used in the mocap system are typical spherical in shape, causing difficulties in labeling and matching the markers during the computation of the 3D marker positions. However, for a more demanding biomechanics analysis that requires accurate geometric parameters such as the mass, centroid, moment of inertia, and the principle axes of each body segment, the fictitious model cannot precisely represent the real body. It is a time consuming process and a very complicated task to compute real human body dynamics. Up to now, there is no short cut available for body dynamic analysis. An automatic process that integrates real body model creation and motion capturing capabilities into one system is of greatest needed.

Inverse dynamics has been wildly applied in the field of robotics and biomechanics. The purpose of inverse dynamics is to obtain joint reaction force/ moment from a given body model and motion data. In human motion analysis, the inverse dynamics method is normally employed to estimate joint force and moments during activities involving both lower extremities and upper body, such as impact force analysis[1], balance control[2, 3] and gait analysis [4, 5]

3D body models are now employed in inverse dynamics analysis [6, 7]. Ensminger [8] used a 3D upper extremities model to determine the joint forces/moments during wheelchair propulsion. Liu [9] presented a 3D lower limb kinetic analysis based on a wireless sensor system. Yoshihiko [10] constructed a complex 3D musculoskeletal model to predict the muscle and tendon tensions. Body segment parameters (BSP) are essential inputs in the dynamics analysis of human motion. These parameters include the mass, centroid, moment of inertia (MOI), and principle axes. The conventional use of predictive equations obtained from cadavers or medical scanning of live subjects is one of the methods to calculate the parameters [11, 12]. 3D body segments can be produced by reconstructing the geometry from computer tomography (CT) [13]. Other predictive equations are using magnetic resonance imaging (MRI) [14] and dual energy absorptiometry [15]. Jianjun [16] showed that motion capture data can be used to estimate BSP by building an optimized model. 3D body scanning method [17] is another way to determine the BSP.

In recent years, with the development of motion capture technology [18-20], 3D mocap data can be analyzed by commercial software packages for body dynamics. The authors' laboratory has also dedicated to the research of 3D body technologies. The developed systems include a dual mode body scanning/mocap system [21] and an intelligent body motion processing system (iBMPS) [22]. In this paper, the integrated hardware and software system allows us to estimate the joint forces and moments via inverse dynamics computation using personalized 3D body model and mocap data. Both the body geometric model and the motion captured data are obtained by the same system; in which, BSP can be calculated by the body scanned data automatically, whereas the kinematic data are calculated by an two-phased optimized inverse kinematics algorithm [23].

2   Calculation of the Geometric Parameters

2.1 *Data Structure of the Body Link*

The 3D body model is created from the dual mode optical system. Feature points on the body are recognized by mathematic algorithms according to the anatomic properties [24]. Those feature points are connected into curves that pass through anatomic features of the body and divide the body surface into segments. Each segment is then constructed as a standardized mesh structure. The vertexes of the mesh are called the "structure points." They designate the body geodesic coordinates (BGC). The BCGs are characterized by latitudinal girths and longitudinal curves that similar to that of the Earth; and are normalized in nature that do not subject to dimensions, gender, and age. For example, the left and right burst points arc (40, 10) and (40, 60) respectively. And the waist line is the 22th girth on the torso. The scanned 3D body point cloud is then condensed into a concise body model yet contains all the features, dimensions, and shape of the body that can be easily extracted. In this paper, the 3D body model is called the BGM (Body Geometric Model), and is recorded in the STL (Stereo Lithography) format. Hence, the BSP can be readily computed if the density of the link segment is known. Figure 1 shows a structuralized body model and segmented body links.

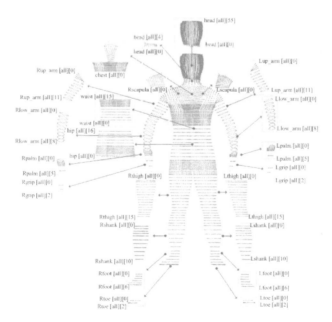

Fig. 1 The structuralized body model

2.2 The Closing Method of the Triangular Mesh of the Human Link

The body STL (Stereo Lithography) data structure is composed of triangular meshes. The mesh encloses the body as a whole. However, when the body mesh is separated into individual segments, each mesh becomes open shape. Therefore, the computation of BSP should close the open sections first. There are two cases of the open sections: one is the structure points of the open section locate at the same girth, as shown in Figure 2. The open sections are located at the first and last girths of the link. Figure 3 shows the second case of an open section, such as the chest link, the structure points of the side open sections are not on the same girth. However, we can always add an auxiliary point P_c on a suitable location to build triangular meshes and close up the open sections.

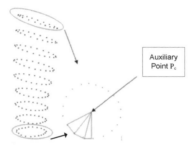

Fig. 2 The construction of the auxiliary point P_c

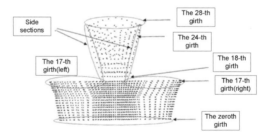

Fig. 3 The structuralized body model

2.3 Calculation of Geometric Parameters

The surface of each link is composed of triangular meshes readily to calculate the volume of each link, we can connect the triangular mesh to the auxiliary point to form a tetrahedron. By summing up the volumes of all tetrahedron, we can calculate the volume and find the centroid of each link.

$$V_{OABC} = \frac{1}{6}(\overrightarrow{OA} \cdot (\overrightarrow{OB} \times \overrightarrow{OC})) = ((a_{11}a_{22}a_{33} + a_{13}a_{21}a_{32} +$$
$$a_{12}a_{23}a_{31}) - (a_{13}a_{22}a_{31} + a_{12}a_{21}a_{33} + a_{11}a_{23}a_{32})) \tag{1}$$

The MOI and principle axes (the dynamic frame) of this tetrahedron can be computed according to Reference [25]. Using parallel axis theorem, we can get the dynamic frame and MOI of each link segment by summing up all the MOI of the tetrahedrons.

2.4 Verification of Geometric Parameters

The densities of different body link segments for different human races have been extensively studied by many researchers and are available in the literatures [26, 27]. They can be used for computing the mass from the calculated volume. Once the densities of the link segments are known, we can compute the mass of each link and the body total mass to verify the computed result of geometric parameters. An example body model is a student in our lab. Total mass of the body model is 69.97 kg and the real mass of the person is 72.0 kg. The percentage error is 2.81. Table 1 lists the verification of computed results for 21 body segments.

Table 1 The calculation of each body link mass

Body link	kg	Body link	kg
hip	16.03	neck	0.53
waist	13.31	head	5.13
chest	7.94		
L_scapular	2.91	R_scapula	2.99
L_uparm	1.47	R_upArm	1.64
L_lowArm	0.87	R_lowArm	0.99
L_palm	0.38	R_palm	0.31
L_grip	0.15	R_grip	0.16
L_upLeg	4.40	R_upLeg	4.57
L_lowLeg	2.02	R_lowLeg	2.16
L_foot	1.10	R_foot	0.95
Total mass of the model			**69.98**
Real mass of the human body			**72.0**
% error			**2.81**

3   Calculation of Kinematic Parameters

A body kinematic model (BKM) is created by assigning a specific joint type between two adjacent body segments, as shown in Fig. 4(b). Depending on the anatomic location and motion characteristic, each joint has 0 to 4 degrees of freedom to reproduce the relative movements. The BKM has 48 DOFs in total; it contains five kinematic chains: a trunk and four limbs. The trunk has allocated eight DOF, four to the waist, and four to the neck. Eight DOFs are assigned to both arms and legs. For each of scapula joint, 4 DOFs are also assigned. For any 4-DOF joint, the first three are revolute pairs in yaw-pitch-roll order and the last DOF is modeled as a prismatic pair in the direction of the third (roll) axis. Joints with less than four DOF are all revolute. The joint assignment is given in Figure 4. Before motion capturing, each body link segment is affixed at least one marker. During body scanning, the location of the marker with respect to the link is automatic registered. So that the position of each body link can be computed once the attached marker is recognized.

3.1   Kinematic Parameters of the Base Link

The base link of the body model is chosen as the hip link. The motion data representing the location of link N to the origin ($L2O$) is given by a homogeneous transformation matrix:

$$
L2O_N = \begin{bmatrix} r_{11} & r_{12} & r_{13} & d_x \\ r_{21} & r_{22} & r_{23} & d_y \\ r_{31} & r_{32} & r_{33} & d_z \\ 0 & 0 & 0 & 1 \end{bmatrix}
\tag{2}
$$

Where the upper left 3x3 sub-matrix $[r_{ij}]$ is the rotational part of the transformation, and $\underline{d}$ is the translation between the link and the origin. First, the fix angles yaw-pitch-roll (ψ, θ, ϕ) are obtained from the rotational matrix [28]:

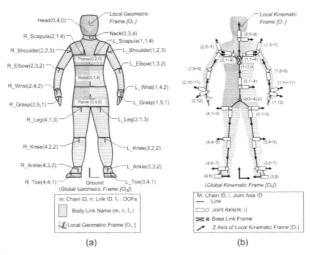

(a)   (b)

Fig. 4 Human body models: (a) Body segments of the BGM. (b) Kinematic chains and joint axes of the BKM.

$$\theta = A \tan 2(-r_{31}, \sqrt{r_{11}^2 + r_{21}^2})$$ (4)

$$\psi = A \tan 2(\frac{r_{21}}{\cos \theta}, \frac{r_{11}}{\cos \theta})$$ (5)

$$\phi = A \tan 2(\frac{r_{32}}{\cos \theta}, \frac{r_{33}}{\cos \theta})$$ (6)

Using numerical analysis, the angular velocity and angular acceleration are calculated by taking the differentials of the fixed angles. The velocity and acceleration of each link can be obtained as well by taking derivatives of the translational part of the matrix. The first order differential is computed from the central difference by Equation (7) [29]. We can use this equation again to get the second order differentials.

$$f'(\theta_0) = \frac{1}{12h}(-f_2 + 8f_1 - 8f_{-1} + f_{-2}) + O(h^4)$$ (7)

$O(h^4)$ is the error term, and h is the step length. Where f_1 is defined by:

$$f_1 \equiv f(\theta_0 + h)$$ (8)

3.2   Kinematic Parameters of All Joints

Figure 5 shows the relation between the kinematic frame and dynamic frame. The definitions of the D&H parameters are listed in Table 2. In this paper, we use the convention that the Nth coordinate system in the kinematic frame is located on the Nth axis with Z_n lies along the axis and X_n lies along the common normal of axes N and $N+1$. The transformation between the dynamic frame to the kinematic frame of the link is given by L_n.

Table 2 Definitions of the D&H parameters

D&H parameters	Definitions
a_N	Link length
α_N	Twist angle
r_N	Offset
θ_N	Joint angle

Now we can derive the angular velocity of the axis N in terms of the joint angle θ_N, which is obtained from the inverse kinematics process by [23].

$$\omega_N = \omega_{N-1} + \dot{\theta}_N w_N$$ (9)

For prismatic joint:

$$\omega_N = \omega_{N-1}$$ (10)

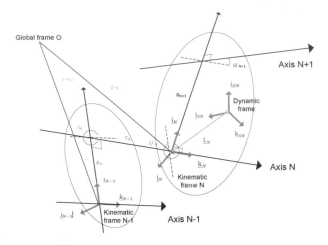

Fig. 5 Definitions of the kinematic and dynamic frames

Where w_N is the unit direction vector of the axis N. Differentiate Eq. (10), we can get angular acceleration $\dot{\omega}_N$ of the link and the linear acceleration a_{GN} of the link centroid by the following equations.

$$\dot{\omega}_N = \frac{d\omega_N}{dt} = \dot{\omega}_{N-1} + \ddot{\theta}_N w_N + \dot{\theta}_N w_{N-1} \times w_N \tag{11}$$

for prismatic joint:

$$\dot{\omega}_N = \dot{\omega}_{N-1} \tag{12}$$

$$a_{ON} = \ddot{p}_N \tag{13}$$

$$a_{GN} = \ddot{p}_N + \dot{\omega}_N \times \ell_N + \omega_N \times (\omega_N \times \ell_N) \tag{14}$$

4   Full Body Inverse Dynamics

This section introduces the procedure to calculate the full body inverse dynamics (FBID). Newton-Euler method is used to calculate the joint force/moment of the body model. Similar to that of Reference [30], the recursive Newton-Euler method is divided into two parts, and each part is composed of two steps.

4.1   Part I of the Newton-Euler Method

Part I(a) of the Newton-Euler method is to find the kinematic parameters of each links by beginning with the base link and work out to the end-effector and load. The kinematic parameters obtained previously were referred to the global frame. For recursive computation purpose, they should be expressed in N^{th} kinematic frame.

Referring to Figure 5, K_N is the transformation from frames N to N-1:.

$$K_N = \begin{bmatrix} \cos\theta_N & -\sin\theta_N & 0 & \alpha_N \\ \sin\theta_N\cos\alpha_N & \cos\theta_N\cos\alpha_N & -\sin\alpha_N & -r_N\sin\alpha_N \\ \sin\theta_N\sin\alpha_N & \cos\theta_N\sin\alpha_N & \cos\alpha_N & r_N\cos\alpha_N \\ 0 & 0 & 0 & 1 \end{bmatrix} \quad (15)$$

We can separate the transformation matrix into a 3×3 rotation matrix and a 3×1 translation vector. The above equation can be rewritten as:

$$K_N = \begin{bmatrix} V_N U_N & V_N S_N \\ 0 & 1 \end{bmatrix} \quad (16)$$

$$V_N = \begin{bmatrix} 1 & 0 & 0 \\ 0 & \cos\alpha_N & -\sin\alpha_N \\ 0 & \sin\alpha_N & \cos\alpha_N \end{bmatrix}, \text{ and}$$

$$U_N = \begin{bmatrix} \cos\theta_N & -\sin\theta_N & 0 \\ \sin\theta_N & \cos\theta_N & 0 \\ 0 & 0 & 1 \end{bmatrix}, \text{ and } S_N = \begin{bmatrix} a_N \\ 0 \\ r_N \end{bmatrix}$$

Where V_N and U_N represent the rotations from frames N to N-1 along axis k_N and i_{N-1} respectively. S_N is the translational vector between the frames.

Now all the kinematic parameters should be referred to the N kinematic frame. Use subscript to denote the parameter of this axis, and superscript for the frame it refers to. Since axis N lies along the z-axis of the kinematic frame N, we have

$$w_N^N = k \quad (17)$$

$$w_{N-1}^N = U_N^T V_N^T k \quad (18)$$

$$\omega_N^N = \omega_{N-1}^N + \dot{\theta}_N k \quad (19)$$

$$\omega_{N-1}^N = U_N^T V_N^T \omega_{N-1}^{N-1} \quad (20)$$

The superscript "T" means the transpose (and hence, the inverse) of the matrix. Similarly for the angular acceleration:

$$\dot{\omega}_N^N = \dot{\omega}_{N-1}^N + \ddot{\theta}_N k + \dot{\theta}_{N-1} w_{N-1}^N \times k \quad (21)$$

$$\dot{\omega}_{N-1}^N = U_N^T V_N^T \dot{\omega}_{N-1}^{N-1} \quad (22)$$

for prismatic joint

$$\omega_N^N = U_N^T V_N^T \omega_{N-1}^{N-1} \quad (23)$$

$$\dot{\omega}_N^N = U_N^T V_N^T \dot{\omega}_{N-1}^{N-1} \quad (24)$$

Acceleration of center of mass

$$a_{GN}^N = \ddot{p}_N^N + \dot{\omega}_N^N \times \ell_N^N + \omega_N^N \times (\omega_N^N \times \ell_N^N) \qquad (25)$$

Part I(b) of the Newton-Euler Method is to find the resultant forces and moments on each link from the given kinematic parameters. Newton-Euler's equation is given by:

$$\begin{cases} \sum F = ma_G \\ \\ \sum M = I_G \dot{\omega} + \omega \times I_G \omega \end{cases} \qquad (26)$$

where F, M are the external force and moment, respectively; I_G and m are the inertia matrix and mass of the link respectively.

Euler's equations are written relative to the dynamic frame. In Part I(a), we transfer the kinematic parameters from global frame O to the kinematic frame N. If we want to use these equations, we need to modify Eq. (26) so that it can be used in the kinematic frame.

The transformation from dynamic frame to kinematic frame is:

$$P = L_n P' + \ell_N \qquad (27)$$

Where P' is the position of a point in the kinetic frame and P is the position of that point in the kinematic frame.

So we can modify Equation (26) to be referred to the N kinematic frame.

$$\begin{cases} F_N^N = m_N a_{GN}^N \\ \\ M_N^N = \sum L_N M_N^G = J_G^N \dot{\omega}_N^N + \omega_N^N \times J_G^N \omega_N^N \end{cases} \qquad (28)$$

where J_G^N is the inertia matrix transformed to the kinematic frame:

$$J_G^N = L_N I_G L_N^T \qquad (29)$$

4.2   Part Two of the Newton-Euler Method

Having been modified and referred the Newton- Euler equations to the kinematic frame, we can apply the equations for the inverse dynamics problems. Step II(a) begins with the end-effector and external load and works back to the base link, then does a free body analysis on each link until all forces and moments of every link in the chain are known. Step II (b) determines joint force or moment for each link segment.

The end-effectors of the body model are the head, both hands, and both feet. We can solve the forces and moments by balancing if any external loads applied on the end-effectors are known. For a typical body motion, the unknown force/moments are the reactions from the ground. If two feet are all touch the ground, we have more unknowns than available equations, so that the problem cannot be solved without further assistants or assumptions. Nevertheless, when one (or no) foot touches the ground, the balance force of the lower limbs can be solved if the forces/moments of upper limb chains are known. The flow chart of FBID, according to the four ground-touching cases, is depicted in Figure 6.

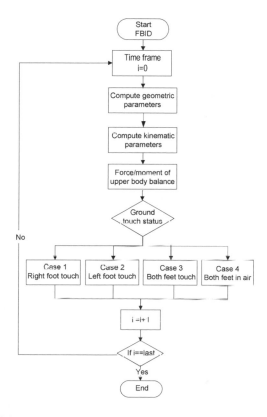

Fig. 6 Flowchart of the FBID

For case 1, the right foot touches the ground, the FBID starts from either of the two hands, head, and left foot to the hip and finally to the right foot. The second case happens when right foot lifts up. We can start the FBID from two hands, head, and right foot to the hip and finally to the left foot. For case 3, both feet touch the ground, we apply the smooth transition assumption (STA) [31] to the walking motion. The STA was proposed to solve the indeterminate problem of double support phase in a walking motion. In the last case, for which the subject jumps into the air, we can start the FBID from all end-effectors of the body model and find the resultant force of the base link (the hip).

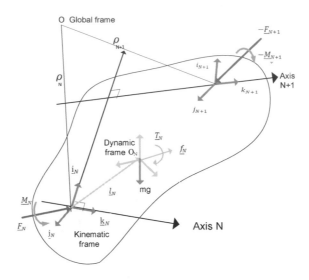

Fig. 7 The free body diagram of our two force link with kinematic frames

For a two force link as an example shown in Figure 7, the Newton/Euler equations of based on the kinematic frame are derived as:

$$\underline{F}_N^N = \underline{F}_{N+1}^N + m_N \boldsymbol{a}_{GN}^N + m_N \underline{g}_N^N \tag{30}$$

$$
\begin{aligned}
M_N^N = M_{N+1}^N + \ell_N^N \times F_N^N - (\boldsymbol{\rho}_N - \boldsymbol{\rho}_{N+1} + \ell_N)^N \times F_{N+1}^N \\
+ J_G^N \dot{\omega}_N^N + \omega_N^N \times J_G^N \omega_N^N
\end{aligned}
\tag{31}
$$

4.3 Motion Tracking Experiments

To get the body geometric parameters, we first scan the person and then create a personalized body model. For the kinematic data, our system can also track the markers for the motion data. Figure 8 shows a person wearing a leotard with markers on the body. Figure 9(a) shows the actor is under scanning, and (b) is under motion capturing by using the dual mode D2000 system.

4.3.1 Lifting Feet Experiment

During the experiment, the person walks slowly by lifts his feet alternatively. We calculate the ankle joint force and compare the results from literatures. The total weight which one foot withstand during normal walking speed is about 1.5 times the body weight and 2 to 3 times the body weight during running [32]. We can calculate his right (left) ankle joint force when he lifts his left (right) foot highest in the air.

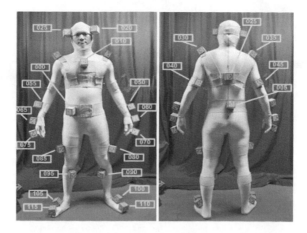

Fig. 8 (a) Front view of the actor, (b) Rear view of the actor wearing leotard with markers

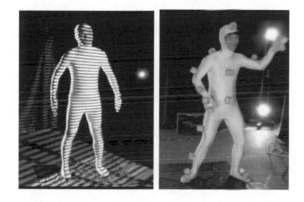

Fig. 9 (a) The actor is under body scanning, (b) under motion capturing

Table 3 lists the resultant force in the z component of both ankles on their kinematic frames. The force magnitudes are shown in Newton and kgf. The third column lists the ratio of forces divided by the personal body weight for comparison. The ratios of both feet are around 1.5. This is quite compliance to the values in the literatures.

Table 3 The joint force of ankle in z-direction

Body link	Newton (kgf)	ratio
L_ankle_z	1021.0 (104.13)	1.45
R_ankle_z	1047.5 (106.83)	1.48

4.3.2   Walking Experiment

Now we will apply the smooth transition function to the walking motion. By using the STA assumption, we obtained the ankle joint force in a complete gait cycle as shown in the left side of Figures 10 to 12. The calculated moments of ankle joint are shown in the left side of Figures 13 to 15. For comparison, the right side of the figures shows the results from Reference [31].

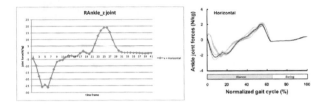

Fig. 10 Ankle joint force in x direction

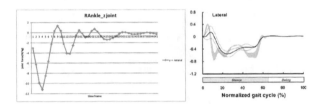

Fig. 11 Ankle joint force in y direction

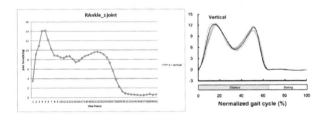

Fig. 12 Ankle joint force in z direction

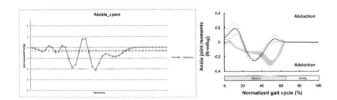

Fig. 13 Ankle joint moment in x direction

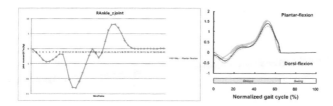

Fig. 14 Ankle joint moment in y direction

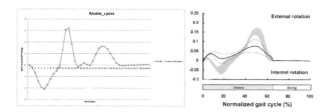

Fig. 15 Ankle joint moment in z direction

4.3.3   Jumping Experiment

For the case of both feet in the air, we still can calculate the force and moments of the hip link from upper body and lower body. Figure 16 shows the z-component of the Hip position in a jumping motion. The person jumped in the air at time frames 57 to 71.

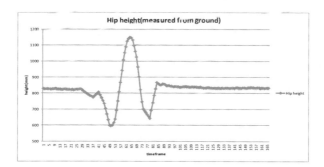

Fig. 16 The height of the Hip

Figure 17 shows the resultant forces and moments, respectively, of the Hip during the jump-in-air time frames. The non-zero values illustrated the Hip didn't balance, and the inverse dynamics computation gave incorrect results.

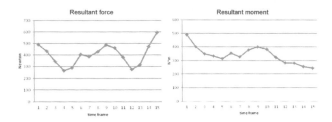

Fig. 17 Resultant force and moment of the Hip link

5   Discussion

In this study, the joint forces and moments of the full body can be estimated by the motion capture data, using personal BSP in an efficient and fully automatic approach. However, there are still some problems needs to discuss.

1. The estimated joint forces and moments of the full body dynamics in this study are found good match to the results of literatures in the sagittal plane only. They have less accuracy in other two coordination planes. Since the result calculated in reference [31] also have quite different patterns in other planes when comparing with the force plate data, this indicates that the smooth transition assumption didn't suitable for everybody and may need modification according to individuals.

2. The non-zero values of resultant force/moment on the Hip during jump-in-air are incorrect, since they should be zero for balancing. There might be some reasons leading to such errors. First, the raw motion data have not yet been properly filtered and smoothed; this would introduce error during numerical differentiations; and great care should be taken on applying numerical method. Second, the frame rate of the capturing system is 30Hz; it might not be high enough for fast movement like jumping. Third, the dynamic modeling in this study only considered the external forces. The internal effects, such as the joint actuation torque, compliance, and damping produced by the muscles are not considered. The person jumped into the air and may activate muscles to change his postures. Those factors may influence the resultant joint torque to be balanced at the instant. It is suggested a more rigorous modeling technology should be applied to accurately estimate the internal joint force. And a more elaborated method is of greatly anticipated for the evaluation of the internal actuations. For example, if the Hip link should be balanced by the internal actuations, does it offer a clue to estimate the dynamic parameters of muscular stiffness and damping coefficient by the given inertia and motions?

6   Conclusion

Traditional FBID method needs two sets of apparatuses, one for body scanning to create body model, the other is for body motion tracking. Engineers should struggle with laborious works of labeling the markers, registering markers on the body, and assigning BSP, etc. In this study, the developed software integrates a dual mode 3D body scanning/motion capturing capabilities that can create 3D personalized body model, capture body motion, and analyze the body's motion data. The system can solve inverse dynamics problem fully automatically without manual intervention. With our standardized body model, geometric parameters that are required for biomechanics analysis can be computed accurately without tedious measurement and assignment.

After all, this preliminary study is just a pilot run on the newly developed system; it still requires further rigorous testing and verification. However, this study postulates a convenient technique for FBID analysis and provides a novel tool for analyzing the 3D body motion data. It is believed these features would quickly become a trend and be accepted by the fields of dancing and athletics, biomechanics, apparel design animation, and rigorous amusement in which the body model and motion data should be unified and generated from the player himself.

References

1. Bisseling, R.W., Hof, A.L.: Handling of impact forces in inverse dynamics. Journal of Biomechanics 39, 2438–2444
2. O'Kane, F.W., McGibbon, C.A., Krebs, D.E.: Kinetic analysis of planned gait termination in healthy subjects and patients with balance disorders. Gait & Posture 17, 170–179
3. Robert, T., Chèze, L., Dumas, R., Verriest, J.P.: Validation of net joint loads calculated by inverse dynamics in case of complex movements: Application to balance recovery movements. Journal of Biomechanics 40, 2450–2456
4. Forner-Cordero, A., Koopman, H.J.F.M., van der Helm, F.C.T.: Inverse dynamics calculations during gait with restricted ground reaction force information from pressure insoles. Gait & Posture 23, 189–199
5. Ren, L., Howard, D., Boonpratatong, A.: Three-dimensional whole-body gait analysis: an investigation into the fundamental problem of inverse dynamics. Gait & Posture 30(suppl. 2), S78–S79
6. MacKinnon, C.D., Winter, D.A.: Control of whole body balance in the frontal plane during human walking. Journal of Biomechanics 26, 633–644
7. Ren, L., Jones, R.K., Howard, D.: Whole body inverse dynamics over a complete gait cycle based only on measured kinematics. Journal of Biomechanics 41, 2750–2759
8. Ensminger, G.J., Robertson, R.N., Cooper, R.A.: A model for determining 3-D upper extremity net joint forces and moments during wheelchair propulsion. In: IEEE 17th Annual Conference on Engineering in Medicine and Biology Society, vol. 2, pp. 1179–1180 (1995)
9. Liu, T., Inoue, Y., Shibata, K., Shiojima, K.: Three-dimensional lower limb kinematic and kinetic analysis based on a wireless sensor system. In: 2011 IEEE International Conference on Robotics and Automation (ICRA), pp. 842–847 (2011)
10. Nakamura, Y., Yamane, K., Fujita, Y., Suzuki, I.: Somatosensory computation for man-machine interface from motion-capture data and musculoskeletal human model. IEEE Transactions on Robotics 21, 58–66

11. de Leva, P.: Adjustments to Zatsiorsky-Seluyanov's segment inertia parameters. Journal of Biomechanics 29, 1223–1230
12. Yeadon, M., Morlock, M.: The appropriate use of regression equations for the estimation of segmental inertia parameters. Journal of Biomechanics 22, 683–689
13. Pearsall, D., Reid, J., Livingston, L.: Segmental inertial parameters of the human trunk as determined from computed tomography. Annals of Biomedical Engineering 24, 198–210
14. Pearsall, D., Reid, J., Ross, R.: Inertial properties of the human trunk of males determined from magnetic resonance imaging. Annals of Biomedical Engineering 22, 692–706
15. Lee, M.K., Koh, M., Fang, A.C., Le, S.N., Balasekaran, G.: Estimation of Body Segment Parameters Using Dual Energy Absorptiometry and 3-D Exterior Geometry. In: Lim, C.T., Goh, J.C.H. (eds.) ICMBE 2008. IFMBE, vol. 23, pp. 1777–1780. Springer, Heidelberg (2009)
16. Jianjun, Z., Yi, W., Shihong, X., Zhaoqi, W.: Estimating human body segment parameters using motion capture data. In: 2010 4th International Universal Communication Symposium (IUCS), pp. 243–249 (2010)
17. Norton, J., Donaldson, N., Dekker, L.: 3D whole body scanning to determine mass properties of legs. Journal of Biomechanics 35, 81–86
18. Vicon Motion Systems, http://www.vicon.com/
19. OptiTrack, http://www.naturalpoint.com/optitrack/
20. Qualysis, http://www.qualisys.com/
21. Tsai, M.J., Lee, H.W., Lung, H.Y.: Dual-Mode Optical Measurement Apparatus and System. U.S. Patent Pending (13/188,724)
22. Tsai, M.J., Lung, H.Y.: Development of an Intelligent Body Motion Processing System. Paper submitted to ASME Journal of Mechanisms and Robotics (2012)
23. Tsai, M.J., Lung, H.Y.: Two-Phase Optimized Inverse Kinematics for Motion Replication of Real Human Models. Paper submitted to International Journal of Robotics (2012)
24. Tsai, M.J., Fang, J.J.: A Feature Based Data Structure for Computer Manikin (2007)
25. Chang, C.H.: Numerical Simulations for a 3D System Composed of Polyhedral Blocks-Dissection of Polyhedral Blocks. Master Thesis of Dept. of Civil Engineering, National Central University, Zhongli (2006)
26. Zatsiorsky, V.: Kinetics of Human Motion. Sheridan Books (2002)
27. Zheng, S.Y.: Modern Sports Biomechanics, 2nd edn. National Defense Industry Press, Beijing (2007)
28. Waldron, K., Schmiedeler, J.: In: Cisiliano, B., Khatib, O. (eds.) Handbook of Robotics, ch. 1. Springer, Berlin (2008)
29. Richard, J.D.F., Burden, L.: Numerical Analysis, 8th edn. Brooks/Cole (2005)
30. Tsai, L.-W.: Robot Analysis, ch. 9. John wiley & Sons, Inc., New York (1999)
31. Ren, L., Jones, R.K., Howard, D.: Whole Body Inverse Dynamics Over a Complete Gait Cycle Based Only on Measured Kinematics. Journal of Biomechanics 41, 2750–2759 (2008)
32. Farley, C.T., Ferris, D.P.: Biomechanics of Walking and Running: Center of Mass Movements to Muscle. Department of Integrative Biology, University of California, Berkeley, USA: Ferris DP. (1998)

25 Lower-Limb Muscle Function in Human Running

Anthony G. Schache, Tim W. Dorn, and Marcus G. Pandy

Abstract. This paper provides a brief summary of work completed to date in our research laboratory investigating lower-limb muscle function during human running. Muscle function has been evaluated using a variety of methods, including muscle electromyography, inverse dynamics, and computational musculoskeletal modeling. It is evident that the coordination amongst the major lower-limb muscles changes considerably when running speed is progressed from jogging through to maximum sprinting. The ankle plantarflexor muscles appear to have a dominant role up to running speeds of around 7 ms^{-1}. For running speeds beyond 7 ms^{-1}, the hip flexor and extensor muscles become far more critical. These findings provide insight into the strategies used by the lower-limb muscles to maximize running performance and have implications for the design of injury prevention programs.

1 Introduction

Running is a fundamental skill. It is a critical requirement for almost all sporting activities. Understanding the biomechanical function of the major lower-limb muscle groups during running is important for improving current knowledge regarding human high performance as well as identifying potential factors that might be related to injury. There are a variety of methodological approaches that can be taken to study lower-limb muscle function during running, including: (a) the measurement of muscle electromyographic activity [1, 2]; (b) the use of inverse dynamics to determine lower-limb joint moments of force (or torques), net joint powers, and work [3,5]; and (c) the use of computational musculoskeletal modeling to calculate certain parameters that cannot be directly measured via non-invasive experiments [6,9]. Our research group has applied a combination of these approaches to address several research questions of primary interest, for

Anthony G. Schache · Tim W. Dorn · Marcus G. Pandy
Department of Mechanical Engineering, University of Melbourne,
Parkville, Victoria 3010, Australia
e-mail: pandym@unimelb.edu.au

V. Kumar et al. (Eds.): *Adv. in Mech., Rob. & Des. Educ. & Res.*, MMS 14, pp. 323–327.
DOI: 10.1007/978-3-319-00398-6_25   © Springer International Publishing Switzerland 2013

example: Which lower-limb muscles are most important for increasing running speed? and How does faster running influence the mechanics of certain biarticular muscles, such as the hamstrings? Herein we provide a summary of some of the main findings from work completed to date.

2   Overview of Experimental Methods

Data were collected from nine healthy adult runners (five males, four females; age, 27.7 ± 8.0 years; body mass, 73.1 ± 8.6 kg, height, 176 ± 7 cm). Each participant ran on an indoor synthetic running track at four discrete steady-state running speeds: slow running at 3.5 ms^{-1} (N=9), medium-paced running at 5.0 ms^{-1} (N=9), fast running at 7.0 ms^{-1} (N=8) and maximal sprinting at 8.0 ms^{-1} or greater (N=7). Small reflective markers were mounted at specific locations on the trunk, legs and arms and the marker trajectories were recorded using a three-dimensional motion capture system (VICON, Oxford Metrics Ltd., Oxford, UK). Ground reaction forces were measured using eight force plates (Kistler Instrument Corp., Amherst, NY, USA). Lower-limb muscle electromyographic data were acquired using a telemetered system (Noraxon Telemyo 2400T G2, Noraxon USA Inc., Scottsdale, AZ, USA).

3   Lower-Limb Muscle Function with Increasing Running Speed

To determine which muscles are most important for increasing running speed, we initially applied an inverse-dynamics approach to calculate the torques, net powers and work done at the lower-limb joints [4]. The most substantial increases in magnitude were displayed by the sagittal-plane torques, net powers, and work done at the hip and knee joints during the terminal swing phase of the stride cycle. For example, when running speed changed from 3.5 ms^{-1} to 9.0 ms^{-1}, the peak hip joint power generation and the peak knee joint power absorption during terminal swing increased 12.1-fold and 8.1-fold, respectively. In contrast, the work done at the knee joint during the stance phase of the stride cycle was not affected by running speed, whereas the work done at the ankle joint during stance increased when running speed changed from 3.5 ms^{-1} to 5.0 ms^{-1}, but plateaued thereafter. In terms of lower-limb muscle function, this study revealed that in order to progress running speed towards maximal sprinting the increase in biomechanical load generated by the hip flexor and extensor muscles during the swing phase of the stride cycle was substantially greater than that generated by the knee extensor and ankle plantarflexor muscles during stance.

While our inverse-dynamics analysis generated some important and interesting observations, the ability of this approach to quantify the biomechanical load experienced by an individual lower-limb muscle is limited by the mechanical redundancy of the human musculoskeletal system. In other words, because many muscles cross each lower-limb joint, a net joint torque can be satisfied by an infinite combination of muscle forces. It is therefore not possible to discern the actions of individual muscles from net joint moments alone [10]. Hence, our next step was to apply computational musculoskeletal modeling [6]. Such an approach allows the contributions of individual lower-limb muscles to joint and center-of-mass accelerations to be determined, information which perhaps best describes the functional role of a muscle.

When running speed progressed from slow to fast, stride length displayed a greater percentage increase in magnitude than stride frequency; however, beyond 7.0 ms^{-1} the opposite occurred. This result is consistent with what has been reported by others [11]. Thus, faster running is initially achieved by increasing stride length at a greater rate than stride frequency, but eventually a threshold is reached and a shift in strategy occurs whereby the progression to maximum running speed is achieved by increasing stride frequency at a greater rate than stride length. Our computational modeling indicated that the ankle plantarflexor muscles were primarily responsible for this strategy shift. Specifically, for running speeds up to 7 ms^{-1}, the ankle plantarflexors (i.e., the gastrocnemius and soleus muscles) provided a significant contribution to vertical support and hence increases in stride length. For speeds beyond 7 ms, these muscles likely shortened at relatively high velocities and had less time to generate forces needed for support. Consequently, running speed was progressed to maximum by having the hip flexors and extensors (i.e., the iliopsoas, gluteus maximus, and hamstring muscles) accelerate the hip and knee joints more vigorously during swing, thus increasing stride frequency. These findings offer insight into the strategies used by the lower-limb muscles to maximize running performance. The function of the ankle plantarflexor muscles appears critical, and it could be theorized that a key difference between elite and sub-elite sprinting athletes relates to the rate at which these muscles produce maximum force [12]. Compared to their sub-elite counterparts, elite sprinters may have the capacity to produce maximum force from the ankle plantarflexor muscles in a much shorter period of time, thus allowing them to reach higher speeds of running before needing to shift strategies.

4   Hamstring Muscle Function during Sprinting

One consequence of the switch to a hip-dominant strategy as running speed approaches maximum sprinting is that the magnitude of the forces (gravity and centrifugal) acting about the hip and knee joints during the terminal swing phase of the stride cycle increase dramatically. Large 'external' hip flexor and knee extensor torques are produced, which are primarily opposed by the hamstring muscles. It is therefore not surprising that the majority of hamstring muscle strain-type injuries occur when running at maximal or close to maximal speeds [13]. In order to aid in the development of rehabilitation and prevention strategies that are specific to the mechanism of injury, musculoskeletal modeling has been used to understand the mechanics of the human hamstring muscles during sprinting by our research group [14] and others [15, 16]. The consistent finding from all studies completed to date is that during the terminal swing phase of the stride cycle the hamstrings reach peak muscle-tendon unit stretch, produce peak force, and perform much negative work (energy absorption). It has therefore been proposed that for sprinting the biarticular hamstrings are at greatest risk of injury during terminal swing when they are contracting eccentrically [17]. Such a proposal is consistent with what has been concluded from two independent case studies that analyzed biomechanical data collected during an acute *in vivo* hamstring muscle strain-type injury [18, 19]. The main implication is that interventions for rehabilitating and preventing hamstring strain injuries should be biased towards fast eccentric contractions performed at long muscle-tendon unit lengths.

5   Summary and Future Directions

This review has provided a brief summary of work completed to date in our research laboratory investigating lower-limb muscle function during human running. Evidence has been presented demonstrating that the coordination amongst the major lower-limb muscles changes considerably when running speed is progressed from jogging through to maximum sprinting. Our future work is directed at investigating lower-limb muscle function for continuous maximal accelerations and comparing results to those already obtained by analyzing a spectrum of discrete steady-state speeds. We are currently also using ultrasound imaging to directly evaluate muscle-fiber strain, and endeavour to integrate such measurements with a computational modeling approach to generate a more complete understanding of lower-limb muscle function during running.

References

1. Cappellini, G., Ivanenko, Y.P., Poppele, R.E., Lacquaniti, F.: Motor patterns in human walking and running. J. Neurophysiol. 95, 3426–3437 (2006)
2. Kyröläinen, H., Avela, J., Komi, P.V.: Changes in muscle activity with increasing running speed. J. Sports Sci. 23, 1101–1109 (2005)
3. Belli, A., Kyrolainen, H., Komi, P.V.: Moment and power of lower limb joints in running. Int. J. Sports Med. 23, 136–141 (2002)
4. Schache, A.G., Blanch, P.D., Dorn, T.W., Brown, N.A.T., Rosemond, D., Pandy, M.G.: Effect of running speed on lower limb joint kinetics. Med. Sci. Sports Exerc. 43, 525–532 (2011)
5. Winter, D.A.: Moments of force and mechanical power in jogging. J. Biomech. 16, 91–97 (1983)
6. Dorn, T.W., Schache, A.G., Pandy, M.G.: Muscular strategy shift in human running: dependence of running speed on hip and ankle muscle performance. J. Exp. Biol. 215, 1944–1956 (2012)
7. Hamner, S.R., Seth, A., Delp, S.L.: Muscle contributions to propulsion and support during running. J. Biomech. 43, 2709–2716 (2010)
8. Glitsch, U., Baumann, W.: The three-dimensional determination of internal loads in the lower extremity. J. Biomech. 30, 1123–1131 (1997)

9. Sasaki, K., Neptune, R.R.: Muscle mechanical work and elastic energy utilization during walking and running near the preferred gait transition speed. Gait Posture 23, 383–390 (2006)

10. Pandy, M.G., Andriacchi, T.P.: Muscle and joint function in human locomotion. Annu. Rev. Biomed. Eng. 12, 401–433 (2010)

11. Weyand, P.G., Sternlight, D.B., Bellizzi, M.J., Wright, S.: Faster top running speeds are achieved with greater ground forces not more rapid leg movements. J. Appl. Physiol. 89, 1991–1999 (2000)

12. Weyand, P.G., Sandell, R.F., Prime, D.N.L., Bundle, M.W.: The biological limits to running speed are imposed from the ground up. J. Appl. Physiol. 108, 950–961 (2010)

13. Askling, C.M., Tengvar, M., Saartok, T., Thorstensson, A.: Acute first-time hamstring strains during high-speed running. Am. J. Sports Med. 35, 197–206 (2007)

14. Schache, A.G., Dorn, T.W., Blanch, P.D., Brown, N.A.T., Pandy, M.G.: Mechanics of the human hamstring muscles during sprinting. Med. Sci. Sports Exerc. 44, 647–658 (2012)

15. Chumanov, E.S., Heiderscheit, B.C., Thelen, D.G.: Hamstring musculotendon dynamics during stance and swing phases of high-speed running. Med. Sci. Sports Exerc. 43, 525–532 (2011)

16. Thelen, D.G., Chumanov, E.S., Sherry, M.A., Heiderscheit, B.C.: Neuromusculoskeletal models provide insights into the mechanisms and rehabilitation of hamstring strains. Exerc. Sport Sci. Rev. 34, 135–141 (2006)

17. Chumanov, E.S., Schache, A.G., Heiderscheit, B.C., Thelen, D.G.: Hamstrings are most susceptible to injury during the late swing phase of sprinting. Br. J. Sports Med. 46, 90 (2012)

18. Heiderscheit, B.C., Hoerth, D.M., Chumanov, E.S., Swanson, S.C., Thelen, B.J., Thelen, D.G.: Identifying the time of occurrence of a hamstring strain injury during treadmill running: A case study. Clin. Biomech. 20, 1072–1078 (2005)

19. Schache, A.G., Wrigley, T.V., Baker, R., Pandy, M.G.: Biomechanical response to hamstring muscle strain injury. Gait Posture 29, 332–338 (2009)

26 Toward Broader Education in Control System Design for Mechanical Engineers: Actuation System Selection and Controller Co-design

K. Srinivasan

Abstract. Control system design education for mechanical engineers, in its current form, focuses primarily on control algorithm design. We argue here that control system design is performed best when it is broadened to include requirements development and actuation system design, to be performed jointly with design of the control algorithm. We review practices in specifying control system performance and capabilities of actuation technologies for control applications. An existing unifying framework for representing actuation system capabilities and their selection for applications of interest is presented and assessed. Developments needed for an improved methodology for actuation system selection are enumerated. First, actuator comparison must be extended to include system-level characterization of performance. Second, mechanical actuation applications should be classified in more generic terms and application requirements framed accordingly. Third, compilation of performance characteristics for actuators and actuation systems need to be more comprehensive and better linked to underlying technological limitations.

1 Motivation and Introduction

Control system design education for mechanical engineers, as currently implemented, focuses primarily on control algorithm design and underemphasizes the selection and design of mechanical actuation systems, as well as the formulation and significance of performance requirements for different

K. Srinivasan
Department of Mechanical and Aerospace Engineering
The Ohio State University
Columbus, OH 43210
e-mail: Srinivasan.3@osu.edu

V. Kumar et al. (Eds.): *Adv. in Mech., Rob. & Des. Educ. & Res.*, MMS 14, pp. 329–347.
DOI: 10.1007/978-3-319-00398-6_26   © Springer International Publishing Switzerland 2013

applications. We focus here on applications requiring high performance and closed loop control because of the unique associated knowledge base. In typical undergraduate controls curricula, students are 'given' the mechanical actuation system and load, and system performance requirements ostensibly suited to the application, and the bulk of their efforts is directed at modeling and analysis of the given physical system, and feedback control algorithm design to best meet the performance specifications. Issues related to the selection of the actuation system, both the technology of actuation and the sizing of the actuation system components, are essentially underemphasized and, in most cases, avoided. In graduate controls courses, as a rule, the emphasis on control algorithm design is even greater, since the control applications studied are chosen intentionally to be more demanding in terms of performance in order to illustrate the need for more sophisticated control algorithms. As in undergraduate courses, the selection/design of the mechanical actuation system is usually not considered. Actuators and actuator selection are usually considered in classes on mechatronics, but the emphasis in these courses is on familiarizing students with multiple actuation technologies and quantitative understanding of the static and dynamic performance of these technologies. Issues such as rational procedures for selection of actuation technologies based on desired performance specifications are rarely addressed, and considerable weight is placed on conventional practice in performing the task. The more narrowly defined task of sizing actuation system components is usually structured so as to meet basic performance requirements such as maximum loads, accelerations or velocities appropriate for the application, rather than a more complete set of performance requirements characterizing an acceptable level of dynamic performance.

In the current educational context of interest then, to take an example, if we are considering the feedback control of slide position on a machine tool, *determination of* control system *performance requirements* based on machine tool use, and *selection* of the appropriate actuation technology, e.g. hydraulic power versus electrical power, based on the performance requirements, would be rarely addressed by the student. Instead, the actuator for the application would be specified, usually reflecting prevalent engineering practice, and would be considered as part of the plant being controlled. The emphasis in most cases would then be on modeling and analysis of the given plant's dynamic response, and synthesis of the appropriate control algorithm to achieve the specified closed loop performance. In mechatronics courses, selection of the actuation technology is considered as a task performed prior to and largely independently of the rest of the control system design. Sizing of the actuator itself based on performance specifications related to needed force/torque/power/motion is relatively straightforward once the actuator type has been chosen, and is usually treated adequately.

The focus of control design being on control algorithm design, and the perspective on actuator selection as a function undertaken prior to controller design, reflects current industrial practice where actuation system selection is viewed as part of the mechanical design of the system. The role of the control system designer is seen as beginning after mechanical design decisions have been made, and the task of the control system designer is seen as getting the best performance from the chosen design hardware, rather than helping select the actuation type and hardware as well to best match the desired system performance. The current emphasis on control algorithm design reflects appropriately the knowledge-intensive nature of the control algorithm design task, the largely manual performance of that task in current practice, and hence the need for practitioner and student competence in this skill.

The central premise of this paper is that control system design is performed best when specification of the desired performance of the control system, selection of the actuation technology and sizing of the actuator, and design of the control algorithm are carried out jointly, with full awareness of the interactions between different aspects of the overall design. Such an approach offers potentially greater returns in terms of achievable system performance, and the corresponding skill would be valued as a higher-level skill by employers of controls engineers. We argue here that education in the broader control system design function, involving performance requirements development for applications and mechanical actuation technology and actuator selection/design to best match the application performance requirements, will also help better develop the student's problem-solving and synthesis skills and appreciation for engineering applications. Especially in a context of evolving actuation technology and newer applications, such skills would be highly valued, as there is little by way of documented engineering practice to guide actuation technology selection and actuator design in such emerging areas. Consequently, developing the higher level synthesis skills inherent in the broader approach to control system design is a way for controls engineers to continue to provide value in complex design environments involving emerging technologies as well.

This perspective on control system design education may also be viewed as being responsive to some of the stated requirements for successful industry deployments of new control technology across multiple application domains identified in a recent state of the art review titled "The Impact of Control Technology - Overview, Success Stories, and Research Challenges" (Samad and Annaswamy, 2011), and published by the IEEE Control Systems Society. Quoting from this study: "Despite its maturity as a discipline, control engineering is often a technology that is considered only after the plant has been designed. The design of a plant such that it can be effectively controlled is still rare in many applications." Emerging needs identified include "co-design of plant, sensors, actuators, and control for desired closed-loop performance". The improved awareness on the part of the control engineer of the context within which control is to be exercised that would result from such an approach will also enable development of control algorithms more likely to lead to effective implementations. While the cited study stated the need as common across multiple application domains, we argue here specifically that broadening control system design education to include such co-design of the actuator and the plant, in the application domain of mechanical system control, is important for mechanical engineering students and would allow them to bring more value to their function as control engineers. We consider the status and capabilities of tools for such co-design in the following section.

Practices in specifying control system performance and capabilities of actuation technologies for control applications are reviewed in the following section. An existing unifying framework for representing actuation system capabilities and their selection for applications of interest that is being used with some success in recent years is then presented and assessed. Developments needed for an improved methodology for actuation system selection are enumerated in the final section, and constitute our goals for ongoing work in this area.

2   Practices in Performance Requirements Development and Assessment of Actuation Technologies

Development of detailed control systems performance requirements for specific applications is usually the province of application engineers, requires considerable awareness of all aspects of these applications, and is very much a context-dependent task that varies significantly from one application to another. Documentation resulting from such work is usually information proprietary to the organizations performing the work. As such, this function is best viewed as one that is appropriately learnt on the job and in the context of organization-specific practices. Consequently, the fact that there is little by way of open technical literature on methodologies for performing this function is only to be expected, and the fact that the topic is usually not part of academic curricula is not a shortcoming that we see the need to address here. It is useful however to see how capabilities of competing actuation technologies are assessed or represented in practice, and to see what implications such practices have for the education of control engineers.

Figure 1 represents the relative capabilities of two established actuation technologies, electrohydraulic (EH) and electromechanical (EM), in terms of the power level and speed of response of actuation, as seen by Moog, an aerospace control systems vendor (Maskrey and Thayer, 1978). The lower boundary represented the limits on the capabilities of electromechanical actuation at the time, whereas the upper boundary represented limits on the capabilities of electrohydraulic actuation. The aerospace control applications noted on the figure were amenable therefore primarily to electrohydraulic actuation at the time of the publication of the cited reference. The measures used to represent power level and speed of response were stated therefore in terms meaningful only for electrohydraulic actuation, more specifically, a servovalve or a pump (for hydrostatic drives) controlling fluid flow to an actuator which powered the load motion. Power level was presented in terms of horsepower corresponding to a 3000 psi pressure drop across the valve or pump, or in terms of flow rate in gpm for a valve pressure drop of 1000 psi. The speed of response was presented in terms of the frequency corresponding to which the servovalve frequency response had a 90° phase lag. It should be noted in particular that the axes are in log scale, with multiple decades along each axis.

The association of any named application on the figure with the corresponding power level - speed combination has more significance on how this application compares with another application, rather than the narrowly defined absolute region on the space the application occupies. So, for instance, missile fin position control usually requires much higher speeds of actuator response than aircraft primary flight control and usually involves much lower power levels. The ranges of response speeds and power levels of flight control systems can themselves be wide. While Figure 1 did present useful information on the relative capabilities of different actuation technologies and the relative actuation needs of a variety of applications, its principal utility in the manner presented was in relating applications to the capabilities of the electrohydraulic actuation hardware that would be required for these applications. The perspective was also one of informing the user of these systems, and hence the customer of the control systems vendor, of the appropriateness of the actuation systems for the applications noted. The reliance upon open loop specifications in terms such as flow rates and servovalve phase lag that are related only to electrohydraulic actuation limits the utility of the results in situations requiring evaluation of alternative actuation technologies. In fact, actuation needs of applications are really better stated in terms of closed loop performance specifications on the speed of response rather than open loop specifications, as such specifications can be independent of actuation technology. Such a specification on the actuation closed loop speed of response is used in Figure 2, also reflecting the same control systems vendor perspective (Thayer, 1988) and this representation is therefore more useful. The variety of actuation technologies considered is also broader, and includes electropneumatic (EP) actuation, and electropneumohydraulic (EPH) actuation. The representation serves the purpose of showing that, at the time of the publication, electrohydraulic actuation continued to have significantly more capability than electromechanical actuation for aerospace applications.

Figure 2 is more useful than Figure 1, but it shares a critical lack of transparency on how the limiting envelopes shown for any of the actuation technologies were determined. Since the limiting envelopes are really closed loop system-level characterizations, they depend on the characteristics of all of the system components as well as the control algorithms for the closed actuation loops. Since the component characteristics, and the analysis and design methods used to determine the closed loop control algorithms corresponding to the performance envelopes, are not specified by the hardware/systems vendor, there is a resulting lack of transparency. Taking valve controlled electrohydraulic actuation as an example, the closed loop actuation system characteristic (actuated load dynamics in Figure 2) depends on the characteristics of the servovalve, actuator (e.g. piston or fluid motor), sensor(s), motion converter if used, (e.g. ball screw), and the control algorithm used for closed loop control, and hence depends on a large number of variables related to the actuation that are not specified. Since transparency is important in educational settings, the issue of how best to represent system-level capabilities of any actuation technology meaningfully for a variety of applications, while maintaining transparency, needs to be resolved. We note, however, that the characterization of actuation technology capabilities at the system level, as in Figures 1 and 2, is important and should be retained in any representation of actuation technologies that is used for design support.

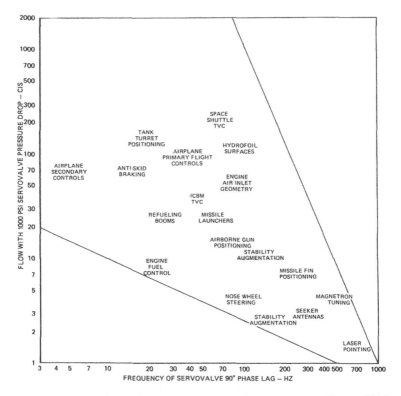

Fig. 1 Aerospace application performance requirements (Maskrey and Thayer, 1978)

Yet another consideration in representing the capabilities of actuation technologies is that different applications require emphasis on different measures of performance. While power levels and speeds of response are highlighted in Figures 1 and 2, applications may emphasize other aspects of performance, such as weight, size, cost, efficiency, duty cycle, reliability etc, and may also allow effective comparison of actuation technologies. Environmental considerations may emphasize other requirements such as temperature, vibration/shock limits, nuclear hardening, EMI etc. Business and support issue considerations such as service requirements, reusability, environmental impact, and ownership are important as well, though they may not all be amenable to quantification. It is important therefore that comparison of actuation technologies accommodate a variety of performance measures.

Other comparisons of different actuation technologies on a set of common performance measures have also been reported, either in the context of a specific application such as robotic actuation (Hollerbach et al., 1992), a specific type of actuation such as microactuation (Fukuda and Menz, 1998) or MEMS actuation (Bell et al., 2005), or by way of comparing one or more emerging actuation technologies with other actuation technologies (Kornbluh et al., 1998). Hollerbach et al. (1992) compare the performance of a variety of technologies for macrorobotic applications including established technologies such as hydraulic (electrohydraulic), electromagnetic (electromechanical), and pneumatic (electropneumatic), and emerging technologies such as piezoelectric, shape memory alloy (SMA), polymeric, and magnetostrictive actuation. The different actuation technologies are considered with a view to understanding the source of limitations on actuation technologies and industrial design and manufacturing practices, and comparisons are done both at the actuator level and at the actuation system level. Typical of the insights into actuation technologies are observations such as i) an electric motor's torque/mass ratio depends on electromagnetic design whereas its power/mass depends also on the limitations of the power electronics and ii) electric motor currents are limited by the motor's ability to dissipate the heat generated at the windings. Representative actuator level comparisons of the different technologies in terms of stress, strain, strain rate (S.R.) and mechanical efficiency (M.E.) are listed in Table 1, the numbers being derived from a combination of sources such as performance data from experimental prototypes, technological limitations, and behavior intended to be 'representative'.

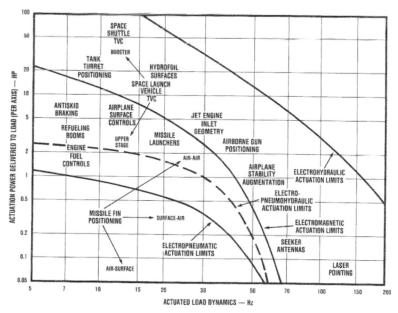

Fig. 2 Aerospace application performance requirements and actuation technology limits (Thayer, 1988)

As compared to the actuation technology comparisons in Figures 1 and 2, there is more transparency behind Table 1. Polymeric actuators have reasonable strain levels as compared to the established macroactuation technologies such as hydraulic, electromagnetic, and pneumatic, whereas shape memory alloys, piezoelectric and magnetostrictive actuation have comparatively good stress levels but lower or much lower strain levels. Shape memory alloy actuation is particularly poor in terms of its efficiency. While these comparisons do provide insight into the relative capabilities of actuation technologies, the comparisons do not extend very readily to more direct and general comparison of their usefulness for specific robot actuation tasks. Torque/mass and power/mass are listed in Table 2 for the different actuation technologies based upon commercial products or prototypes and represent system-level capabilities. Unfortunately, the manner of determination of these system level capabilities does not provide any transparency into the underlying factors and limits their utility in the educational settings we seek to support. Methodologies leading to the compilations of actuation capabilities in Table 1 and 2, and other similar compilations (Kornbluh et al., 1998), while serving well their intended use of comparing actuation technologies broadly for a single application, fall short of providing support for a broader framework for control system design.

Roadmaps for control technology developments are platforms allowing for the collective development of performance requirements for control systems in the application areas of interest. While this task is very much context-dependent, there is some transparency on how performance requirements are developed and more documentation of the underlying methodologies for identifying candidate actuation technologies. Moreover, roadmaps for technology developments perforce address emerging technologies along with established technologies as appropriate, and emphasize system-level capabilities so as to be able to gauge impact on the application area of interest. A recent such roadmap on actuators for gas turbine engines (Webster, 2009) by the Research and Technology Organization of the North American Treaty Organization (NATO) considered actuator requirements for gas path control within the engine in terms of airflow manipulation, flow switching, flow control, and geometry control. Given the mechanical nature of the actuation, actuation technologies were viewed in terms of their technical capabilities such as maximum force, energy density, stroke, response speed or repetition rate, input energy type, resolution/controllability support system requirements, and environment limitations, for example, temperature or pressure. Starting with generic actuation requirements for the gas path control functions noted above, capabilities of established and emerging mechanical actuation technologies were surveyed first, and we'll return later to the topic of how such comparison was performed here. Then, based on a consideration of more specific application-related requirements and developments, an actuation technology roadmap was developed. For each of the control applications considered, significant environmental conditions, requirements on actuator operation constituting a partial list of performance specifications, candidate actuator technologies, current technology readiness level (TRL) and projected timeframe for TRL 6, and actuation challenges were identified. Taking as an example compressor blade tip clearance control, specifications for actuation stroke, velocity, displacement resolution, and force levels, along with frequency of actuation were determined. Pressures and temperatures of operation were also noted. A variety of actuation technologies were identified as candidates, including pneumatic, hydraulic, electromagnetic or electromechanical, and piezoelectric, along with the challenges for each actuation technology. Weight was identified as a potential challenge for hydraulic and electromagnetic actuation, stiffness for pneumatic actuation, temperature and strain level for piezoelectric actuation, seals for hydraulic actuation, and effectiveness of control for electromagnetic and pneumatic actuation. The study also noted that cost was a concern for all of the actuation technologies for compressor clearance control as compared to the more technically challenging problem of turbine blade tip clearance control, as the performance benefits realizable from the former were lower. We argue that awareness of the broader issues involved in comparing actuation technologies in the manner described here would enable control engineers to play a more significant role in the mechanical design tasks related to actuation system design that precede control algorithm design. The goal of the broader control system design education that we advocate is to cultivate such awareness in the control engineer.

Table 1 Typical performance characteristics of actuators (Hollerbach et al., 1992)

Actuator	Stress (MPa)	Strain	S.R. (s^{-1})	M.E.
Electromagnetic	0.02	0.5	10	0.9
Hydraulic	20	0.5	2	0.8
Pneumatic	0.7	0.5	10	0.9
NiTi SMA	200	0.1	3	0.03
Polymeric	0.3	0.5	5	0.3
Piezoelectric	35	0.001	2	0.5
Magnetostrictive	10	0.002	2	0.8
Muscle	0.35	0.2	2	0.3

Table 2 Comparison of actuator characteristics (Hollerbach et al., 1992)

Actuator	Torque/mass	Power/mass
McGill/MIT EM Motor	15 N·m/kg	200 W/kg
Sarcos Dexterous Arm electro hydraulic rotary actuator	120 N·m/kg	600 W/kg
Utah/MIT Dexterous Hand electropneumatic servovalve	20 N·m/kg	200 W/kg
NiTi SMA (Hirose *et al.*, 1989)	1 N·m/kg	6 W/kg
PVA-PAA polymeric actuator (Caldwell, 1990)	17 N·m/kg	6 W/kg
Burleigh Instruments inchworm piezoelectric motor	3 N·m/kg	0.1 W/kg
Magnetoelastic (magnetostrictive) wave motor (Kiesewetter, 1988)	500 N·m/kg	5 W/kg
Human biceps muscle	20 N·m/kg	50 W/kg

3   Assessment of a Unifying Framework for Actuation System Selection

In evaluating the capabilities of different actuation technologies for gas turbine engines (Webster, 2009), the author relied upon a methodology for selection of mechanical actuators developed and reported on by Huber et al. (1997), Zupan et al. (2002), and Bell et al. (2005). We summarize that methodology here. A detailed list of actuator performance characteristics or measures is compiled and shown in Table 3 and used as the basis for actuator evaluation. Ranges of achievable performance characteristics are then estimated for different actuation technologies, based upon manufacturers' data and simple models of how actuator performance is limited fundamentally by basic phenomena such as resonance and thermal response. While tabulation of these ranges of performance characteristics is given by Huber et al. (1997) and is informative, graphical representations of pairs of these performance characteristics such as Figures 3 and 4 may be seen to be more effective visually.

Figure 3 shows actuation limits for different types of actuators in terms of bounds on actuation stress and actuation strain, these performance characteristics and others being defined in Table 3. Logarithmic scales are used on both axes in order to cover multiple decades of the performance characteristics. The boldfaced lines displayed are really the upper right corners of the corresponding actuation limits. Applications requiring high stroke would normally require actuators toward the right of the figure whereas those requiring high actuation stress levels would

Table 3 Definitions of actuator performance characteristics (Huber et al., 1997)

performance characteristic	definition
actuation stress (σ)	The applied force per unit cross-sectional area of an actuator.
maximum actuation stress (σ_{max})	The maximum value of actuation stress in a single stroke which produces maximum work output.
actuation strain (ϵ)	The nominal strain produced by an actuator; an actuator of initial length L extends to a total length of $(1 + \epsilon)L$.
maximum actuation strain (ϵ_{max})	The maximum value of actuation strain in a single stroke which produces maximum work output.
actuator density (ρ)	The ratio of mass to initial volume of an actuator. (We neglect the contribution to mass from power supplies, external fixtures and peripheral devices. For example, in the mass of a hydraulic cylinder, we include the working fluid and the cylinder, but neglect the compressor, servo-valve, cooling system and mounting fixtures.)
actuator modulus (E)	The ratio of a small increment in σ to the corresponding small increment in ϵ when the control signal to an actuator is held constant. (In general this differs from the measured modulus $d\sigma/d\epsilon$ which depends upon the control signal.)
volumetric power (p)	The mechanical power output per unit initial volume in sustainable cyclic operation.
efficiency (η)	The ratio of mechanical work output to energy input during a complete cycle in cyclic operation.
strain resolution (ϵ_{min})	The smallest step increment of ϵ (order of magnitude approximations are given).

use actuators toward the top of the figure. Applications requiring high energy densities would use actuators toward the upper right corners of the figure as they correspond to higher values of the product $\sigma\epsilon$, constant contours for which are displayed as discontinuous lines with a slope of -1. The discontinuous lines with a slope of +1 represent contours of constant σ/ϵ, a modulus or stiffness-like quantity, the latter being denoted by E, Table 3. Actuators with their upper right boundaries in regions of high σ/ϵ also have high modulus values E, and are better suited for open loop applications, whereas those in regions of low σ/ϵ have low modulus values and require closed loop control in order to achieve accuracy. Inferences based on Figure 3 such as those stated above are valid if actuator sizes are comparable. Figure 4 shows plots of actuation limits in terms of actuation power volumetric density and frequency of actuation, the latter being an open loop measure in this case, and again underscoring the advantages of hydraulic actuation over pneumatic and electromechanical actuation as in Figure 2. There are important differences between Figures 2 and 4, and we note these differences when we later address the question of the utility of these representations for making actuator design decisions as part of the control system design process.

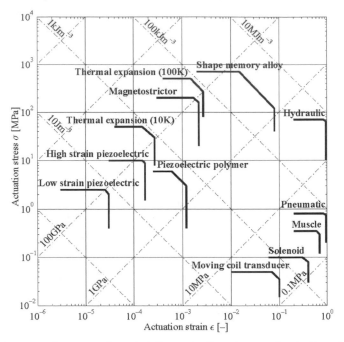

Fig. 3 Actuator performance characteristics: actuation stress and actuation strain (Huber et al., 1997)

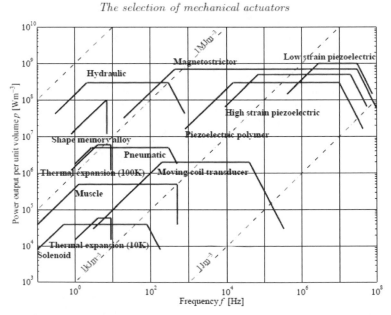

Fig. 4 Actuator performance characteristics: power density and frequency range (Huber et al., 1997)

The utility of the representation of mechanical actuator performance developed by Huber et al. (1997) is that it allows broad comparison of families of actuators based on a variety of performance characteristics, and identification of feasible actuation methods early in the design stage. The framework and associated procedure for actuator selection as it stands currently is however limited in its ability to effectively support design for closed loop control. These limitations, and the enhancements needed to overcome them, are the following. First, closed loop system capabilities depend on more than just actuator performance characteristics. For example, the characteristics of other actuation system components, such as power transmission components and the control algorithm, are relevant as well in determining closed loop system performance. Therefore, the actuator comparison must be extended to performance characteristics of actuation systems so that system level capabilities, preferably closed loop, may be captured in the comparison. Second, we need to classify mechanical actuation applications in more generic terms, and to capture more application requirements in such classification, so that application requirements may map better and more broadly to actuation system performance characteristics. Third, the compilation of performance characteristics for actuators and actuation systems must be more comprehensive in order to incorporate recent and continuing developments in actuation technology. It is also important that better linkages be established to underlying technological limitations so that the compilations have more lasting value. The utility of the resulting enhanced framework would seem to be better suited to evaluation of fewer candidate actuation technologies at a time, suggesting that the enhanced framework might be appropriate for a later stage of the design process after early design decisions narrowing the choice of actuation technologies have been made based on the methodology proposed by Huber et al. (1997) or others of a similar nature. The control system design education context of primary interest here is compatible with such a later design stage and would require the enhancements listed here. We elaborate on these enhancements below.

4   Toward an Improved Methodology for Actuation System Selection and Design

The need for accommodating system level considerations in selecting actuation technologies for different applications, and the limitations of looking only at actuator characteristics for this purpose, has also been noted by Huber et al. (1997) and by Webster (2009), and is the first of the enhancements proposed here. For example, as noted in connection with the definition of 'actuator density' in Table 3, the term excludes system components other than the actuator, an omission that would be significant for weight sensitive applications when considering hydraulic actuation since such actuation requires components other than the actuator, such as the servovalve and hydraulic power supply. The latter is also a resource usually shared by multiple actuators and presumably best dealt with qualitatively in most cases. More importantly, some performance characteristics such as speed of response and resolution are more meaningful for actuation selection when they refer to closed loop system-level characteristics rather than actuator or open loop characteristics. While this poses challenges since control algorithm design decisions are yet to be made, the limitation caused by omission of such system-level considerations has been noted by Webster (2009) in connection with the actuator roadmap for gas turbines, and by Granosik and Borenstein (2005) in their evaluation of actuators for a serpentine robot. The latter reference also augments the actuator characteristics in Figures 3 to include data for pneumatic bellows and electric motors with leadscrew transmissions to convert rotary motion to linear motion, two candidate actuators of interest for the application. Figure 5 shows the result. Actuation system compliance, which is a closed loop system-level performance characteristic, is a significant criterion for the application, and its implications for choosing between electric and pneumatic actuation for this application are dealt with entirely qualitatively in the cited reference, with the final decision to choose pneumatic bellows for actuation being based on the natural compliance of pneumatic actuators.

While this approach may have been appropriate for this application, there are many applications where it is important to be more quantitative in evaluating alternate actuation systems and it is therefore important to formulate a methodology for actuation system selection that can accommodate closed loop system-level performance characteristics more quantitatively. For instance, many of the aerospace applications in Figure 2 were amenable to more than one actuation system solution more than two decades ago as indicated by their

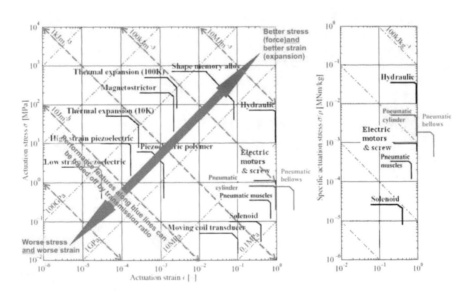

Fig. 5 Actuation system evaluation for serpentine robot (Granosik and Borenstein, 2005)

relationship to the corresponding actuation technology boundaries, and many more are expected to be so today as a result of developments in actuation technologies. The contributions of components to system-level performance in applications of interest therefore need to be accommodated in a manner that supports co-design of the actuation system and control algorithm. It is well known for electrohydraulic servomechanisms, for instance, that the hydraulic resonance resulting from hydraulic fluid compliance and the mass or mass moment of inertia of the load is a limiting factor on the performance of closed loop control systems (Merritt, 1967). In another application domain, machine tool slide motion control, it is similarly known that the axial stiffness of the ball screw transmission system limits the performance of slide motion control systems (Younkin, 2003). It would be immensely valuable for the effectiveness of proposed control solutions therefore, if these limitations on control system performance could be captured in generic form and considered as part of the actuation system selection process. Clearly, doing so would require more considered evaluation of the mechanical actuation context within which the control system is to be function. In an educational setting, the student exposed to such design practices would have a broader and hence more beneficial exposure to control system design.

The second enhancement noted above, and one needed to better support actuation system selection for applications, is to classify mechanical actuation applications in more generic terms so that application requirements may be mapped better to actuation system performance characteristics. Huber et al. (1997) consider a few highly simplified actuation applications amenable to analytical procedures for actuator selection, an example being the selection of an actuator for cyclic oscillation of a mass m at frequency f and motion amplitude X while minimizing actuator volume. The same research group developed actuator selection procedures (Zupan et al., 2002) that used a data base of actuators and their performance characteristics such as those in Table 3, to select a set of feasible actuators based upon threshold values of performance criteria such as actuator weight, actuation frequency, force, stroke, or simple combinations of these performance criteria. The actuator selection from this reduced set of actuators was then performed using analytical procedures to optimize an additional criterion. Both sets of examples applications are highly simplified and not very representative of broad variety of actuation applications of sufficient engineering interest. Other variations on the idea of investing simple contours on the space of actuator performance characteristics with significance for classes of applications were also noted by Huber et al. (1997). For instance, contours of straight lines with slope of -1 on Figure 3 may be seen to correspond to regions of constant work per unit volume, and consequently actuators with their boundary regions in the upper right corner of the figure are appropriate for energy-intensive actuation tasks requiring compactness of actuators. There is considerable scope for improving the extent of mapping of application requirements to actuation system performance characteristics.

We propose to start with consideration of a limited set of actuation applications instead and classify them broadly and generically in terms of their requirements.

Our expectation is that these classifications of application characteristics would either identify new actuator performance characteristics of interest or, alternatively, different combinations of actuator performance characteristics that would be significant for different application categories. To cite a simple example that is known already, applications where inertial loads are dominant would favor actuators with high values of force/mass or torque/inertia as for robotic manipulators. Applications where, in addition, there is a volumetric constraint on the packaging of the actuator, would benefit from actuators with high values of torque/inertia per unit volume. It is important that such a classification effort begin with a consideration of application requirements. In doing this task, we plan to work with application engineers with experience in developing performance specifications in the selected application domains. We had stated earlier that the development of detailed performance requirements for specific applications is a context-dependent task that is currently left to application development engineers. We propose here instead to classify mechanical actuation applications in terms generic enough for the educational context, and discriminating enough in their requirements on actuation to allow the formulation of procedures for actuation system selection and control system design.

The third and final enhancement needed to better support actuation system selection for applications is to enlarge the compilations of performance characteristics by the references noted above, to include more classes of actuators and actuation systems, and to better link their limitations to underlying technological limitations. An example of the proposed extension to other actuators is the addition, by Granosik and Borenstein (2005), of performance characteristics for motor - leadscrew transmissions and pneumatic bellows to the data compiled by Huber et al. (1997) as shown in Figure 5. Motor-leadscrew transmissions, and actuators using linkages to achieve mechanical advantage, need to be included in the data base of actuators because of their prevalence in practice. While compiling larger data bases of available actuators is an appropriate way to do this, it is also important, where possible, to identify technological limitations on actuator performance. By doing so, the capabilities of the corresponding type of actuation may be explicitly bounded with less effort as compared to relying upon an extensive data base to implicitly represent the bound. For example, Huber et al. (1997) note that the operating frequency of piezoelectric and magnetostrictive actuators is limited by the lowest structural resonance frequency. For shape memory alloy and thermal expansion actuators, both of which depend upon temperature change for actuation, the operating frequency is limited by convective heat transfer coefficients. In both cases, the smallest available size of commercially available actuators has been used to determine the maximum operating frequency. For hydraulic and pneumatic cylinders, the maximum sliding speeds that can be tolerated by the seals are limited and, when combined with lower limits on actuator lengths, results in upper limits on the power per unit volume. Upper limits are also imposed upon the pressure in hydraulic and pneumatic cylinders based on practice and considerations of safe high pressure containment. Hollerbach et al. (1992) have noted that motor torque/mass ratios are limited by electromagnetic design limits such as the maximum magnetic flux density and by heat dissipation capabilities of motors which in turn limit motor currents, while motor power/mass ratios are limited further by the volt-amp rating of the power electronics. The same may be expected to be true of other forms of electromagnetic actuation such as solenoids and moving coil actuators (Gomis-Bellmunt et al., 2007).

Recent developments in macroactuator technology have included hybrid forms of actuation such as electrohydrostatic actuators (Frischmeier, 1997) for aircraft flight control surface actuation that are competitive against more conventional forms of actuation. These actuators rely upon electric motors near the control surfaces powering hydraulic motors that in turn power hydraulic cylinders moving the control surfaces. Closed hydraulic circuits at the control surfaces are used for actuation. The reliance upon aircraft-wide electric power transmission (Power-by-Wire) in such systems eliminates the aircraft-wide hydraulic power transmission employed by systems that rely upon electrically controlled (Fly-by-Wire) servovalve-cylinder combinations at the control surfaces, and hence reduces system weight and complexity. One instance of such an actuator consists of a DC motor powering a fixed displacement pump - cylinder combination, the motor speed being varied under closed loop control of the flight control surface. Another consists of a variable displacement pump - cylinder combination at the control surface and driven by a constant speed AC motor, the pump displacement being varied under electrohydraulic closed loop control. Characterization of the performance of such hybrid actuators in terms comparable to more conventional forms of actuation will therefore be of considerable use in control system design for this class of applications, and is included as part of the enhanced compilation of actuator performance proposed here.

In order to demonstrate the benefits of our approach, we propose to develop actuation system selection procedure enhancements and procedures for co-design of actuators and control algorithms for two classes of applications that satisfy the following criteria: they should be established and prevalent enough for a knowledge base of performance specifications as well as multiple commercially supported candidate actuation technologies to be available, and they should be sufficiently demanding of performance for closed loop system solutions to be necessary. We consider aircraft flight control surface actuation (Gee, 1984: Ravenscraft, 2000) and machine tool and robot control (Srinivasan and Tsao, 1997) as two classes of applications that satisfy the criteria. We expect that detailed performance requirements for specific applications are probably not well-documented in the open technical literature, but that the relevant knowledge base can be compiled from practicing controls engineers involved in application development. The resulting methodologies have the potential to broaden control system design education in the manner envisaged, and to enhance the value of our graduates in control engineering tasks in these established application domains. We expect also that, once the benefits of the proposed approach are demonstrated here, the methodologies developed here may be applied to broader classes of applications involving newer forms of actuation, such as microactuation (Fukuda and Menz, 1998) and MEMS actuation (Bell et al., 2005). Both these categories of applications lack the accumulated knowledge base resulting from established practices in industry, and offer the potential for future industrial practices to benefit greatly from the methodologies developed here.

References

1. Bell, D.J., Lu, T.J., Fleck, N.A., Spearing, S.M.: MEMS actuators and sensors: observations on their performance and selection for purpose. Journal of Micromechanics and Microengineering 15, 153–164 (2005)
2. Frischmeier, S.: Electrohydrostatic Actuators for Aircraft Primary Flight Control – Types, Modelling, and Evaluation. Technical Report, Technical University of Hamburg (1997)
3. Fukuda, T., Menz, W.: Micro Mechanical Systems, Principles and Technology. In: Handbook of Sensors and Actuators, vol. 6. Elsevier, Amsterdam (1998)
4. Gee, B.: The Implications of the Control Surface Actuation System on Flight Control System Design. In: Billings, S.A., Gray, J.O., Owens, D.H. (eds.) Nonlinear System Design. IEE Control Engineering Series, vol. 25, ch. 10. Peter Peregrinus Ltd., London (1984)
5. Gomis-Bellmunt, O., Galceran-Arellano, S., Sudria-Andreu, A., Montesinos-Miracle, D., Campanile, L.F.: Linear Electromagnetic Actuator Modeling for Optimization of Mechatronic and Adaptronic Systems. Mechatronics 17, 153–163 (2007)
6. Granosik, G., Borenstein, J.: Pneumatic actuators for serpentine robot. In: 8th International Conference on Walking and Climbing Robots, London, U.K., pp. 719–726 (September 2005)
7. Hollerbach, J.M., Hunter, I.W., Ballantyne, J.: Robotics Review 2 - A Comparative Analysis of Actuator Technologies for Robotics, pp. 299–342. MIT Press (1992)
8. Huber, J.E., Fleck, N.A., Ashby, M.F.: The Selection of Mechanical Actuators Based on Performance Indices. Proceedings of the Royal Society of London, A 453, 2185–2205
9. Kornbluh, R., Pelrine, R., Eckerle, J., Joseph, J.: Electrostrictive polymer artificial muscle actuators. In: Proceedings of the IEEE International Conference on Robotics & Automation, Leuven, Belgium, pp. 2147–2154 (May 1998)
10. Merritt, H.E.: Hydraulic Control Systems. John Wiley & Sons Inc., New York (1967)
11. Ravenscroft, S.: Actuation Systems. In: Pratt, R.W. (ed.) Flight Control Systems, Practical Issues in Design and Implementation. IEE Control Engineering Series, vol. 57, ch. 3. IEE, UK, AIAA, USA
12. Samad, T., Annaswamy, A.: The Impact of Control Technology - Overview, Success Stories, and Research Challenges. IEEE Control Systems Society (2011)
13. Srinivasan, K., Tsao, T.-C.: Machine Tool Feed Drives and Their Control - A Survey of the State of the Art. 75th Anniversary Issue of the ASME Journal of Manufacturing Science and Engineering 119, 743–748 (1997)

14. Thayer, W.J.: Electropneumatic Servoactuation – An Alternative to Hydraulics for Some Low Power Applications. Technical Bulletin 151. Moog Inc., East Aurora (1988)
15. Webster, J.: Actuator Requirements and Roadmaps. In: More Intelligent Gas Turbine Engines, ch. 7, RTO Technical Report TR-AVT-128, North Atlantic Treaty Organization, Neuilly-sur-Seine Cedex, France (2009)
16. Younkin, G.W.: Industrial Servo Control Systems: Fundamentals and Applications, 2nd edn. Marcel Dekker Inc., New York (2003)
17. Zupan, M., Ashby, M.F., Fleck, N.A.: Actuator Classification and Selection – The Development of a Database. Advanced Engineering Materials 4(12), 933–940 (2002)

27 Biologically Oriented Subjects in European Master on Advanced Robotics – Sharing the Teaching Experience

Teresa Zielinska and Krzysztof Kedzior

Abstract. We share our experience gained during realization of the world-wide European Master on Advanced Robotics - EMARO (Erasmus Mundus program coordinated by Ecole Centrale de Nantes - France). Besides of classic subjects like robot dynamics and kinematics, control theory, artificial intelligence or computer vision, in EMARO we teach the biomechanics and bio-robotics. Our aim is to supply the students with complimentary knowledge stimulating their creativity for designing and prototyping non-conventional robots. Gained knowledge from the area of biomechanics allows the EMARO graduated to take a job not only in robotics but also in biomedical or biomechanical fields. Five year experience with two years EMARO program confirmed that our graduated are not only continuing their carrier in robotics but many of them are working in biomedical centers designing active prosthesis, testing the human gait properties and so on. There are also EMARO graduated developing walking machines, designing some novel constructions, or working in area of occupational biomechanics with ergonomics, focusing on human - machine interactions.

1 Introduction

Analysis of educational offers around the world shows that there are not many master programs devoted to robotics, moreover the biorobotics is often taught through laboratory projects and not by regular lectures. The topics are here focusing on selected problems like swimming robots, walking machines or flying robots (i.e. see [1]). According to our knowledge there is no biorobotic textbook offering general and comprehensive fundamentals, availabke books are introducing selected projects

Teresa Zielinska · Krzysztof Kedzior
Faculty of Power and Aeronautical Engineering Warsaw University of Technology,
Warsaw, Poland
e-mail: {teresaz,kkedzior}@meil.pw.edu.pl

V. Kumar et al. (Eds.): *Adv. in Mech., Rob. & Des. Educ. & Res.*, MMS 14, pp. 349–354.
DOI: 10.1007/978-3-319-00398-6_27   © Springer International Publishing Switzerland 2013

or problems – i.e. [4] (robotic fish, and snake, study of human blood cells and brain). The book [2] published in 1989 by S.M.Song and K.Waldron can be considered as one of the first textbooks in area of walking machines explaining fundamental knowledge about the design, control and motion planning of statically stable walking machines. This knowledge is still actual and is used by many students when making prototyping and creating the control systems for simple multilegged machines. Such kind of activities excellently validate the gained knowledge and develop engineering skills. Biorobotics course is located in the third semester of four-semesters EMARO program. Concurrently to this subject students attend Biomechanics. The Biorobotics module consists of 20 lecturing hours and 20 hours devoted to project. The lecture expands students knowledge concerning the biological locomotion mechanisms, and ways of its utilization in the world of technology. The students learn how to develop a simple walking machine. The following problems are discussed: – classification of on-land animals from the point of view of locomotion, – basic features of animal gaits, – biological fundamentals of motion control, – biologically inspired walking machines and its design solutions focusing on leg structures. In this part we apply our study results on animal motion properties [3], we explain the functions and structures of biological Central Pattern Generators (CPG) and refer to our results on CPG based motion generation in bipeds [5]. The gait synthesis methods together with gait diagrams, and generation of leg-end trajectories are explained. Different structures of control systems including hardware and software are also summarized. We use our experience gained during development of different walking machines. Some of our works in this field were preformed in cooperation with K.Waldron team from Ohio State University within Maria Curie-Sklodowska Funds. Postural stabilization methods in animals and walking machines, reaction forces in biology and force control methods applied in walking machines are also explained to the students. They learn basic principles of aerodynamics and hydrodynamics for the purpose of flying and swimming robots design. Aim of our lecture is to stimulate the creativity concering the robots deveopment with increased or dedicated (to specific environment) mobility. Students posses the knowledge concerning the motion principles, sensing and motion control in biology, they know how the animals are adapted to their living conditions. We expect that this knowledge will be applied to search for proper mechanical structure. During the project students must elaborate the concept of biologically inspired robot being able to perform some selected task in some specific environment.

2   First Steps towards Biorobotics

Before we started, considered as prestigious, international EMARO program we included the elements of biorobotics and biomechanics in regular study program offered to local students. Fig.1 a) – top shows six-legged walking machine Mentor which was designed and built together with control system during undergraduate studies as the diploma project (2005-2006). This picture is showing the machine moving by tripod gait, which is the fastest gait observed in the insects

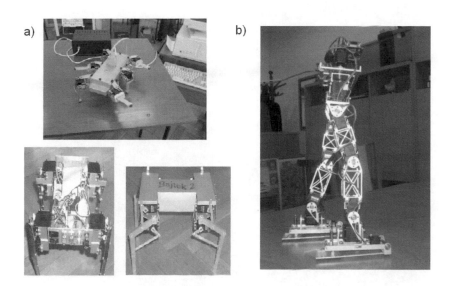

Fig. 1 Robots developed by students: six-legged and four-legged – a), biped – b)

world. Four-legged machine Bajtek (Fig.1 a) – bottom) was also developed for undergraduate   and later, as the continuation, for graduate diploma work (2005-2008). The legs of this device are inspired by pantograph structure applied in OSU hexapod [2]. In Bajtek student implemented quadruped crawl which is the slowest gait observed during statically stable walk of four-legged animals.

Our research on properties of human locomotion resulted in the development of small biped with the leg dimensions proportional to the human legs - Fig.1 – b). This prototype together with its control system was made by undergraduate student, master student developed gait generation method using genetic algorithm [5]. A human gait pattern was here applied as the reference signal when developing the formal model of Central Pattern Generator.

3   Selection of Student Projects

During Biorobotics course, after gaining the knowledge about animal motion principles with neuro-biological fundamentals of locomotion and with overview of some biologically inspired robots, students select the robot task and propose the mechanical structure. Proposition is described and presented for group critical review during seminar. The most promising proposition will end with technical design done under supervision of a specialist in mechanical design. This proposition is next prototyped using 3D printing technique. Bellow is presented selection of student works [6]. Fig.2 a) shows the fruit picking robot with the design inspired by giraffe. Robot has four supporting legs, fruit container - trunk, and telescopic pipe - neck (prismatic arm) delivering cut by automatic scissor fruit to the container. Robot illustrated in

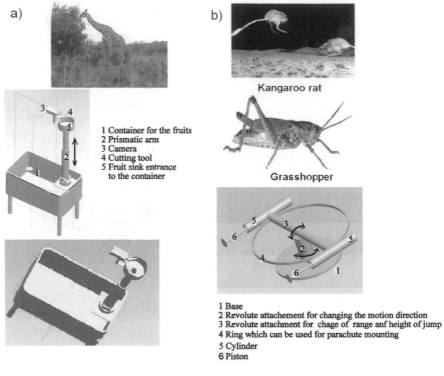

a)

b)

Kangaroo rat

1 Container for the fruits
2 Prismatic arm
3 Camera
4 Cutting tool
5 Fruit sink entrance
  to the container

Grasshopper

1 Base
2 Revolute attachement for changing the motion direction
3 Revolute attachment for chage of range anf height of jump
4 Ring which can be used for parachute mounting
5 Cylinder
6 Piston

Fig. 2 Inspired by giraffe fruit picking robot – a), hopping robot inspired by kangaroo rat or grasshopper – b)

Fig.2 b) was inspired by grasshopper and kangaroo rat. It is small, lightweight robot dedicated for exploration. Upper ring (3) orientation can by modified (moving it up or down) what influences the height and range of jump, by side rotation of the ring (2) the jump direction is modified. Heavier base (1) makes the stable landing support. The motion is obtained by fast release of the pistons (6) mounted in the cylinders (5) those producing the jet impulse. With those two listed above projects students referred to body build and locomotion methods of vertebrates.

Robot inspired by scorpion (Fig.3) was proposed for search of the of natural disaster area. Robot will be able to dig in the rubble and explore underground passages. Trunk bending offers better mobility comparing to robots with one stiff trunk. This project refers to the knowledge about invertebrates body build and motion principles

Interesting worm-like robot was proposed for exploration of rocks and caves (cave climber). Robot consists of several expandable segments connected by revolute joints. Both ends of the robots are equipped with heads containing batteries and control equipment. Four cameras are used for environment recognition – Fig.4. Body shape adaptation to the surface is obtained due to the revolute joints applied between segments. The grasp (fixing to surface) is made by spikes (or hooks) located by both ends of the body. The spikes are hidden during crawling motion (when the

Fig. 3 Inspired by scorpion robot for rubble exploration

segment expands) and its releasing is obtained by ejection spring, thuse affixing the body. The segments shorthening is obtained due to the electromagnetc forces produced by electromagnet. Robot applies peristaltic motion expanding and contracting its segments like the earthworm.

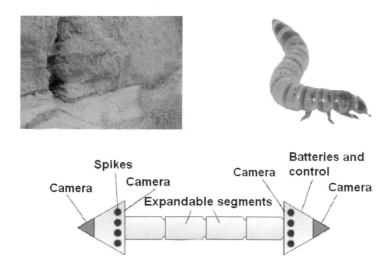

Fig. 4 Inspired by earthworm cave climbing robot

4 Conclusion

Study of biological patterns for the purpose of technical world does not mean that the best design ideas can be obtained by copying and imitating. Biology through natural selection delivers the solutions which are just satisfactory in current living conditions. Biological patterns shall be used as the suggestion for the search of effective technical design, they can be applied, or modified or even rejected. Moreover the biological world which was an inspiration for the design of diverse automatons, and in consequence the robots, now stimulates the development of many other aspects of engineering. *Animats* (short for artificial animals) and *biorobots* (biologically inspired robots) have been created and it is just the question of time when they will assist us in our everyday life. This will be achieved thanks to the work of many generations of invertors, engineers and scientists, K.Waldron and his team had made here also a significant contribution.

References

1. Metta, G., Nori, F., Pattacini, U., Fumagalli, M.: Anthropomorphic robotics website A course for the bioengineering curricula,
   `http://www.liralab.it/teaching/ROBOTICA/`
2. Song, S.-M., Waldron, K.J.: Machines that Walk. The MIT Press (1989)
3. Zielinska, T.: Biological Aspects of Locomotion. In: Pfeiffer, F., Zielinska, T. (eds.) Walking: Biological and Technological Aspects. CISM Courses and Lectures, vol. 467, pp. 1–30. Springer (2004)
4. Liu, Y., Sun, D.: Biologically Inspired Robotics. Taylor and Francis Group (2012)
5. Zielinska, T., Chew, C.-M., Kryczka, P., Jargio, T.: Robot gait synthesis using the scheme of human motion skills development. Mechanism and Machine Theory 44(3), 541–558 (2009)
6. Zielinska, T. (supervisor): Biorobotics projects by students. Warsaw University of Technology (November 2010)

28 From the Tool Room to the Class Room

Blaine W. Lilly

Abstract. In this brief memoir, the author re–visits his undergraduate and gradu-ate years in the Department of Mechanical Engineering at Ohio State University, as well as his years as a tool and die maker apprentice at the General Motors Fish-er Body plant in Columbus, Ohio. The author's experience as an apprentice and as a young engineer confronted by the task of building a trotting quadruped have had a clear effect on his later work in helping to create a totally new course for second–year students in mechanical engineering.

1   Early Days

My work with Ken Waldron began while I was an undergraduate student in Me-chanical Engineering at Ohio State University in 1982. I returned to Ohio State to pursue an engineering degree while working forty hours a week as a tool and die maker apprentice at the Columbus General Motors Fisher Body plant on West Broad Street. After receiving a B.A. in English from OSU in 1971, I briefly pur-sued an M.A. in English at the University of Colorado before dropping out and spending the next three years thoroughly enjoying myself (and wasting precious time) in Boulder during the Seventies.

I returned to Columbus in late 1976, and started working on the production line at Fisher Body on April 4, 1977. I realized quickly that the only part of factory life that appealed to me was the skilled trades, specifically the tool room where my father and cousin worked as die makers, and so began the long process of becom-ing a tool and die maker in late 1977. Neither the company nor the union particu-larly wanted to see me in the apprenticeship program: the company considered

Blaine W. Lilly
Department of Mechanical and Aerospace Engineering
The Ohio State University
Columbus, OH
e-mail: lilly.2@osu.edu

V. Kumar et al. (Eds.): *Adv. in Mech., Rob. & Des. Educ. & Res.*, MMS 14, pp. 355–370.
DOI: 10.1007/978-3-319-00398-6_28   © Springer International Publishing Switzerland 2013

me to be too old at 28, and the UAW local had no great love for my father or his brother (who was a general foreman in the tool room). The upshot was that I was required to jump through several extra hoops before I was accepted into the program. Although I had completed a B.A. degree from Ohio State, I was told to take the GED (high school equivalency) exam, which I did. Then I was told that I needed shop training, so I went downtown and enrolled in a machine shop class at Columbus State. Finally, as the union local was running out of reasons to keep me out of the program, I was told that I really needed additional math classes, since I had had no math since leaving high school. This led me to enroll at Ohio State in the summer of 1978. I believe I'm most likely the only member of the OSU engineering faculty who began his academic career in Math 102!

To my surprise, I found that I enjoyed being back in the classroom, and after completing Math 102 I jumped into pre–calculus, and then into the first calculus sequence. Looking around for another class to take, I chose Chemistry 121, and found that I really enjoyed it as well. At about this time I was finally allowed to enter the apprenticeship program, which meant that I could count on working second shift, from 3:30 PM to midnight, forty hours a week, for the next several years. This was great news, since by then I had made the decision to pursue an engineering degree. For the next several years I attended classes every morning, left campus at 2 PM, changed my clothes along with my personality, and showed up in the tool room at 3:30; leaving the plant at midnight, I went home, read the morning paper, and went to bed, up again in five hours to head to OSU.

For the next two years, I worked my way steadily through the prerequisites for entering the mechanical engineering department. During this time I had essentially no contact with the College of Engineering at all, as I was enrolled in Continuing Education, so I simply got a copy of "Book Nine" and started plowing through courses one after another. My only 'advice' came from a counselor in the College of Engineering who, when I said that I wasn't sure whether to go for electrical or mechanical engineering, replied that it didn't really matter which major I chose, since I was very unlikely to finish either one! This encounter just increased my determination, and by 1981 I was enrolled in the mechanical engineering major and well on my way to my first engineering degree, which took nineteen quarters to complete.

I've often thought that I was extremely lucky to have had the opportunity to work in the tool room at Fisher Body while taking engineering classes at OSU. In those years I was surrounded by gifted teachers, both in school and on the job. The apprenticeship gave me the opportunity to learn to work with my hands, and that experience has been very important to my career. In my years at Ohio State, I believe I developed a reputation as someone to go to for advice on design and fabrication. For myself, I learned that not all knowledge can be put into words, let alone a computer program. Humans learn through their hands just as much as they learn with their eyes and ears. Having the opportunity to become a precision worker in metals has been invaluable to me.

Another lasting lesson I took away from my experience in the tool room was an appreciation for how effective the apprenticeship method is for learning a skill. For forty hours every week I was working closely with highly skilled tool and die makers, most of whom were dedicated to teaching their craft. Of course it's impossible to exactly replicate the interaction between apprentice and master craftsman at the university, but as I'll explain below, the apprenticeship experience has been a very strong influence in my own development as a teacher of engineering. One goal that I've tried to achieve in the new course that I describe below is to offer our students, as closely and as often as possible, the one–on–one transmission of skill and knowledge that I experienced as an apprentice.

My first real encounter with the Mechanical Engineering department was through Ernie Doebelin's second control systems course, which in those days was ME 382. I well remember my first conversation with Ernie – I had studied hard for the first midterm, but only managed to pull a C+. I went to see Ernie in his office on the ground floor of Robinson Lab (next door to what would become the first "Walker Project" office), and told him that I was thinking of dropping out of the major, since I clearly wasn't capable of doing the work. Ernie quickly informed me that my C+ was one of the highest grades in the class, and under no circumstances should I consider leaving! I soon learned that Professor Doebelin's grading alphabet typically started with the letter B and proceeded from there. Years later Ernie would always go out of his way to mentor me when I became a member of the engineering faculty. After Ernie passed away in late 2010, I heard that Professor Mike Moran had gone to visit him in hospice during his last week, and found him working on the proofs of his final book, a professor to the very end.

I entered the department just a year or so after two young assistant professors by the names of Kinzel and Srinivasan joined the faculty. Gary Kinzel, as always, was quick to spot the student who seemed to be having trouble fitting in: in this case it was the older guy at the back of the room who disappeared every afternoon. It helped that I loved his kinematics course, ME 553, and did well in it. Gary quickly became the mentor I had been looking for, and in fact he's continued to play that role for me up to the present day. I've always believed that the fact that we're both 'non–observant Appalachians' played a role in Gary's willingness to guide me through my very unconventional career, but in truth Gary has always been the go–to guy for students who don't quite fit the mold for whatever reason.

When I became unemployed during the recession of 1982, Gary urged me to apply for a position with the ME department as a student machinist, and interceded (I believe) with the department to ensure that I was hired. In any event, I soon found myself working with Vince Vohnout in the initial stages of the Adaptive Suspension Vehicle Project. Of course, as I quickly learned, in those days no one worked 'with' Vince, you worked 'for' Vince. He was (as always) quite demanding, but apparently I had learned enough at my apprenticeship to keep him happy, and I stayed on as a machinist while finally finishing the B.S.M.E. in March of 1983. It was through this work on the early stages of the ASV that I first became acquainted with Ken, who had joined the department in 1980.

2   Graduate Work: The Quadruped Project

I would not be called back to work at General Motors until after the recession ended in 1984; following graduation I immediately began to work on my M.S. in Mechanical Engineering with Ken Waldron as my thesis advisor. I had originally assumed that Gary Kinzel would be my advisor, but apparently Ken and Gary had noticed that I was at least an adequate machinist, and had other plans. The task I was given as my thesis topic seemed simple enough at the time, but as it turned out, took me almost three full years to complete – much to my advisor's consternation!

Waldron's charge was very clear: design and build a four–legged, completely mechanical device that would trot. The goal of the thesis was to prove that a machine could be made to run, not walk, using a fixed gait without a high level of control. Ken carefully explained to me exactly what he meant by *trot*: a well–defined gait with support provided by alternating diagonal pairs of feet, with each support phase separated by a phase during which no feet would be on the ground. He also made it clear that he wanted the machine to function with the absolute lowest level of control possible. After that, it was up to me to make it work.

In retrospect, I was both flattered and overwhelmed by the challenge I'd been handed. Although I was ten years older than my fellow grad students, I still considered myself (accurately) to be a rank beginner as an engineer. After a few months of reading Alexander and McMahon, pondering cats with severed spinal cords that were capable of running on treadmills, and many long discussions with Vince Vohnout, John Gardner, and Tom Ward, I came up with a design that looked very much like a small card table with legs that appeared to be collapsing. My design used the same basic pantograph leg that the ASV team had designed, only at a much smaller scale. In this case, each leg was driven by a pair of cams, one to control vertical motion, the other horizontal. All four legs used exactly the same set of cam profiles, with diagonal pairs timed to move in unison. All eight cams were driven through chain drives by a single electric motor suspended centrally beneath the frame, which was powered through an umbilical cord that would trail behind the machine while in motion. I took great care to ensure that the machine was balanced both right and left, and fore and aft.

Ken and his team approved the design without many changes, as I recall, and then began the long period of fabrication. Gary Gardner, who coincidentally had been an apprentice to my father at Fisher Body, was in charge of the ASV machine shop and allowed me to use the shop after hours. In those days the Rockwell Aviation plant was still in operation out at Port Columbus airport, and I tagged along when Vince or one of his crew would go out to the Rockwell scrap yard looking for cheap aluminum for the ASV. So it happened that I decided to build my machine using 7075 aluminum, not knowing what that entailed – I'd had no experience working with aluminum at Fisher Body. Of course, 7075 is aircraft grade aluminum, very tough and very difficult to machine. So my machining skills improved along with my engineering skills.

The most difficult part of the project was designing and fabricating the cams that would control the motion of each leg. The pantograph leg design required that the cam followers would both extend and retract under load, so I decided to capture the followers in grooves, rather than rely on springs to keep the followers in contact with the cam surface. Once I had arrived at what I thought would be a feasible cam profile, I had to find some way to machine the complex curves into the cam plates. At that time in the mid–1980s the only CNC milling machine on the OSU campus was a Cincinnati–Milacron T–10 in the ISE department shop. My apprenticeship at GM had not prepared me to program CNC machinery at all, and I was at a loss how to proceed. I eventually went to talk with the professor who oversaw the machine for ISE, who promised that he'd start on my cams immediately. Several months later they were still untouched on his office floor, so I quietly rescued them, and with some help from Cedric Sze and Al Miller in ISE, was able to figure out on my own how to program and run the T–10. As it turned out, the cams worked as planned, with no further modification necessary.

If memory serves, the machine was complete within a year, but then began the long process of trying to get it to actually run. The problem, as we always assumed it would be, was in starting the machine from a standstill. I was confident that if I could somehow bring the machine into contact with the ground while the legs were already in motion, it would run. But exactly how to achieve that took quite awhile to determine. I recall many long nights out at the Walker building on Kinnear Road (I had been called back to work at GM by then), with Jerry Kingzett and others, trying to find some way to start the machine from a dead stop. At this point I must be honest and admit that the solution was suggested to me by Ken Ku, one of the undergraduates who worked for Vince in the drafting shop. Ken became aware of bearings that would only turn in one direction, and suggested that I try them on the feet of the machine. This turned out to be the solution. I encased the outer shell of each bearing in an elastomer, and mounted them so that the bearing would turn if the leg were moving forward, but lock if the direction was reversed. This scheme allowed the machine to start from a dead stop on the floor: the rear legs were able to roll to the forward position while the forward pair propelled the machine forward and slightly upward.

Shortly after this breakthrough came a day that I will never forget, though I've wished many times that it had turned out differently. One sunny morning in the late summer of 1986 we rolled the machine out on the shop floor with its new feet, plugged it in, and – off it went, trotting perfectly across the shop. Success! Even better, a success that was captured on video, which as it turned out allowed me to graduate. Because immediately after this first test, I decided (with considerable urging from the witnesses) to take the machine out to the parking lot and test it over a longer distance. I plugged it into a much longer extension cord, the machine again took a few steps, and then stepped into a slight depression I had overlooked in the parking lot and broke a leg. By this point, three years after assigning me the project, Ken was more than happy to declare the initial test a success, and be done with me and my interminable trotting project. But that few seconds of trotting captured on video was a lifesaver to me: I finally graduated in Autumn 1986 with an M.S. in Mechanical Engineering.

Ken later told me that he showed the short video at a conference, and thus was able to prove his point that running could be accomplished without a high level of control. Although I never considered publishing the results of the thesis, intending to do further work after repairing the machine, the work was later cited by a few other researchers in the field. Professor Jim Schmiedeler was kind enough to send me a list of papers that reference the quadruped, including his own Ph.D. dissertation which he completed in 2001 under Ken's guidance. I've included these papers as references [1–4]. To quote from Jim's dissertation [5]:

> Lilly…proved that high-level control is not required for symmetrical, dynamic locomotion with his quadruped trotting machine that used a series of cams for passive coordination.

What happened to the machine after that was regrettable. I had quit my job at Fisher Body the previous year in order to finish my Masters degree. By 1985, I had finished the apprenticeship and as the lowest seniority person in the tool room found myself working a different shift every week. Under those conditions it was almost impossible to finish my work on the trotting machine, so I applied for a six–month leave of absence to finish things up at OSU. One might think that the management in the tool room would see some benefit in having a tool and die maker with an M.S. in mechanical engineering, but they didn't see it that way. They insisted that they couldn't do without me, so I walked out the door, much to my family's chagrin, and never looked back. By this point I had been married for three years, and was supporting my first wife while she finished her Ph.D. in Electrical Engineering. The need to get a job was paramount, so with some intercession (again) from Gary Kinzel, I quickly found myself employed as a staff engineer for the newly established ERC for Net Shape Manufacturing, working under Professor Taylan Altan.

My intent was to get back to the trotting machine as quickly as possible, rebuild and strengthen the legs, which I had clearly under–designed, and run further tests. As it turned out, I delayed getting back to the lab on Kinnear Road, and in the meantime a group of undergraduates working for Ken discovered the machine and cannibalized it for spare parts for a class project. By the time I discovered this, it was too late, and three years of my life were gone. Figure 1 shows the machine in its supporting frame, along with its very somber, but relieved, creator.

Looking back at this experience, I can say now that it was without question one of the formative experiences of my life as an engineer. One of the reasons I left Fisher Body after finishing the apprenticeship was the lack of interesting work. The tool room at the Columbus plant was devoted almost exclusively to die repair, not the construction of new dies. It was clear to me that it would be years before I was senior enough to be allowed to actually build a new die, and my brief run–in with management convinced me that I was also not in line for a position on the engineering staff. The quadruped, on the other hand, was my own work; I had considerable advice and help from the rest of the ASV team while building it, but it was very much my own concept, and that was a new experience for me. Confronting the problems involved in turning a concept into a real design, and then turning that design into an actual engineered artifact, taught me a lot about engineering and about myself. There were many weeks and months when it seemed unlikely that the quadruped would ever be completed, but in the end, I took away from the project a real sense of accomplishment.

Fig. 1 The author and his quadruped, 1986

3   Transition to the Classroom

I worked for the ERC under Taylan Altan for three years, assisting graduate students in designing and building the tooling they needed for their research in net shape manufacturing. At this time I first became acquainted with CAD in the form of CATIA Version 1, and taught myself how to program CNC equipment. In July of 1989 my wife and I moved to Germany for a year, where she had secured a NATO post–doc position at the RWTH in Aachen. Midway through our year there, our marriage came apart, and in late 1990 I found myself back in Columbus with no idea what to do with myself. My plans for the future had disintegrated along with my marriage, my parents were both in very poor health, and my once–promising future as an engineer seemed to have disappeared overnight.

In those years Robert Frost's famous line from 'Death of the Hired Man', *"Home is the place where, when you have to go there, they have to take you in..."* was a constant refrain in my head. Taylan Altan generously took me back into the ERC, and I decided to start work on a Ph.D. My original intent was to complete a one–of–a–kind program in the history of technology, a subject that had always been of interest to me. I quickly discovered that this topic was of interest to no one in OSU's Department of History, so after two quarters of trying in vain to find an advisor, I began to pursue a Ph.D in Industrial & Systems Engineering, with Professor Jerry Brevick as my advisor. Through my work with the ERC, I had gotten to know Professor Al Miller quite well, and had expressed an interest in teaching at the university level to him. My first chance came in 1992, when through Al's intercession I taught a basic course in engineering methods at the Central Ohio Technical College in Newark.

My big break came in 1994, when Al Miller, by then the Chair of ISE, asked if I would be interested in helping Professor Kos Ishii with the very popular class he had developed in Mechanical Engineering, ME 682. Kos had been recruited to OSU from Stanford by Ken, who by this point was the Chair of Mechanical Engineering. I had been working with Kos through the ERC for two years by then, helping him now and again with CAD and CNC issues as they arose, and found him very charismatic and easy to work with. Kos had created two new courses in the ME curriculum, ME 682 and ME 683. The courses were very closely coupled, with ME 682 providing the background for ME 683, which was essentially a course in programming table–top CNC milling machines. The demand for the course soon got to the point at which Kos could do nothing else, so Ken and Al agreed to find someone to teach the course one quarter a year to free up his time for research.

I quickly agreed to take on the task of teaching ME 682 once a year, but before I had the chance to start, the situation changed drastically. Kos's advisor at Stanford, Phil Barkan contracted leukemia, and Stanford asked Kos to return there and take over for him. Kos's sudden departure from OSU meant that the ME department had no one to teach their most popular technical elective, so before I quite knew what was happening, I had a new appointment as a Lecturer in both Mechanical Engineering and Industrial & Systems Engineering, and began a four–year long stint of teaching full time while finishing the Ph.D. I very clearly recall a long talk about my future I had with Kos before he left Ohio State, during which he gave me a piece of advice that I've never forgotten: "Make yourself indispensable". That was it. I've tried very hard to follow that advice ever since. Tragically, Kos died unexpectedly at the age of 51 in 2009, but without question he had an important role in setting me onto the path I've followed ever since then.

In 1998, after four years of teaching eight classes a year, I finally finished my dissertation in ISE. Those four years were among the hardest years of my life, as I tried to complete a dissertation while creating a new technical elective in CAD for ME and ISE students, evolving Kos's original ME 682 in the direction of a design for manufacturing course, and revising and teaching a course in CNC machining. Thanks to the strong support of Ken in ME and Al in ISE, I secured a joint position as Assistant Professor in both departments in April of 1998, with ISE as my tenure initiating unit. During my time as an Assistant Professor I devoted more time than I should have to teaching, continuing to pour most of my effort into 682, which by 2004 had developed into a full–blown product engineering course. The course by this point had become one of the most popular courses in the College of Engineering, and was no doubt a major factor in the College's somewhat surprising decision to grant tenure and promote me to Associate Professor.

In my first several years at OSU, I saw myself as being an integral part of the manufacturing faculty in the ISE department. As time went by, I found myself growing less interested in manufacturing engineering and more interested in product design engineering. This transition was helped along when Professor Julie Higle, Al Miller's successor as Chair in ISE, made the decision to de–emphasize manufacturing in ISE, and as a consequence eliminated my course in machining (ISE 622) from the curriculum. As a result of this and other factors, I began to migrate away from ISE and toward mechanical engineering. Returning from a year's sabbatical in Hungary in 2007–2008, I changed the terms of my joint appointment, and currently hold a 75% appointment in Mechanical Engineering, a 25% appointment in ISE, and an adjunct appointment in the Department of Design.

4   Reworking the Mechanical Engineering Curriculum

For the past three years, as Ohio State University moved from an academic calendar based on the quarter system to one based on semesters, I found myself deeply involved in the discussions initiated by Cheena Srinivasan, as Chair of Mechanical Engineering, to re–think and rebuild the mechanical engineering curriculum from the ground up. Following extensive discussions with industry, alumni, students, and faculty at Olin College, the University of Texas, and MIT, as well as Ohio State, we elected to devote a much larger fraction of the curriculum to experiential learning. Three major changes resulted from this decision:

- the required capstone course in the senior year was expanded to a two–semester sequence, with several different tracks made available to students with varying interests. These tracks ranged from conventional mechanical engineering problems to motorsports, biomedical device design, product design, and an interdisciplinary experience in conjunction with the College of Engineering.
- the manufacturing processes course that traditionally was taught in the second year was moved to the senior year and totally re–designed to better fit the needs of mechanical engineers. The new course, which has been developed by Professor Jose Castro in ISE, assumes a background in heat transfer and machine design, and emphasizes experiential learning in the lab coupled with extensive analysis and simulation of manufacturing processes.
- an entirely new course, "Introduction to Design in Mechanical Engineering" is now required of all students entering the major, in the second semester of the sophomore year. This course is intended to provide students with skills they will need to successfully complete subsequent design courses in the major, and to make them effective engineers more quickly when they begin their careers.

The thrust of the new course was to introduce second–year students to hands–on skills that they will need to be successful engineering designers. This entails a need to provide extensive experience in the machine shop, the CAD studio, and the electronics laboratory. At Ohio State, given the large number of students who are enrolled in the major, our primary constraint in the design of this course was how to provide a truly rich experience to our students while simultaneously coping with burgeoning enrollments and limited resources. Cheena Srinivasan entrusted the development of this course to Professor Lisa Abrams and me, with extensive support and creative input from Joe West, the long–time electronics wizard of the ME department, and Chad Bivens, a very talented mold maker who had recently joined the department as the machine shop supervisor. Our task was made much easier by working alongside three very talented, dedicated, and hard–working graduate students: Michael Neal, Ryan Kay, and Angelica Liu.

The undergraduate mechanical engineering program at Ohio State University is currently one of the largest in the country. Over the past seven years, the program has seen a significant increase in enrollment, from 865 undergraduates in 2005 to 1337 in 2011. The ME program is now the largest undergraduate major in the College of Engineering at Ohio State, and one of the largest is the USA. In planning a course that would rely so heavily on experiential learning, the primary consideration was simply the logistics of moving large numbers of students through a structured set of exercises while maximizing the pedagogical benefit to the student.

Beyond introducing mechanical engineering to entering students in an experiential context, an important secondary goal was to create a device – a 'teaching platform' – that could be integrated readily into many other courses in the curriculum. Like many other mechanical engineering undergraduate programs, our curriculum is essentially divided into four distinct sections: mechanics, machine design, thermal and fluid systems, and dynamic systems. In creating this course, our intent was to show the students how these seemingly disparate parts of the curriculum fit together in practice. Hence we needed to create a platform that was rich enough to support its integration with subsequent courses over the remaining four semesters, and wide enough to span the entire discipline.

From the beginning, our intent was that the course itself would be quite consciously built around the fabrication and testing of this teaching platform. Students would attend two 1.5 hour lectures per week, in addition to a two–hour lab period, which would be spent in either the machine shop or the electronics lab. Lectures would be structured to provide needed information 'just in time' for students to apply it in the shop or lab. Our belief was that by immediately applying concepts introduced in lecture to a real device, students would internalize and retain the material more effectively.

Thus our first task was to create a suitable teaching platform: the mechanical device that the students would construct and test. Because of the centrality of the device to this course and most likely subsequent courses, this decision was not an easy one. The device had to be simple enough to be constructed by second–year students with very limited hands–on skills, but at the same time, it needed to be complex enough to satisfy the requirements outlined above. In summary, the device needed to:

- Challenge students to develop real proficiency in the machine shop and electronics lab;
- Offer interesting control scenarios, the better to integrate with the microprocessor;
- Provide a rich enough experience to cover a wide range of mechanical engineering problems;
- Integrate well with subsequent courses in the curriculum.

After much discussion, and based on our review of the literature, we elected to have the students work in small teams to fabricate a fairly complex inline two–cylinder compressed air motor. The idea for using a compressed air motor as the 'teaching platform' came from similar projects at the US Coast Guard Academy and Cornell University. While researching several air motor designs, we discovered that a group of students in the OSU mechanical engineering program had designed and built a two–cylinder version of a motor for their capstone project during the previous year. This motor became the basis of our new design.

Fig. 2 The initial OSU air motor design

During the summer of 2011, a graduate student in our program, Angelica Liu, working closely with Chad Bivens, re–designed the original capstone air motor to create a more robust design that was also quite a bit more challenging to fabricate. This motor became the prototype for the first pilot class in Autumn Quarter, 2011; one of these motors is shown in Figure 2. The device was assembled from over thirty separate components, the majority of which were machined on lathes and milling machines by the students. The tolerances required for robust performance were considerable, but the students were able to machine parts from aluminum, bronze, and steel to one–thousandth of an inch or better. In this version of the motor the flow of air to the cylinders is controlled by a cam and valves; because cam timing was crucial, the cam was machined on a CNC milling machine by the lab supervisor.

Much tuning and tweaking of the motors was necessary to achieve optimal performance, but by the end of the first pilot, all of the student teams succeeded in building fully functioning motors. One very clear result of this effort was the sense of accomplishment felt by the students. They put in long hours outside of class, and were frankly surprised when their efforts resulted in success. Another result was also quickly apparent: this initial design of the motor, while fulfilling for the students, was simply too complex to undertake given the very large number (~200) of students who would be enrolled in the course in Autumn Semester, 2012. With over thirty separate components, and with the level of finesse needed to successfully assemble the motor, it was clear that this design simply was not feasible for the scaled–up course.

In addition, it was still not clear to us how to incorporate a microprocessor into the design of the motor. Our initial thinking had been that the student teams would work on their own to design a mechanical 'load' which would be driven by the air motor, with the load controlled by the Arduino© microprocessor. However, after finishing the initial pilot course, and coming to grips with the number of hours required for the students to become proficient at machining, we realized that it would be impractical for the students to attempt any project in the machine shop that was not highly structured. We simply did not have the resources, either in staff or in the physical plant, to support over two hundred students working simultaneously on independent designs. While we remained enthusiastic about having the students fabricate and assemble an air motor, it was clear that we needed a much simpler motor design that would better incorporate the Arduino© microprocessor, and thus eliminate the need for a cam and valve system.

With the completion of the first pilot in December, 2011, we stepped back and re-evaluated the assumptions we had been working under for much of the previous year. This process led to a total re–thinking of the course objectives, and a re-ordering of the course constraints. It was clear that our original idea of following the air motor project with a more open–ended design/build experience was totally impractical, given the number of students who would be taking the class. It was also clear that the cost of running this course could be a strong disincentive to its sustainability over the long term, unless we were able to find ways to reduce the costs significantly.

With this in mind, we arrived at a new set of constraints to guide the further development of the course:

- a less complex motor design, probably by eliminating the need for a cam and valves;
- better integration between the motor and the Arduino microcontroller;
- a more flexible teaching platform that could easily accommodate various numbers of students working together.

Luckily for us, at this time, and quite on his own, Joe West was experimenting with a very small replica of the WWII–vintage Wright Cyclone radial engine. We were able to quickly adapt this design during the month of December, and have it ready for the second pilot course, which began in January, 2012. While the first version of the motor required that all of the components be in place before it would function correctly, the radial design gave us much more flexibility in this regard.

Fig. 3 The second iteration has fewer unique parts, with air flow controlled by a microprocessor

Although all of the teams in the initial pilot course were able to complete their motors on time, it was clear that we would need to plan for the inevitable situation in which some members of a team did not finish. The first version of the air motor was not at all flexible in this regard: it simply would not run unless it was complete. The new radial design, however, could be constructed to operate with any number of cylinders, from one to six. With this design, each team member would be responsible only for completing a single cylinder sub–assembly (cylinder block and cap, piston, and connecting rod). If one student did not finish, the remainder of the team would still have a functioning motor. This design is also better at accommodating the large number of students involved, as it allows six students to assemble a single motor, which reduces the cost of running the course considerably.

Figure 3 shows the six–cylinder version of the radial engine. The frames on which the cylinders are assembled are machined by our shop personnel and provided to the students. Each student then assembles his or her cylinder sub–assembly to the frame, which is then assembled to one of nine test rigs which incorporates the microprocessor, twelve solenoid valves, and the air lines. This design also greatly reduces the number of solenoid valves needed for the class, which in fact are the most expensive component of the motor. The solenoids are used in common by all students for testing.

The Arduino© microprocessors are used to control the set of solenoid valves which send air to each cylinder. The Arduino© concept was originally developed at the Interaction Design Institute Ivrea, in Italy, for use in teaching design and engineering students how to program a relatively simple but sophisticated microprocessor[16]. These devices have evolved into an extremely useful tool for students, designers, and engineers looking for a low–cost but capable controller, and are extremely popular among the 'maker' community. For the course, each student is required to purchase an Arduino© kit rather than an expensive textbook. Students spend roughly half of the semester in the electronics lab, learning to program the microprocessor and use it with various sensors and actuators. Each student is then required to write and test their own version of code for controlling the flow of air to the assembled motor.

Fig. 4 Some of the thirty–three air motors completed during Autumn Semester, 2012

In 2014, I will have been teaching in the College of Engineering at Ohio State for twenty years. I think it's fair to say that during that time I've established a reputation as one of the 'teachers' in the College, and find myself filling the role of teaching mentor frequently. When I look back over the past two decades at the various courses I've created, modified, and taught, it's clear to me now that my early experiences in the tool room and with the quadruped have had a strong influence on how I teach my courses. In truth, these experiences are so deeply embedded in my 'professor DNA' that until I was asked to contribute a short article to this *festschrift*, I had never given much thought to just how important they were. I fully realize that I've been incredibly fortunate to have had these experiences, and I've done my best over the past two decades to fulfill Kos Ishii's advice to 'make myself indispensable'. My academic career at Ohio State University will draw to a close in another few years, but I hope the new course will live on for some time to come.

References

1. Schmiedeler, J.P., Waldron, K.J.: Leg Stiffness and Articulated Leg Design for Dynamic Locomotion. In: ASME Design Engineering Techical Conference, Montreal, DETC2002/MECH–34331 (2002)
2. Wan, X., Song, S.-M.: A Cam–Controlled, Single–Actuator–Driven Leg Mechanism for Legged Vehicles. In: ASME International Mechanical Engineering Congress and Exposition, Anaheim, CA, IMECE2004–62266 (2004)
3. McKendry, J., Brown, B., Westervelt, E., et al.: Design and Analysis of a Class of Planar Biped Robots Mechanically Coordinated by a Single Degree of Freedom. ASME J. Mech. Des. 130, 1–8 (2008)
4. Fisler, J.: Maximum High Jump with a Robotic Leg. M.S. Thesis, Swiss Federal Institute of Technology, Zurich (2008)
5. Schmiedeler, J.P.: The Mechanics of and Robotic Design for Quadrupedal Galloping. Ph.D. Dissertation, The Ohio State University, p. 16 (2001)

29 Project-Based Learning in Engineering through Street Performance Robot Contest

Shigeo Hirose, Gen Endo, and Edwardo F. Fukushima

Abstract. This paper introduces the outline of a creativity education course "Machine creation" that has been conducted at the Department of Mechano-Aerospace Engineering, Tokyo Institute of Technology, for 22 years. The course is designed to provide students with extensive opportunities to work with real objects to bridge classroom lectures and hands-on experiences. The students work as a team within a time limitation under budget to create a "Street Performance Robot" which can entertain audience. We assume that students can efficiently learn about a process of product development through this course from planning of the robot to the final presentation of the robot. To evaluate this assumption, we carried out questionnaire survey for the current students and alumni who are currently working as engineers in the real world. The results suggest that this course can provide a valuable experience for students and many alumni agree with our educational methodology.

1 Introduction

In today's world, where value systems are diversifying and technology is being commoditized, sustained growth of a company cannot be achieved simply by improving the performance of already existing products. Rather, what is required of today's engineers is to combine basic technologies and propose methods of use based on unprecedented, new ideas, and thereby create new value in people's lives.

Therefore, systematic nurturing of creativity is a key issue for educational institutions producing human resources. The importance of "education on Monodzukuri (hands-on experience of design and manufacturing)" and "nurturing creativity" in engineering has been recognized by many universities in recent years, but the Department of Mechano-Aerospace Engineering of Tokyo Institute of Technology has been addressing this issue since 1990, 23 years ago. The department has developed

Shigeo Hirose · Gen Endo · Edwardo F. Fukushima
Tokyo Institute of Technology, 2-12-1 Ookayama, Meguro-ku, Tokyo, Japan
e-mail: {hirose,gendo,fukusima}@mes.titech.ac.jp

V. Kumar et al. (Eds.): *Adv. in Mech., Rob. & Des. Educ. & Res.*, MMS 14, pp. 371–385.
DOI: 10.1007/978-3-319-00398-6_29   © Springer International Publishing Switzerland 2013

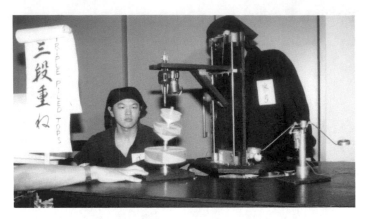

Fig. 1 Street performance robot spinning a top (1994)

a curriculum for nurturing creativity, and has provided education stressing hands-on experience and fabrication [1].

This paper has two main purposes. First, it provides an overview of a course called "Machine Creation," which the authors proposed and have been teaching for 23 years, and its contribution to the systematization of education using robots. In this course, students are given an extremely vague assignment of "making a robot which can entertain people." Groups of 4-5 students each fabricate a robot over a period of about 4 months, and at the end of that process they present their robots to an audience at the "Street Performance Robot Contest" (Figure 1). From the planning stage, where groups decide what kind of robots to make, through the subsequent stages of machining, fabrication, control and presentation at the contest, the intention is for students to gain a comprehensive experience, in a short time, of the entire development process experienced by engineers in the real world.

Secondly, this paper conducts a questionnaire survey of the educational effectiveness of Machine Creation, and discusses the results. Various engineering education programs have been proposed using robots as subject matter, but it is difficult to objectively assess their educational effectiveness. Many of the references administer questionnaire surveys to students who have taken the courses [2]-[6], or assess course effectiveness by examining where students are employed at graduation [7][8].

However, the authors feel it is difficult to determine whether the education is truly effective simply by surveying students immediately after they have taken a course (or, within 1-2 years at the latest). The reason why is that, even if students believe they have gained useful knowledge and experience from a course, the benefits may be transient, and soon forgotten as short term memory fades. Our true intent is not to develop the skill of overnight cramming to boost test scores. The true purpose of engineering education is for students, through their courses, to transform the knowledge and experience they have acquired into wisdom, and then to continually make use of those assets going forward.

Therefore, for this paper, a questionnaire survey was administered to alumni who took Machine Creation in the past, and are currently working on the front lines in the real world. In this follow-up survey, the alumni were asked to reflect on the course, and indicate whether it is truly useful as engineering education, and what sort of impression they have of it today. At the same, a questionnaire survey was administered to current students, to compare survey results and thereby verify the effectiveness of improvements in course operation which have been made every year.

2 Machine Creation

2.1 Overview of Course

"Machine Creation" is a required course (3 units, 3 classes/week) in the second term of the third year of the Department of Mechano-Aerospace Engineering, and in the 2011 academic year it was held for a total of 15 weeks. Machines are designed and fabricated with the aim of presenting them at the Street Performance Robot Contest held in week 14. The assignment given to students is basically only to "make a street performance robot which can entertain people," and the exact nature of the robot's performance is left entirely up to the imagination of the students. In this course, students fabricate robots in teams made up of 4-5 students. Students are assigned to teams by the instructor, based on their grades in the course, and their responses to a questionnaire administered prior to the course asking them what sort of robot they want to fabricate, and what role they want to play on their team (e.g., project manager, or mechanical/electrical/program engineer). (Assignment to teams is discussed afterward.) In the 2011 academic year, 49 students took the course, and they were divided into 12 teams.

The integrated creation studio for carrying out fabrication is equipped with a drilling machine, lathe, milling machine and other machine tools, and the students do their own machining based on prior training and guidance by the technical staff. The workshop also has a price list for items such as structural materials, mechanical parts, electrical parts, actuators and other components, and each team fabricates their robot within a budget of $225.

The necessary parts can be purchased separately by individual students or groups at locations such as Akihabara, but students must pay themselves for any excess over the budget. (It is assumed that expensive devices such as power supply units, motor driver circuits, air cylinders, solenoid valves, and air compressors will be rented, and these are not included in the budget.)

To check the progress of fabrication, review meetings are held about every 3 weeks, and each group makes a presentation for about 5 minutes in front of the instructors. After each review meeting, students are instructed, as an assignment, to submit a progress report (four A4 pages). In week 14 of the course, the Street Performance Robot Contest is held before an audience of the instructors and second year students from the Department of Mechano-Aerospace Engineering. To determine student grades, basic points for each group are calculated based on scores for each review meeting and the Street Performance Robot Contest, and individual students are comprehensively evaluated, taking into account their attitude toward coursework, and record of tardiness and absences. (For details on how the course is operated, please see Reference [9].)

2.2   Features of the Street Performance Robot Contest

At present, almost all of the robot contests held widely throughout Japan are the oppositional type, but for this assignment, the authors selected a performance type competition, where robots must be judged subjectively. The reasons for this are as follows.

1) The assignment given to students is vague

If the competition is the oppositional type and the rules are clearly specified, the students can focus on the technical issues of complying with the rules, and achieving victory. However, if winning and losing are clear-cut, then after competitions over many years, the approach to winning tends to converge on an orthodox robot form which operates with high reliability, and the scope for demonstrating creativity is narrowed. Also, if the robot forms are the same, then what determines victory may not be ideas, but simply differences in fabrication capabilities. On the other hand, there are many ways to entertain people, and a wide variety of possible ideas, so there is broad leeway for demonstrating creativity. In addition, students have to show originality from the planning stage, where they decide of what sort of robot to make. This makes it possible to foster creativity beyond just fabricating an item, and includes planning skills for deciding on what to make.

2) Students compete to appeal to an audience

Since the assignment is to make something which will entertain people, the robots come in all descriptions, and the audience can view them without getting bored. The students give a presentation in front of a large audience, and the reception-laughter, cheers or awkward silence-provides the students with direct feedback.

This leaves a strong and lasting impression on the students as direct experience. Also, in this contest, a technically superior robot will not be rated highly if its superiority is hard to convey to a general audience. For serious students who are apt to fixate on technical aspects, the authors believe it is a valuable experience to understand the importance of communicating with people, and adopting the third-person perspective of an ordinary person. However, it is difficult to objectively evaluate the entertainment value and appeal of a performance. Evaluation of students' work in this course is multi-faceted, and based not only the assessment of audience members who saw the contest, but also on technical evaluation of interim review meetings and report content.

3) The contest can be held continuously

If the robots do performances, then the contest can be repeated every year with the same rules, and there is no need to change the competition every time, as is frequently done with oppositional robot competitions. This greatly reduces the burden on the instructors who manage the contest. In fact, as of this year, the Street Performance Robot Contest has been held 21 times with almost the same rules. The fact that the rules are the same has a number of advantages. For example, actuators, controllers, power supplies and other equipment can be reused over multiple years, and the costs of holding the course can be reduced.

2.3 Street Performance Robot Contest

The Street Performance Robot Contest is held as indicated below. Each team does a presentation/performance in less than 3 minutes, and the instructor and audience cast votes to rank the robots in order from most to least interesting. For voting, each audience member is given 2 ping-pong balls, and the instructor is given 10 ping-pong balls. At the end of the contest the number of votes is counted, and the ranking is announced. The audience members can freely view and vote without limitations, especially the second year students in the Mechano-Aerospace Engineering Department, who will take Machine Creation in the following year, are encouraged to join voting. The aim is to have them experience the atmosphere of the contest, and show them their goal for the following year, and thereby stimulate their thinking.

Figure 2 shows the "ExciteBicycle" robot by Team 9, which was the course winner in the 2011 academic year. This is a bicycle simulator, and if the operator pedals it and turns the handlebars, he or she can propel and steer a small tricycle. This is a machine with the qualities of a game, where players compete on whether they can finish a specified course (with features such as tunnels and bridges) within a certain time. The rotation speed of the pedals driven by the player is measured with a student-built optical encoder employing a reflective photo-interrupter. The rotation speed value is sent to the tricycle robot via Xbee, and used to rotate the drive wheels. A camera is mounted on the tricycle robot, and "you are there" operation is achieved by projecting the image from that camera with a projector.

Fig. 2 Bicycle simulator "ExciteBicycle" by Team 9 (2011)

In recent years, it has become possible to perform PC-linked information processing and comprehensive staging using video cameras, projectors and audio input/output. Some robots have employed visual feedback using OpenCV, and some have used Speech SDK, a voice recognition and synthesis API from Microsoft. This shows how students devise ideas in accordance with the current technical environment. In this way, students freely develop ideas, including final staging, and there is leeway for them to address novel technical issues. This approach is a notable advantage of this contest.

A truly diverse range of robots has appeared at the past 21 contests, but they can be roughly classified into the following 8 categories. (Examples are given in parentheses.)

1. Street performance (juggling, plate spinning)
2. Playing musical instruments (guitar, keyboard, violin)
3. Achieving specific capabilities (batting)
4. Substituting for personal tasks (automatic tooth brushing, cosmetics)
5. Preparing and serving foods or drinks (making ramen)
6. Competing against and interacting with people (target game)
7. Story-based performances (fairy tales, marionettes)
8. Movement coordinated with video (video jumps out into reality)

To put it another way, even if a robot is independently invented by students in the new academic year, in most cases it will fall under one of the above categories, and there will have been similar robots in the past. However, even if there have been similar robots, we believe there is no problem provided that the new robot is not a perfect imitation of a past robot.

One reason why is that the primary emphasis of this contest is on the process in which students themselves generate their own ideas, conduct prototyping and experiments, and discover and solve problems. Another reason is that the evaluation criteria differ from the novelty required in academic research. This contest does not aim to discover "world firsts." Its primary emphasis is giving students their first experience.

3   Questionnaire-Based Evaluation

This section reports on questionnaire surveys administered to students in the 2011 academic year who took the Machine Creation course (referred to below as "current students") and currently working alumni who took the course in academic year 1994 (referred to below as "alumni").

In order to verify the persistence and effectiveness of the proposed education, ideally it would be best to base the analysis on a questionnaire survey administered to students at various stages (immediately after graduation, 3 years later, 5 years later, and 10 years later), but contacting many alumni and receiving valid questionnaire results from them would require a tremendous amount time and labor. Therefore, as a first step toward evaluation by course alumni, we administered a questionnaire

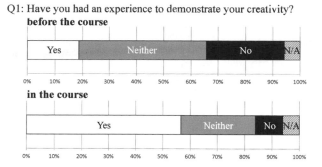

Q1: Have you had an experience to demonstrate your creativity?

Fig. 3 Questionnaire for current students

to alumni from the 1994 academic year, who were easy to recruit for this survey because they were classmates of one of the authors.

One of the authors himself took the class as a student in the 1994 academic year, and thus I remember, as a student, how the course was run at that time, and can discuss it in comparison with the current approach.

After describing the questions and results for current students and alumni, respectively, we compare the responses of current students and alumni to the same questions.

3.1 Questionnaire for Current Students

A questionnaire survey relating to nurturing creativity was administered to 49 students who took this course in academic year 2011, and responses were received from 48 of them. In the pre-survey for assigning students to teams, which is administered prior to the start of the course, there is a question asking "Have you ever previously demonstrated creativity?" By asking this same question in the post-questionnaire after the course is finished, it is possible to determine whether students had an experience of demonstrating creativity through their participation in the Machine Creation course.

Figure 3 shows the results of the pre- and post-questionnaires. The figure shows that, while 19% of students responded that they were able to demonstrate their creativity in the pre-questionnaire, this percentage increased dramatically to 56% in the post-questionnaire. When students were asked for specific examples they experienced during Machine Creation, there were a total of 25 responses, and of these 9 related to ideas in the planning stage, and 8 were specific designs for mechanism structure. However, students also gave various other specific examples such as electrical circuits, programming and solutions when confronted with problems. For this reason, it is conjectured that this course provides opportunities to demonstrate creativity in a variety of areas.

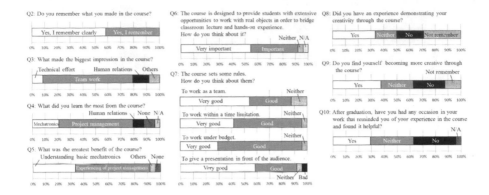

Fig. 4 Questionnaire survey about "Robot Creation" for alumni

3.2 Alumni Questionnaire

A questionnaire form was sent to the 36 alumni whose contact information could be obtained from among the 48 students who took Machine Creation in the 1994 academic year. Responses were obtained from 24 alumni. The collection rate was 67% (50% of all students who took the course).

3.2.1 Alumni Attributes

Since alumni are likely to have a different understanding of the course depending on their job category and position in the real world, the first items in the survey were designed to determine the attributes of the alumni subject to the survey. In terms of job category, the alumni were divided roughly 50-50 between manufacturing and non-manufacturing industries. Within the non-manufacturing industries, almost all of the alumni were working as engineers in fields such as electric power and transportation. In terms of position, alumni at the subsection chief or chief level (with 1 or 2 subordinates) accounted for 70% of the total. In terms of duties, 80% of the alumni responded that they were engaged more in technical work relating to R&D than in management work. Their ages were 38-39. For the above reasons, it is likely that the alumni who responded are a generation of engineers who are actively carrying out engineering work.

3.2.2 Evaluation of Machine Creation

The left and center columns of Figure 4 show the results of the alumni questionnaire for questions designed to evaluate the significance and rules of the Machine Creation course. First of all, when alumni were asked if they remembered what kind of robot they made in the course they took 17 years before, 58% responded that they remember clearly, and 32% that they remember[1].

Next were the questions: Q3: What made the biggest impression in the course?, Q4: What did you learn most from the course?, and Q5: What was the greatest benefit of the course? In the results for all these questions, the most frequent response was accomplishing a project through joint work as a team. The percentage of students who replied that they had project experience prior to taking the course was about 10%, and thus for almost all of the students this was their first project-type joint work. Therefore it likely left a lasting impression and was very educational. For this reason, it is likely that the significance of this course lies in accomplishing a project through joint work as a group.

Next, the alumni were asked how they felt about the fact that the course is designed to provide students with extensive opportunities to work with real objects in order to bridge the gap between classroom lectures and hands-on experience. The responses showed that 54% thought such efforts were very important, and 38% felt they were important. These results can be regarded as supporting the importance of education by working with real-world objects and not just class lectures.

Alumni were also asked their opinions about the following rules, which were adopted as part of operating the Machine Creation course: To work as a team, To work within a time limitation, To work under a budget, and To give a presentation in front of the audience. The responses in all cases were very positive, and thus this framework can be regarded as appropriate.

The above results show that engineers working in the real world have a high opinion of the importance of manufacturing technology, and the associated education methodology, which have previously been advocated by the authors.

3.2.3 Nurturing Creativity

The right hand column in Figure 4 shows survey results regarding nurturing of creativity. The questions were: Q8: Did you have an experience demonstrating your creativity through the course?, Q9: Did you find that you became more creative through the course?, Q10: After graduation, have you had any occasion in your work that reminded you of your experience in the course, and that made you find the course helpful? The results show that roughly 30% of the alumni responded that they demonstrated creativity, remembered the course in work, and found the course helpful. In the questionnaire for current students in the lower part of Figure 3, roughly 60% responded that they were able to demonstrate creativity, and thus the rate for alumni was about half of that. Opinions may differ on whether to regard this 30% as a large or small percentage, but considering that the students took the course 17 years before, and that experiences tend to be forgotten as time passes, this result can be regarded as showing that the effect of this education approach in nurturing creativity is maintained. Among the alumni who responded that they demonstrated

---

[1] Since it is possible that alumni who did not remember the course did not return the questionnaire, it is not possible to extrapolate from these results to results for the entire group. However, at the very least it is evident that alumni who did respond remembered the course.

creativity and found the course helpful, there were some who responded in the free comment portion of the questionnaire that the experience went beyond just a course, and had a very powerful impact on them, e.g., influencing them to choose a career as an engineer, or serving as a foundation for the projects they carry out in their current work.

3.3   Comparison of Overall Impression of Current Students and Alumni

As was pointed out in Section 2.3, the course operation and equipment used are reviewed every year. Therefore, between the alumni who took the course in the 1994 academic year and the current students in the 2011 academic year, improvements were made every year in the method of running the class, even though robots were made based on the same assignment every year. In order to actually ascertain the effectiveness of these improvements, this section asks the same questions to both current students and alumni regarding their overall impression of the course, and then compares the results. The biggest difference in the operation method between the 1994 and 2011 academic years was the method of group assignment, and the fabrication time on campus outside of course time. Whereas, in the 1994 academic year, group assignment was done by mechanically assigning students in the order of their attendance numbers, in the 2011 academic year, group assignment was done at the discretion of the instructor, after determining the desires of each student by administering a questionnaire, while taking into account student grades.

Also, in the 1994 academic year, fabrication could not be finished during, and went on long after, course time, and the usual approach was, right before the contest, to work independently every day until the last train in the university workshop, and in some cases students worked through the night over consecutive days in the workshop.

Working until late at night in this way may be a good memory for students, but it is not really the best approach considering the safety of students and the burden on the instructor running the course. Therefore, in recent years, methods have been devised to improve the efficiency of fabrication so that machining work is finished, as far as possible, during normal course time. More specifically, fabrication work has been divided into highly dangerous work using machine tools, and safe work such as concept design, circuit fabrication and programming. In the former case, the work is done during course time under the supervision of the instructor, technical staff or TA, and students are given explicit instructions to "concentrate during class time on work which cannot be done without machine tools." On the other hand, the latter type of safe work requires no safety supervision and can be carried out by each group and individual outside of course time. In particular, an approach was devised to permit students to take home some equipment-such as power supplies, motor drivers and microprocessors-so they can continue circuit fabrication and programming work on their while at home, during long breaks such as the winter vacation. In recent years, this approach has eliminated the situation where students work all night in the

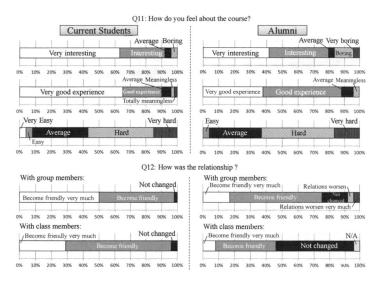

Fig. 5 Comparison between current students and alumni about a general impression of "Robot Creation"

workshop on the day before the contest, and this significantly reduces the burden on the instructor and TAs.

Figure 5 shows the results regarding general impressions of the Machine Creation course, and how relationships with other students changed during the course. The column on the left shows the results for current students, and the column on the right shows the results for alumni.

Q11 on general impressions asked about 3 aspects: whether the course was interesting, whether it was a good experience, and whether it was easy or hard. The general trend of the responses was almost the same for both current students and alumni, with more than 80% indicating that the course was interesting and a good experience. For the question about whether the class was easy or hard, more than 60% of both groups indicated that the class was hard. These results indicate that the course presented many difficulties and was definitely not easy, but most of the students had an interesting, fulfilling and valuable experience.

Incidentally, 17% of alumni, but only 4% of current students, responded that the course was boring or very boring. In the free comment space, almost all of the alumni who responded that the course was boring said that their relationships with group members worsened, and thus their motivation to do the work declined. This point also shows up in the results of Q12 which asked how relationships with group and class members changed due to the course. Whereas there were no negative responses from current students, 8% of alumni responded that their relationships worsened. It is conjectured that the infrequency of dissatisfaction among current students is due to the instructor taking great care when forming groups to ascertain student desires and take into account student grades, and thereby prevent differences from arising between groups in technical skills and motivation. This suggests that the current method of operation is advantageous.

When students were asked what percent of their maximum effort they devoted to the course, and what percent of total satisfaction they derived from the course, the median value in both cases was 80%, and no clear difference was evident between current students and alumni. Finally, when students were asked if they would recommend the course to younger students if it were not required, 94% of current students, and 100% of alumni, said they would, and thus they rate the course extremely highly. This shows that many students have earnestly worked at this course, and derived a high degree of satisfaction from it. In addition, almost no difference was seen between current students and alumni, and this suggests that even when course operation is made more efficient, the quality of the course is maintained.

4 Impressions of Alumni Regarding Machine Creation (Free Comment)

A space for free comment was provided at the end of the questionnaire for alumni, and the alumni were asked to describe the events or experiences which left the greatest impression, and the impressions, improvements or advice they had regarding the course. This section presents and discusses some representative responses.

First, the positive experiences and impressions were as follows.

- The experience I had-of encountering various troubles in determining how to structure a team and different roles, in a situation where members have different strengths-is useful when heading up a team in my current work.
- I feel this is a very significant course for improving project management and presentation skills. If possible, I would like you to say that this course is useful even for non-manufacturing work.
- If hardware and software are not smoothly integrated, a machine will not operate properly; and to make a good robot, it is essential to rely on people, and to help each other out. I learned these two things at the time of the class, and they are the basis of my work today as an engineer.
- In the end, no matter how good the product is, it will be meaningless if the price is too high. No matter how great the technology is, it will be meaningless if the item isn't fun. No matter how good the product is, people won't understand it if you can't convey the advantages to them. These are some of the many common points with my life today, and the fact that I was able to learn these lessons while having fun during my student days was a great experience. The experience was enjoyable, and thus rather then join a manufacturer and continue researching one thing, I wanted to join a company on the consumer side where I could gather exciting technologies from all over, and make them into a product that would make people happy.

The above comments show that this course was not just a transient, fun experience; it provided a valuable experience on running a project as a leader, or working within a project as an engineer. On the other hand, negative experiences and impressions were as follows.

- I remember having a terrible experience due to bad chemistry with my fellow group members. [...] Even today, I'm not very good at mechatronics, and perhaps that is the reason.
- I had no knowledge of mechanical or electrical technology, no ability to use my hands, and no ability to advance the discussion. I keenly felt my inability to do anything.
- When I was taking the course, I was extremely immature, and unfortunately I wasn't able to get anything from the course or enjoy it. As a suggestion, I thought it would be best to just provide technical support as best I could. [...] I don't think it is important for students to experience thinking everything up by themselves. A university is an educational institution, and therefore it should cultivate and nurture the sprouts of learning.

The above impressions show that if personal relationships do not go well with other group members, the student may be left with a strong feeling of ineptness. They also show that there were some students who did not know how to proceed due to unbalanced technical skills between groups. We believe this shows we can reduce negative experiences by devising a new approach to group assignment (like that used now), and having the instructor watch carefully monitor student progress and provide appropriate advice. When current students were actually surveyed via questionnaire about group assignment and instructor support, less than 5% of students indicated dissatisfaction or problems, and this shows that adequate improvement has been achieved.

5 Conclusion

This paper has provided an overview of the "Machine Creation" course held in the Department of Mechano-Aerospace Engineering of Tokyo Institute of Technology. A questionnaire survey was administered, to alumni currently active as engineers in the real world, regarding the significance and educational effectiveness of this course. The results showed that many alumni have favorable opinions of the course significance and operation method. A comparison of current students with alumni suggests that course operation has been improved.

In the free comment section of the alumni questionnaire, some alumni responded that the course was their foundation as an engineer, or that they made use of their experiences in Machine Creation even today. This showed that the course is more than just another course, and indeed can be a vital experience which affects the student's subsequent view of work. For the authors who run this course, this was a very encouraging result which will advance education using robots.

As issues for the future, we would like to administer the same questionnaire to alumni from various academic years, determine how well the educational effect of the course persists, and examine how to contribute to the growth of engineers in the real world. Furthermore, we would like to continuously improve course operation based on the survey results.

In the free comment section of the alumni questionnaires, we received many extremely valuable comments for making specific improvements in course operation-such as comments requesting more active involvement by the instructor and TAs, or requesting guidance so that tasks such as presentations can be experienced by everyone, and not done only by students who are good at them. By effectively incorporating these ideas, we hope, going forward, to contribute to the dissemination and promotion of education to nurture creativity through robotics.

Acknowledgements. The course described in this paper is conducted with tremendous support from the faculty members in the Department of Mechano-Aerospace Engineering of Tokyo Institute of Technology. Many professors and staff members cooperated with actually running the class: Takeshi Yamamoto (technical staff in the Integrated Creation Studio), Katsuhisa Jimbo (former technical staff), and Hideaki Suzuki, Takamichi Koide, Masaru Nishikawa, and Masakazu Yoshii (technical staffs in the Design and Engineering Technology Center of the university). We would like to thank Makoto Ami (former lecturer), Akio Morishima (current associate professor at Chukyo University) and Hiroyuki Kuwahara (Sustinable Robotics). We would also like to express our deep gratitude to Kan Yoneda (current professor at Chiba Institute of Technology), Takeshi Aoki (current associate professor at Chiba Institute of Technology), Atsushi Kawakami (current assistant professor at Aoyama Gakuin University), and Takahiro Tanaka (currently at Mitsubishi Electric). Finally, we would like to sincerely thank all of the alumni of the former Department of Mechanical Science and Engineering, Class of '96, who gave their precious time to respond to the questionnaire for writing this paper.

References

1. Hirose, S.: Creative Education at Tokyo Institute of Technology. International Journal of Engineering Education 17(6), 512–517 (2001)
2. Morishita, T.: Creating Attraction for Technical Education Material and its Educational Benefit (Development of Robotic Education Material Characterized by 3D CAD/CAM and Compact Stereo Vision). Journal of Robotics and Mechatronics 23(5), 665–675 (2011)
3. Iribe, M., Tanaka, H.: An Integrated Hands-on Training Program for Education on Mechatronics. Journal of Robotics and Mechatronics 23(5), 701–708 (2011)
4. Morijiri, S., Ando, Y., Yoshimi, T., Mizukawa, M.: Improvement of Introductory Engineering Education of Mechatronics Based on Outcomes Evaluation by Defining Rubric - Continuous PDCA Cycle Achievement with Reducing Teaching Assistants' Work Load. Journal of Robotics and Mechatronics 23(5), 778–788 (2011)
5. Hayashibara, Y., Nakajima, S., Tomiyama, K., Yoneda, K.: Hands-on Education of Robotics Department for Four Years of College. Journal of Robotics and Mechatronics 23(5), 789–798 (2011)
6. Toda, K., Okumura, Y., Tomiyama, K., Furuta, T.: Hands-on Robotics Instruction Program for Beginners. Journal of Robotics and Mechatronics 23(5), 799–810 (2011)

7. Demura, K., Sakamoto, T., Asano, Y., Matsuishi, M.: Enhancing Student Engineering, Personal, and Interpersonal Skills Through Yumekobo Projects. Journal of Robotics and Mechatronics 23(5), 811–821 (2011)
8. Kumagai, M.: Educating Robot Development in a University Laboratory from First Year - A Trial of a Robotics Club Under Observation at a Laboratory. Journal of Robotics and Mechatronics 23(5), 822–829 (2011)
9. Endo, G., Aoki, T., Suzuki, H., Fukushima, E.F., Hirose, S.: Street Performance Robot Contest as Practice in Creativity Education. Journal of Robotics and Mechatronics 23(5), 768–777 (2011)

30   Research and Science Educations
of Lu Ban's Horse Carriages

Hong-Sen Yan

Abstract. Lu Ban (~ 507-444 BC) was a master carpenter in ancient China. He invented a legendary wooden horse carriage which was first mentioned in the book Lun Heng by Wang Chong (~ 27-97 AD) but was subsequently lost in the long past years. Based on the methodology for the reconstruction design of lost mechanisms developed by the author, his research group at the National Cheng Kung University (NCKU) started to reconstruct various designs of this lost machine in 1993. One of the feasible designs, a planar linkage with 8 bars and 10 joints as the leg mechanism, was further developed into various mechanical horse carriages without and with electrical powers. And, the corresponding hardware and educational kits have been successfully used for science programs at the National Science and Technology Museum and NCKU since 2002.

1   Introduction

Many ingenious mechanical devices were invented in ancient China before the 15th century. However, it was unfortunate that the inventors were unable to preserve the finished objects or keep complete documentation, thus preventing many of the designs to be passed down to latter generations.

Studies and publications on modern walking machines that mimic the motion of animals appeared only in the past hundred years. However, a legendary walking machine named the "Wooden Horse Carriage" might be invented by Lu Ban during the period of ancient China's Era of Spring and Autumn around 500 BC. This design was treated as a novelty, and it can be found in literary records but without surviving hardware.

Hong-Sen Yan
Department of Mechanical Engineering,
National Cheng Kung University, 1, University Road, Tainan 70101, Taiwan ROC
e-mail: hsyan@mail.ncku.edu.tw

V. Kumar et al. (Eds.): *Adv. in Mech., Rob. & Des. Educ. & Res.*, MMS 14, pp. 387–397.
DOI: 10.1007/978-3-319-00398-6_30   © Springer International Publishing Switzerland 2013

This work addresses historical literature regarding Lu Ban's Wooden Horse Carriage first. It follows by introducing the reconstruction synthesis of all feasible leg mechanisms of the wooden horse carriage at the National Cheng Kung University (NCKU, Tainan, Taiwan). Finally, various educational kits, hardware, and activities for K-12 students developed and organized by the National Science and Technology Museum (NSTM, Kaohsiung, Taiwan) and NCKU are presented.

2   Historical Background

Lu Ban (~ 507-444 BC) with Gong-shu as his original family name was a master carpenter and inventor in the Kingdom of Lu during the Era of Spring and Autumn in ancient China. What fascinated people most regarding his numerous inventions were the wooden kite and the wooden horse carriage.

According to legend, Lu Ban built a wooden horse carriage for his aged mother so that she would not tire herself when she went out. The description of this device first appeared in the book Lun Heng by Wang Chong (~ 27-97 AD) in the Eastern Han Dynasty. It states [01]: "It is said that Lu Ban was mourning of the loss of his mother. He built a wooden horse carriage that was well equipped and needed no manual intervention. When his mother rode on, it sped away never to return."

Wang Chong's work was primarily a response to the book Ru Shu. There was a part in it that questioned the credibility of Lu Ban's flying contraption that could stay in the sky for three days. Wang Chong believed that if Lu Ban's carriage could move automatically without stopping, then his kite could also fly for three days without falling. If the wooden carriage could not move on its own, then when Lu Ban's mother was riding in the carriage, the carriage should have stopped moving somewhere, enabling him to find his mother along the 3-day carriage route. According to historical records, however, Lu Ban's mother was never found.

If the wooden bird had existed, then the carriage should also have existed because the design of the flying device should be more difficult than that of the ground carriage. In addition, if the carriage was operated by linkage mechanisms, it would not be a problem for a master carpenter like Lu Ban. And, the correct dimensions and assembly of the parts of the device would certainly be based on experiments done with rich engineering experiences. Furthermore, Lu Ban lived in Dun Huang, a place full of mountainous slopes. This might also suggest that his carriage could move on the rugged terrain possibly based on inertia and the conservation of energy. Therefore, the creation of the wooden horse carriage should be possible. However, this invention was treated as a novelty and quickly disappeared. Nevertheless, it is the earliest story of ancient Chinese walking machines.

3   Reconstruction Designs

In past long years, very few scholars have studied lost ancient Chinese walking machines. Around 1986, Wang Jian of Urumqi in China, built a wooden horse carriage power by an electric motor with batteries based on his ingenious experience and sense of practicality, Fig. 1 [02-04]. This design is composed of a walking mechanism with a leg function and a trailer with balance function. The walking mechanism has four sets of 8-bar linkage with the same configurations.

Fig. 1 Wang Jian's wooden horse carriage

In 1993 the author started a systematic research effort on the lost walking machines at NCKU, especially the lost wooden horse carriage [02-12]. Based on limited historical records and subject to technological constraints of ancient era, all feasible designs of the 4-legged linkage-type walking machines were synthesized, based on "Methodology for the Reconstruction Design of Lost Ancient Mechanisms" developed by the author [04]. It includes the following four steps, Fig. 2:

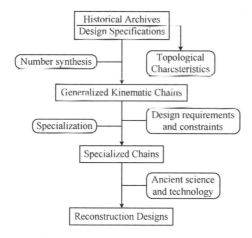

Fig. 2 Process of reconstruction design

Step 1. Design Specifications
Basic design specifications regarding the configuration of the wooden horse carriage are defined as follows according to the study of available historical archives:

1. It is a quadruped walking machine that generates specific gait locomotion, and it mimics the motion of a real horse.
2. Each leg mechanism is a planar linkage with simple revolute joints and one degree of freedom, i.e., the number of links can be 4, 6, 8, 10... etc.
3. A carriage is attached to the back body of the wooden horse to carry riders and also for providing the function of balance.

Step 2. Generalized Kinematic Chains
This step is to obtain or identify the atlas of generalized kinematic chains with the required numbers of links and joints subject to defined design specifications (topological characteristics) by applying the algorithm of number synthesis or simply identifying them from available atlases in reference [04].

Step 3. Specialized Chains
This step is to obtain the atlas of specialized chains with assigned types of links and joints for each generalized kinematic chain obtained in Step 2 through the following sub-steps:

a. For each generalized kinematic chain, identify the thigh link (member 3) and the shank link (member 4) that are adjacent to each other for all possible cases.
b. For each case obtained in sub-step a, identify the ground link (member 1).
c. For each case obtained in sub-step b, identify the crank (member 2).

Then, subject them to the following design requirements and constraints:

1. It has a ground link (frame) as the body.
2. It has a crank not adjacent to the thigh link or the shank link. The crank of the leg mechanism is adjacent to the body, and the fixed pivots of all the four cranks are coaxial.
3. It has a thigh link adjacent to the body and the shank link.
4. It has a shank link not adjacent to the body, but adjacent to the thigh link.
5. The crank, the thigh link, and the shank link must be distinct members.
6. There is a foot point (coupler point) on the shank link to generate a path curve and to contact the ground.

Step 4. Reconstruction Designs
The last step is to obtain the atlas of reconstruction designs from the atlas of specialized chains based on the motion and functional requirements of the lost machine, and by utilizing the mechanical evolution and variation theory to perform a mechanism equivalent transformation. Furthermore, ancient scientific theories and technologies of the subject's time period are applied to find feasible mechanisms that can be considered as the reconstruction designs.

Based on such a methodology, all feasible designs for the leg mechanisms with one degree of freedom and with 6-bar and 8-bar are synthesized.

For a planar 6-bar leg mechanism with simple revolute joints and one degree of freedom, it has 7 joints; and there are two generalized kinematic chains with 6 bars and 7 joints, Fig. 3. Through the process of specialization, there are 32 specialized chains available as shown in Fig. 4. In addition, Figs. 5(a1)-(a15) show the corresponding schematic formats of the leg mechanisms, providing the atlas of all possible reconstruction design concepts for the generalized kinematic chain shown in Fig. 3(a).

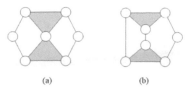

Fig. 3 Atlas of generalized kinematic chains with 6-bar and 7-joint

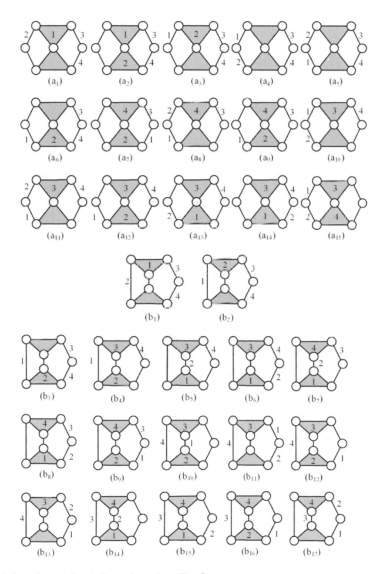

Fig. 4 Atlas of specialized chains based on Fig. 3

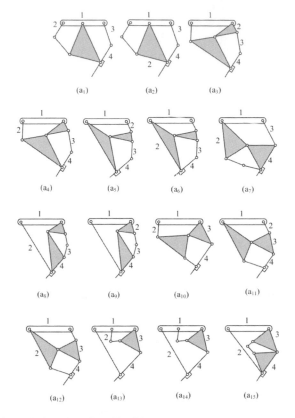

Fig. 5 Atlas of 6-bar designs based on Fig. 3(a)

For a planar 8-bar leg mechanism with simple revolute joints and one degree of freedom, it has 10 joints; and there are sixteen generalized kinematic chains with 8 bars and 10 joints as shown in Fig. 6. By following the same process shown in Fig. 2, there are 117 specialized chains available, and Fig. 7 shows some of the corresponding schematic formats of the 8-bar leg mechanisms.

Fig. 8 shows a physical model developed at NCKU in 1996 [05]. This carriage is pushed to move forward and pulled to move backward. It requires only a small force to push or pull to make it walk up a reasonable slope. When left on a slope around 15 degrees, it moves down without human intervention due to gravity. This might prove that such an invention might really be feasible as indicated in the book Lun Heng by Wang Chong (~ 27-97 AD) in the Eastern Han Dynasty.

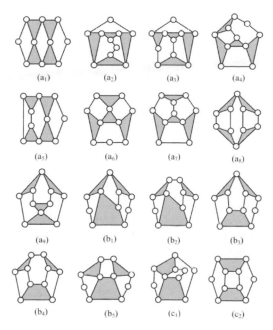

Fig. 6 Atlas of generalized kinematic chains with 8-bar and 10-joint

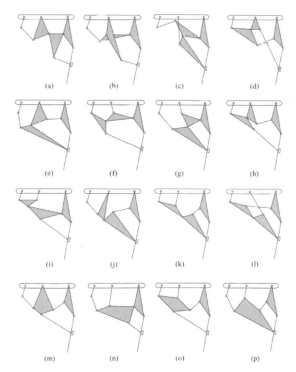

Fig. 7 Some leg mechanisms for the 8-bar designs

Fig. 8 NCKU's 8-bar-type horse carriage

4   Science Educations in Museums

The National Science and Technology Museum (NSTM), inaugurated in November 1997, is the first national museum in applied science in Taiwan, occupying an area of around 19 hectares in site and with 114,000 square meters of floor area. NSTM's educational activities aim to enhance visitor's understanding of science and technology. In addition, its educational program covers all kinds of technology, from the modern to ancient technology and from the traditional to state-of-art technology.

Based on the outcomes of basic research regarding Lu Ban's wooden horse carriage, the author's research group at NCKU further developed the prototype shown in Fig. 8 into some modern horse carriages with an electrical motor, Fig. 9. In addition, the technology of this design was transferred to a local company for commercial production. From 2002, various educational kits, hardware, and activities for K-12 students were developed and organized by NSTM and NCKU.

Fig. 9 NCKU's modern horse carriages

Since each leg mechanism is an 8-bar linkage with 10 joints, educational kits with corrugated plate, thick paper, plastics and metals as materials were designed and manufactured for instructors to illustrate the motion of the linkage mechanism and to allow students to engage in the hand-on assembly, Fig. 10. In addition, a number of physical electrical horse carriages with different sizes and energy sources were also developed, such as a table size, remote control and lead-acid battery design in 2002; several driver's size, manual control and lead-acid battery designs in 2003~2006; and a driver's size, manual control and fuel-cell battery design in 2008~2009, Fig. 11. Furthermore, there were more than 50 mechanical programs and camps with the horse carriages as the main theme during the period of 2003~2008. And, there were 5 special programs for undergraduate and K-12 students held at students' sites organized by NCKU during the same period.

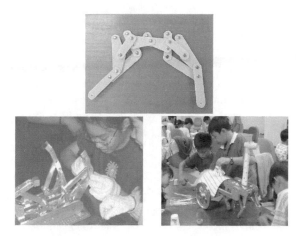

Fig. 10 Educational kits of modern horse carriages

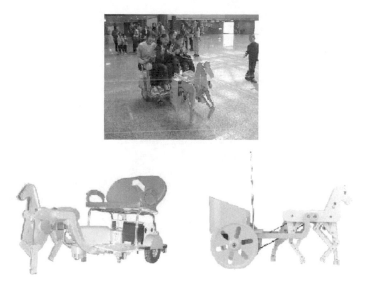

Fig. 11 Products of modern horse carriages

5   Conclusions

Lu Ban's wooden horse carriage is an example of lost ancient Chinese machines with only limited written descriptions, and without any illustrations and surviving evidence. Restoration of this type is more difficult, and some imagination is required. It is believed that this design consists of a four-legged walking machine and a trailer. This work systematically synthesized all feasible linkage-type leg mechanisms of Lu Ban's wooden horse carriage subject to ancient science and technology of the item's time period, based on the methodology for the reconstruction synthesis of lost ancient mechanisms developed by the author.

Based on the built prototype of a reconstruction design with an 8-bar linkage as the leg mechanism as shown in Fig. 8, some modern mechanical horse carriages with an electric motor powered by lead-acid and/or fuel-cell batteries are further developed. And, hardware and educational kits for these carriages have been successfully applied for science programs for K-12 students at the National Science and Technology Museum and National Cheng Kung University since 2002.

As a result, this work provides a solid example of how professors' long-term studies can be a reliable source for transforming basic research into innovative educational programs in science centers and (university) museums.

Acknowledgments. Supports from the Ancient Chinese Machinery Culture Foundation and National Cheng Kung University (Tainan, Taiwan), the National Science and Technology Museum (NSTM) (Kaohsiung, Taiwan), and the Feton Automation Industrial Co., Ltd. (Kaohsiung, Taiwan) are greatly appreciated.

References

1. Chong, W., (Han Dynasty): Lun Heng. Hong Ye Books, Taipei (1983) (in Chinese)
2. Yan, H.S.: A systematic approach for the restoration of the wooden horse carriage of ancient China. In: Proc. of the International Workshop on History of Machines and Mechanisms Science, Moscow, Russian, May16-20, pp. 199–204 (2005)
3. Yan, H.S.: Historical trace and restoration of ancient Chinese walking machines. J. of the Chinese Society of Mechanical Engineers 26(2), 133–137 (2005)
4. Yan, H.S.: Reconstruction Designs of Lost Ancient Chinese Machinery. Springer, Netherlands (2007)
5. Chiu, C.P.: On the design of a wave gait walking horse. Master Thesis, Department of Mechanical Engineering, National Cheng Kung University, Tainan, Taiwan (June 1996)
6. Hwang, K.: On the design of an optimal 8-link type walking horse. Master Thesis, Department of Mechanical Engineering, National Cheng Kung University, Tainan, Taiwan (June 1997)
7. Chen, P.H.: On the mechanism design of 4-link and 6-link types wooden horse carriages. Master Thesis, Department of Mechanical Engineering, National Cheng Kung University, Tainan, Taiwan (June 1998)
8. Yan, H.S.: A design of ancient China's wooden oxen and gliding horse. In: Proceedings of the 10th World Congress on the Theory of Machines and Mechanisms, Oulu, Finland, June 20-24, pp. 57–62 (1999)
9. Shen, H.W.: On the mechanism design of 8-link type walking horses. Master Thesis, Department of Mechanical Engineering, National Cheng Kung University, Tainan, Taiwan (June 1999)

10. Chiang, G.C.: A reconstruction design of Lu-Ban's wooden horse carriage with 10-bar linkages. Master Thesis, Department of Mechanical Engineering, National Cheng Kung University, Tainan, Taiwan (May 2003)
11. Hung, C.C.: On the mechanism design of a hybrid 8-link type walking horse. Master Thesis, Department of Mechanical Engineering, National Cheng Kung University, Tainan, Taiwan (June 2003)
12. Liu, S.H.: On the design of walking horses with 8-link and a sliding pair. Master Thesis, Department of Mechanical Engineering, National Cheng Kung University, Tainan, Taiwan (May 2005)

Author Index

Printed by Printforce, the Netherlands